COMPREHENSIVE SERIES IN PHOTOCHEMISTRY
AND PHOTOBIOLOGY

Series Editors

Donat P. Häder
Professor of Botany

and

Giulio Jori
Professor of Chemistry

European Society for Photobiology

COMPREHENSIVE SERIES IN PHOTOCHEMISTRY
AND PHOTOBIOLOGY

Series Editors: Donat P. Häder and Giulio Jori

*Other Titles in this Series*

COMPREHENSIVE SERIES IN PHOTOCHEMISTRY
AND PHOTOBIOLOGY – VOLUME 6

# Photodynamic Therapy with ALA
## A Clinical Handbook

Editors

Roy Pottier
Department of Chemistry and Chemical Engineering
The Royal Military College of Canada
Kingston, Ontario, Canada

Barbara Krammer
Department of Molecular Biology
University of Salzburg, Salzburg, Austria

Herbert Stepp
Laser Research Institute
Munich, Germany

Reinhold Baumgartner
Laser Research Institute
Munich, Germany

RSCPublishing

ISBN-10: 0-85404-341-1
ISBN-13: 978-0-85404-341-5

A catalogue record for this book is available from the British Library

Published by The Royal Society of Chemistry,
Thomas Graham House, Science Park, Milton Road,
Cambridge CB4 0WF, UK

Registered Charity Number 207890

For further information see our web site at www.rsc.org

Typeset by Macmillan India Ltd, Bangalore, India
Printed by Henry Lings Ltd, Dorchester, Dorset, UK

# Preface for the ESP Series in Photochemical and Photobiological Sciences

"Its not the substance, it's the dose which makes something poisonous!" When Paracelsius, a German physician of the 14th century made this statement he probably did not think about light as one of the most obvious environmental factors. But his statement applies as well to light. While we need light, for example, for vitamin D production too much light might cause skin cancer. The dose makes the difference. These diverse findings of light effects have attracted the attention of scientists for centuries. The photosciences represent a dynamic multidisciplinary field, which includes such diverse subjects as behavioral responses of single cells, cures for certain types of cancer and the protective potential of tanning lotions. It includes photobiology and photochemistry, photomedicine as well as the technology for light production, filtering and measurement. Light is a common theme in all these areas. In recent decades, a more molecular-centered approach changed both the depth and the quality of the theoretical as well as the experimental foundation of photosciences.

An example of the relationship between global environment and the biosphere is the recent discovery of ozone depletion and the resulting increase in high-energy ultraviolet radiation. The hazardous effects of high-energy ultraviolet radiation on all living systems is now well established. This discovery of the result of ozone depletion put photosciences at the center of public interest with the result that, in an unparalleled effort, scientists and politicians worked closely together to come to international agreements to stop the pollution of the atmosphere.

The changed recreational behavior and the correlation with several diseases in which sunlight or artificial light sources play a major role in the causation of clinical conditions (*e.g.*, porphyrias, polymorphic photodermatoses, Xeroderma pigmentosum and skin cancers) have been well documented. As a result, in some countries (*e.g.*, Australia) public services inform people about the potential risk of extended periods of sun exposure every day. The problems are often aggravated by the phototoxic or photoallergic reactions produced by a variety of environmental pollutants, food additives or therapeutic and cosmetic drugs. On the other hand, if properly used, light-stimulated processes can induce important beneficial effects in biological systems, such as the elucidation of several aspects of cell structure and function. Novel developments are centered around photodiagnostic and phototherapeutic modalities for the treatment of cancer, artherosclerosis, several autoimmune diseases, neonatal jaundice and others. In addition, classic research areas such as vision and photosynthesis are still very active. Some of these developments are unique to photobiology, since the peculiar physico-chemical properties of electronically excited biomolecules often lead to the promotion of reactions which are characterized by high levels of selectivity in space and time. Besides the biologically centered areas, technical developments have paved the way for the harnessing of solar energy to produce warm water and electricity or the development of environmentally friendly techniques for addressing problems of large social impact (*e.g.*, the decontamination of polluted waters). While also in use in Western countries, these techniques are of great interest for developing countries.

The European Society for Photobiology (ESP) is an organization for developing and coordinating the very different fields of photosciences in terms of public knowledge and scientific interests. Owing to the ever increasing demand for a comprehensive overview of the photosciences, the ESP decided to initiate an encyclopedic series, the "Comprehensive Series in Photochemical and Photobiological Sciences." This series is intended to give an in-depth coverage over all the very different fields related to light effects. It will allow investigators, physicians, students, industry and laypersons to obtain an updated record of the state of the art in specific fields, including a ready access to the recent literature. Most importantly, such reviews give a critical evaluation of the directions that the field is taking, outline hotly debated or innovative topics and even suggest a redirection if appropriate. It is our intention to produce the monographs at a sufficiently high rate to generate a timely coverage of both well-established and emerging topics. As a rule, the individual volumes are commissioned; however, comments, suggestions or proposals for new subjects are welcome.

Donat-P. Häder and Giulio Jori
Spring 2002

# Volume Preface

Photodynamic therapy, a light-activated process, is now an approved treatment modality for a variety of diseases. Of the various photosensitizers in clinical practice or in the process of development, one stands out as rather unique. Protoprophyrin IX is a natural photosensitizer that can be made by the human body. It is possible to induce the biochemical production of photosensitizing amounts of this natural compound in diseased tissues by the introduction of a small and simple amino acid, namely 5-aminolevulinic acid (ALA). Owing to its small molecular weight, ALA is easily introduced to the body by several routes, including topical (dermal) application.

The application of ALA to a patient in order to induce the biochemical production of excess protoporphyrin IX for the purpose of photodynamic therapy (ALA-PDT) has become a widespread procedure, especially for dermatological purposes. It is easy to use, has a large error margin, and is not dangerous. While it certainly has limitations, this novel approach to photodynamic therapy has proven to be so popular that it deserves to be thoroughly discussed in a separate book.

This monogram on ALA-PDT is primarily aimed at a clinical audience. The introductory chapter is written by the person who first used ALA-PDT on patients. Separate chapters have been written by other clinicians who have considerable and valuable experience in the use of ALA-PDT in their respective fields. Chapter 2, on the other hand, does contain an in-depth coverage of the basic science involved in the process of photosensitization, including an up-to-date review of the possible mechanistic details that are believed to govern all aspects of ALA-PDT. Thus, while most of the chapters are written for the benefit of clinicians who wish to examine ALA-PDT as a possible addition to their treatment arsenal, it also includes one fundamental science chapter so as to hopefully meet the needs of a wide audience.

Roy Pottier
Barbara Krammer
Herbert Stepp
Reinhold Baumgartner

September 2005

# Contents

# Contributors

**H. Barr**, *Cranfield Postgraduate Medical School, Gloucestershire Royal Hospital, Great Western Road, Gloucester, GL1 3NN, United Kingdom*

**R. Baumgartner**, *Laser Research Institute, Klinikum Grosshadern, Ludwig-Maximilians-University, Marchioninistr. 15, 81377 Munich, Germany*

**C. S. Betz**, *Department of Otorhinolaryngology, Head & Neck Surgery, (Head of Department: A. Berghaus, Professor, Dr), Ludwig Maximilians University, Großhadern Medical Center, Marchioninistr. 15, 81377 Munich, Germany*

**A. Curnow**, *Cornwall Dermatology Research, Peninsula Medical School, Knowledge Spa, Royal Cornwall Hospital, Truro, Cornwall, TR1 3HD, UK*

**E. Endlicher**, *Department of Internal Medicine I, University of Regensburg, D-93042 Regensburg, BRD*

**M. Fehr**, *Division of Gynecology, Department of Obstetrics and Gynecology, University Hospital, Frauenklinikstrasse 10, CH-8091 Zurich, Switzerland*

**P. Hillemanns**, *Abteilung I der Frauenklinik, Geburtshilfe und Allgemeine Gynäkologie, Carl-Neuberg-Str. 1, D-30625 Hannover*

**R. Hornung**, *Division of Gynecology, Department of Obstetrics and Gynecology, University Hospital, Frauenklinikstrasse 10, CH-8091 Zurich, Switzerland*

**J. C. Kennedy**, *Department of Chemistry and Chemical Engineering, P.O. Box 17000 Station Forces, Royal Military College of Canada, Kingston, ON K7K 7B4, Canada*

**B. Krammer**, *Department of Molecular Biology, University of Salzburg, Hellbrunnerstr. 34, 5020 Salzburg, Austria*

**A. Leunig**, *Department of Otorhinolaryngology, Head & Neck Surgery, Ludwig Maximilians University, Großhadern Medical Center, Marchioninistr. 15, 81377 Munich, Germany*

**Z. Malik**, *Bar Ilan University, Life Sciences Faculty, Ramat-Gan, 52900, Israel*

**H. Messmann**, *III. Med. Klinik, Klinikum Augsburg, Postfach 101920, 86009 Augsburg*

**R. Pottier**, *Department of Chemistry and Chemical Engineering, P.O. Box 17000 Station Forces, Royal Military College of Canada, Kingston, Ontario K7K 7B4 Canada*

**V. Schleyer**, *Department of Dermatology, Regensburg University Hospital, Franz-Josef-Strauss-Allee 11, D-93053 Regensburg, Germany*

**H. Stepp**, *LIFE-center, Laser-Researchlab, University Hospital of Munich, Marchioninistr. 23, 81377 Munich, Germany*

**W. Stummer**, *Department of Neurosurgery, University of Düsseldorf, Moorenstr. 5, 40225 Düsseldorf, Germany*

**R.-M. Szeimies**, *Department of Dermatology, Regensburg University Hospi-tal, Franz-Josef-Strauss-Allee 11, D-93053 Regensburg, Germany*

**R. Waidelich**, *Department of Urology, University Hospital of Munich, Marchioninistr. 15, 81377 Munich, Germany*

# Acknowledgements

The editors thank Ms. Beverly Kelly for critically reading of all the chapters in this book, and for her helpful and excellent suggestions in making the text more readable. A special thanks also to Professor Brian Pogue for his help in supplementing the suppliers list in the appendix.

*Chapter 1*

# Introduction

## James Cecil Kennedy

**Table of Contents**

## Abstract

This book is intended to be a very practical handbook for physicians who would like to add photodynamic therapy (PDT) to their clinical practice. It is concerned primarily with the specific type of PDT that involves administration of the porphyrin precursor 5-aminolevulinic acid (ALA), and which therefore is commonly referred to as ALA-PDT.

Chapter 1 provides brief descriptions of some basic physicochemical and biological mechanisms that are involved in ALA-PDT, and discusses some of their clinical implications. More detailed discussions are provided in Chapter 2, which is an in-depth coverage of the scientific principles involved in photosensitization. Subsequent chapters discuss the application of ALA-PDT to a variety of anatomical sites and clinical situations. Chapter 2 includes supplementary information of the physics of light delivery, and the Appendix a listing of suppliers of instrumentation used in ALA-PDT.

## 1.1. Outline of the Theory and Technique of ALA-PDT

Light is a form of energy. Molecules of certain chemical compounds (photosensitizers) have the ability to absorb a photon of visible light and then transfer most of their absorbed energy to a molecule of oxygen. This causes a transient increase in the chemical reactivity of the oxygen molecule, and converts it into a relatively strong oxidizing agent known as singlet oxygen (Figure 1). PDT makes use of light-induced singlet oxygen to kill cells by causing lethal oxidative damage to biologically important structures.

The excited states of both the photosensitizer and oxygen have very short half-lives. Consequently, in order to be effective, molecules of both must be in very close proximity to biologically important cellular structures. Moreover, the

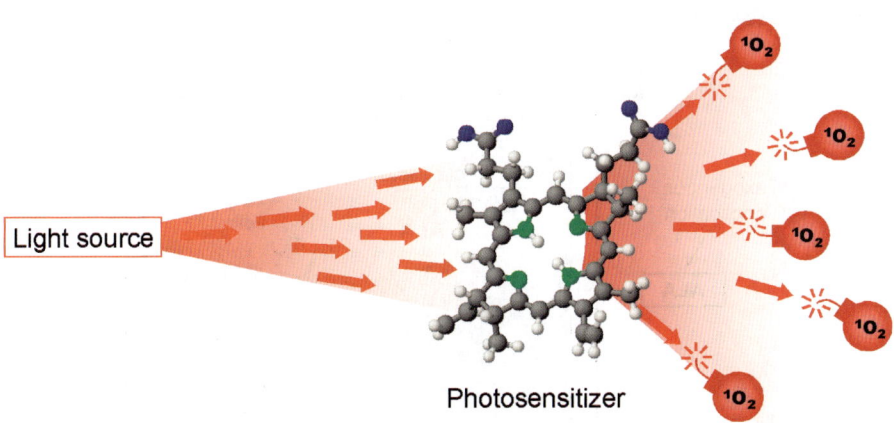

**Figure 1.** A photosensitizer molecule absorbs light of appropriate wavelength and can excite multiple oxygen molecules to a biologically reactive state (singlet-oxygen, $^1O_2$).

primary damage to cells occurs only while they are actually being exposed to the photoactivating light, although lethal cascades that were initiated during such an exposure may continue long after the treatment light has been turned off.

In order to become activated, the photosensitizer must absorb the light. The light therefore must be of wavelengths that lie within the absorption spectrum of the photosensitizer. However, since tissue contains pigments and particulate material that can absorb or scatter light in a wavelength-dependent manner, the particular photoactivating wavelengths that are selected should be ones that are neither absorbed nor scattered strongly by the tissue through which it passes. The choice of a wavelength for PDT usually is a compromise between strong absorption by the photosensitizer and good transmission by the tissue. For very superficial lesions, the major peak in the absorption spectrum works well, but for deeper lesions it is necessary to use light whose wavelength is more toward the red.

Ideally, the phototoxic damage will be restricted to the target tissue, although in practice we often accept a reasonable differential effect. The target tissue therefore must accumulate substantially more of the photosensitizer than does adjacent or underlying or overlying non-target tissue.

### 1.1.1. Tissue Specificity

The photosensitizer used in ALA-PDT is protoporphyrin IX (PpIX), which is synthesized *in situ* from exogenous ALA rather than given to the patient as a preformed molecule (Figure 2). The administration of exogenous ALA bypasses the rate-limiting step in the biosynthesis of heme, and thus forces each step in the pathway to produce its product at the maximum rate possible for that particular step. Since PpIX is the immediate precursor of heme, cells in which the rate of synthesis of PpIX is greater than the rate at which it can be converted into heme, excreted, or otherwise lost to the cell will accumulate PpIX.

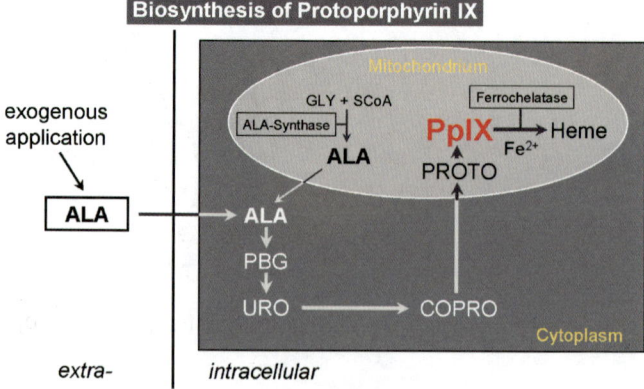

**Figure 2.**   Protoporphyrin IX is synthesized within the intracellular biosynthesis of heme. Upon exogenous delivery of ALA, PpIX is accumulated.

The amount of PpIX that accumulates in malignant, premalignant, and certain other abnormal tissues usually is significantly greater than the amount that accumulates in normal tissues of similar origin. When exposed to exogenous ALA, malignant tissues show a strong tendency to become much more photosensitive than the normal tissues from which they were derived. This is the primary reason for the tissue specificity found with ALA-PDT. Secondary reasons for tissue specificity may involve differences in the capacity of the cells to repair oxidative damage, or differences in the concentration or location of some compound that can function as an anti-oxidant and thus render singlet oxygen harmless.

### 1.1.2. Intracellular Targets

Photosensitizers other than ALA are administered intravenously as preformed molecules. They enter the blood stream, and then enter cells though their plasma membranes. The selective distribution of such photosensitizers depends upon physicochemical differences between the different types of cells, and the phototoxicity that results is a function of the intracellular location and concentration of the photosensitizer. In contrast, PpIX is synthesized by the mitochondria, the primary source of energy for the cell. Oxidative damage to such structures interferes with energy metabolism and can lead to cell death. Once the extra PpIX is produced in the mitochondria, it cannot all be converted into heme. Thus a significant amount will diffuse into the cytoplasm and may eventually find its way into other organelles with the exception of the nucleus. The fact that PpIX was not introduced as a preformed molecule, but rather synthesized *in situ*, means that the final destination of excess ALA-induced PpIX may be quite different from the site of localization of preformed photosensitizers.

### 1.1.3. Fluorescence

As well as being a good photosensitizer, PpIX is strongly fluorescent. It is possible to use ALA-induced fluorescence to locate tiny patches of abnormal tissue. It is possible also to measure the effectiveness of a course of chemotherapy by quantifying changes in the intensity, area, or volume of that fluorescence. For example, fluorescence induced *in vivo* may be used to locate patches of T-cell lymphoma (Figure 3) in the skin and to follow their response to either radiation therapy or chemotherapy. A decrease in fluorescence indicates a decrease in overall metabolic activity (the biosynthesis of PpIX from ALA requires energy), and may correspond to cell death. Again, flow cytometry can be used in conjunction with PpIX fluorescence (induced *in vitro*) to follow the response of leukemic cells to a course of chemotherapy. Failure to observe a decrease in fluorescence intensity of individual leukemic cells might indicate that the chemotherapeutic agent in question is ineffective, or alternatively that only drug-resistant cells are being measured. For such a study, it is very important to use fluorescence intensity standards, and to

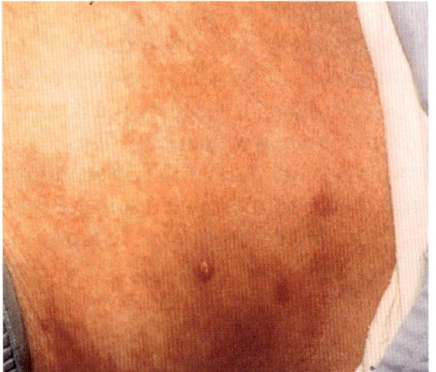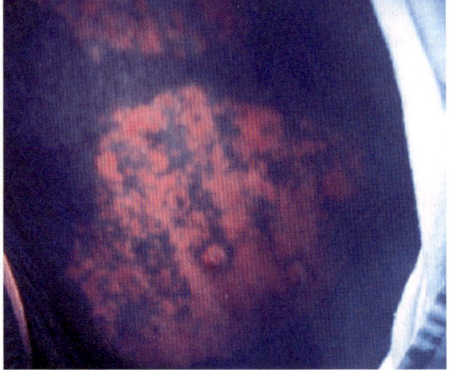

**Figure 3.** Cutaneous T cell lymphoma (mycosis fungoides) (left). ALA-induced PpIX
fluorescence induced in cutaneous T cell lymphoma (right).

distinguish between a decrease in total cell number and a generalized decrease
in their overall metabolic activity.

### 1.1.4. ALA Administration and Approval

Unlike other photosensitizers, ALA can be administered both topically and
systemically. Dissolved in a suitable vehicle, it can pass through intact layers of
keratin. Dissolved in water or in saline, it rapidly passes into the lining of the
digestive, respiratory, and urogenital tracts. Injections (intradermal, intratu-
moral, intravenous, or into the pleural, peritoneal, or pelvic cavities) are
effective also. In clinical studies, systemic application is usually realized
by oral delivery of ALA dissolved in water or juice. Altogether, it is a very
versatile drug.

Topical administration has several advantages. The photosensitized area is
restricted to the site of application, and there are no systemic effects. However,
the maximum depth at which a significant phototoxic effect can be produced is
perhaps not much about 1 mm. Intradermal injection may be helpful at the
periphery of large nodular tumors, and direct injection of tumor nodules may
be effective if the tumor contains areas of devitalized tissue that can function as
slow-release reservoirs for the ALA. A solution of ALA held in the mouth for
15 or 20 min will photosensitize malignant and premalignant tissues of the oral
cavity, and ALA solutions may be injected into the urinary bladder, the uterus,
and the vagina. In all such situations, the effect of ALA is localized primarily to
the site of application, although in some cases there may be a small amount of
leakage systemically.

Both the oral administration of ALA and its intravenous injection may lead
to undesirable systemic effects. These include occasionally reported nausea and
vomiting, tachycardia and hypotension, as well as a general skin photosensitivity for up to 48 h. It is possible to avoid most of these problems by reducing
the dose of ALA. Liver function enzymes may remain elevated for several

weeks, but this elevation is rarely as severe as that produced routinely by chemotherapy.

ALA can be administered as ALA hydrochloride or as ester derivatives. The ester derivatives have different polarities and some derivatives can be useful to treat deeper lesions, since some esters can penetrate deeper into the lesion. The methyl-ester derivative is approved in Europe and Australia for treatment of skin disorders (Actinic keratosis and basal cell carcinoma, Metvix by Photocure, Norway and Galderma, France; see also Chapter 3, 3.1.1.), the hexyl-ester derivative is approved in Europe for endoscopic fluorescence diagnosis of bladder cancer (Hexvix, Photocure). ALA has FDA-approval (USA) also for treatment of actinic keratosis with a blue light source (Levulan Kerastick together with Blu-U light source, DUSA, USA). All other clinical indications are investigational procedures performed in clinical studies.

*1.1.5. ALA/PpIX Clearance*

When ALA is administered by the oral or intravenous route, much of it is cleared from the body as it passes through the liver. The liver has a very large capacity to synthesize PpIX, and most of the ALA-induced PpIX that is synthesized by the liver is excreted via the bile. The kidneys excrete another large fraction. The residual ALA can be converted into PpIX by every cell in the body except mature erythrocytes (which lack the necessary mitochondria). Some normal tissues accumulate relatively large amounts of PpIX, but others accumulate almost none (Figure 4). In the absence of liver or kidney disease,

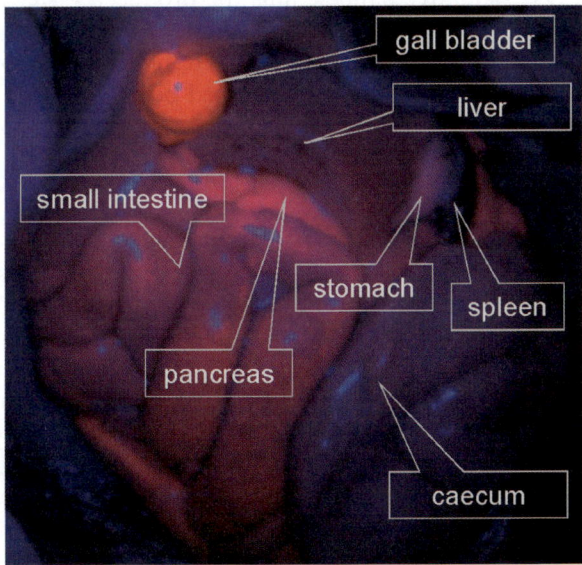

**Figure 4.** Red PpIX-fluorescence 3 h after intraperitoneal injection of 100 mg/kg ALA into a mouse immediately postmortem. Half of the liver had been removed. The bile in the gall bladder shows the strongest fluorescence.

intracellular PpIX returns to normal levels 36–48 h following the systemic administration of the ALA. Until such clearance occurs, the whole body remains at least somewhat photosensitized.

When ALA is administered by topical application, the situation is somewhat different. The stratum corneum, stratum granulosum, and stratum spinosum of the epidermis function as a slow-release reservoir for the ALA, so that instead of being exposed to a relatively short pulse of ALA, the cells of the skin are exposed to ALA over a period of time that may last as long as 12–16 h following its initial application. PpIX continues to be synthesized during this whole time. Consequently, the clearance of PpIX that is synthesized in response to the topical application of ALA is slower than that which follows the systemic administration of ALA, the exact time varying with the thickness of the reservoir tissues. Toenails and the skin on the soles of the feet are relatively slow to lose their PpIX, while the thin skin of the inner arm clears rather quickly. However, in most cases the PpIX concentrations are close to normal 3–4 days after its application. This delayed clearance is usually not a problem, since photobleaching during treatment destroys much of the PpIX.

## 1.2. Factors Affecting Photodynamic Therapy

Many different types of cells are capable of repairing minor oxidative damage. However, if a large number of oxidative events take place within a relatively short period of time, the capacity of a cell to repair itself can be overwhelmed. For effective PDT, oxidative damage must be produced faster than the damaged cells can repair it, until eventually the damage becomes too great for cell survival. The rate at which such damage is produced depends upon a number of interacting factors, some of which are described below.

### 1.2.1. Oxygen

The concentration of oxygen molecules within cells at the target site immediately prior to therapy is important. This concentration decreases during the course of therapy as oxygen molecules are used up in the oxidative reactions. The oxygen can be replenished to some extent so long as there is adequate circulation and diffusion within the target tissue, but the progressive onset of edema during therapy may cause a progressive decrease in this rate of replenishment. It is therefore quite possible for the concentration of oxygen at the target site to limit the effectiveness of PDT.

The total dose of light is the product of the intensity times the duration of exposure. Using lower intensities of photoactivating light over a longer period of time will use up the oxygen more slowly and may permit replenishment during the course of treatment. Dose fractionation (multiple doses of light separated by intervals of dark time) may be helpful also. However, the onset of tissue edema often interferes with the diffusion of oxygen, and it may be necessary to wait as long as 48 h between fractions for the target tissue to become re-oxygenated. It should be noted that at very low light intensities, phototoxic damage may occur so slowly that cellular repair processes negate most of the damage.

### 1.2.2. Photosensitizer

Within limits, the concentration of a photosensitizer in areas immediately adjacent to vital intracellular structures largely determines the effectiveness of the therapy. The half-life of a molecule of photosensitizer that has absorbed the energy of a photon of light is very short, as is the half-life of the resulting singlet oxygen. Light-activated photosensitizer loses its excess energy, and singlet oxygen loses its chemical reactivity without having any therapeutic effect if it is generated too far distant from the intended target.

With other photosensitizers, the usual way to solve this problem is to flood the cell with the photosensitizer. However, ALA-induced protoporphyrin is unique in that it is synthesized within the mitochondria, and therefore is located right at the heart of a cell's primary energy-producing chain of biochemical reactions. A little damage at such a site can have lethal consequences.

Normal tissues may differ in their response to ALA. For example, the normal skin of a child often accumulates significantly more PpIX than does the normal skin of an older person. It is therefore sometimes necessary to adjust the dose of ALA downward in order to obtain a good differential effect. This is particularly important when using ALA-induced PpIX to locate tiny areas of abnormality in the skin, or to follow the response to treatment of subcutaneous nodules that are too small to detect by palpation. It is possible to adjust the ALA concentration to increase the contrast so that the normal skin shows little or no PpIX fluorescence while the abnormal tissues, although definitely less bright than before, are much more visible against the non-fluorescing background.

Protoporphyrin is photobleached very rapidly during the treatment process. As will be explained below, this greatly simplifies the light dosimetry for ALA-PDT.

### 1.2.3. Light

Three factors are important – the wavelength(s) of the photoactivating light, (b) the total dose of light at the target site, and (c) the intensity (dose rate) of the light.

The wavelength(s) must be capable of being absorbed by the photosensitizer molecules. One might predict that the major peak in the photosensitizer absorption spectrum would be the ideal photoactivating wavelength. However, it often happens that these particular wavelengths are strongly absorbed by blood or other body pigments, or are scattered too strongly by cells and other particulate material to permit deep penetration of the tissues. In general, if deep penetration is required, the most effective wavelengths are those at the red end of the photosensitizer absorption spectrum (since blood transmits red wave-lengths). The blue end of the spectrum can be effective for treating more superficial lesions.

The total dose of light is important, but so also is the dose rate (the intensity). Too low a dose rate, and the target tissue may repair much of the phototoxic damage. Too high a rate, and the light is wasted because the concentration of either oxygen or the photosensitizer becomes rate-limiting. Heating of the tissue can become a problem also if the dose rate is too high, especially if red wavelengths are not used, since most of the energy of the non-red wavelengths is absorbed in the superficial layers of the target tissue. In general, light intensity should be reduced if the site is anatomically a poor heat sink (*i.e.*, the ears and the tips of the fingers), or if edema is interfering with removal of excess heat by the blood.

### 1.2.4. Light Dosimetry

Unlike other photosensitizers, protoporphyrin is photobleached very rapidly during treatment. This greatly simplifies dosimetry of the light, since the phototoxic damage can be determined primarily by both the absolute and relative concentrations of photosensitizer that are present in the normal and malignant tissues prior to treatment. Once the photosensitizer in the normal tissue has been photobleached, the addition of a huge excess of light can cause no additional phototoxic damage.

Example – the target lies deep within some normal tissue that we do not wish to destroy. The target tissue contains much more photosensitizer than does the normal, but since the light intensity decreases as we go deeper, giving a lethal dose of light to the target may require us to give a much higher dose to the normal tissue that overlies the target.

If a photosensitizer does not photobleach very rapidly, even a relatively low concentration of that photosensitizer in the superficial tissues could result in serious phototoxic damage if we attempt to give a lethal dose to the target tissue that lies beneath. However, protoporphyrin IX is destroyed by light, and once it has been destroyed in the overlying normal tissues, those tissues may be given any amount of additional light without untoward consequences. It is therefore possible to expose the deeper target tissues to a lethal dose of light while causing only mild and easily repairable damage to more superficial normal tissues, merely by increasing the duration of exposure to the photoac-tivating light.

Example – the target is a lesion on a complex curved surface (for example, the nose and adjacent cheeks). It is almost impossible to expose every part of such a surface to equal intensities of photoactivating light. If we try to give every part of that surface at least the minimum effective dose, then some parts will be given an overdose. Conversely, if we make sure that no part receives an overdose, then some parts will be underdosed.

This would present a difficult problem in light dosimetry if we were using a photosensitizer that did not photobleach readily. However, if we are using

PpIX, then we do not worry about giving too much light, but instead deliberately overdose. The PpIX in the normal tissues will be photobleached long before those tissues experience more than mild damage that is easily repaired.

In summary, when using ALA-PDT we are concerned primarily with giving a curative dose of light to the target. We do not worry about overdosing the normal tissues within the treatment field, since photobleaching occurs and the phototoxic effect is determined primarily by the concentration of photosensitizer that was originally present in the tissues. This assumes that there is a good PpIX concentration differential between the normal and the malignant tissues. If there is not, then reducing the amount of ALA administered may increase that differential, although it also may decrease the degree of photosensitization of the target tissue. Longer exposure times may then be necessary to destroy the target.

## 1.3. Clinical Applications of ALA-PDT

Some of these applications are well-established clinical procedures, others are widely used but still in the process of being accepted, and still others are considered highly experimental. Some are for the treatment of obvious lesions, others for identification of the boundaries of indistinct lesions, and still others for prevention of the further development of occult lesions. A few of the applications are for cosmetic purposes. Examples are given below.

- (i) Premalignant lesions – of the skin (Figure 5), oral mucosa, esophagus, bladder mucosa, vulva, vagina, and cervix.
- (ii) Malignant lesions (all types) – for cure, local control, palliation, prevention, detection, and identification of the boundaries of lesions (malignant glioma).
- (iii) Bacterial infections – Propionibacterium acne.
- (iv) Fungal infections (superficial) – onychomycosis.
- (v) Viral infections – verucca vulgaris and other warts.

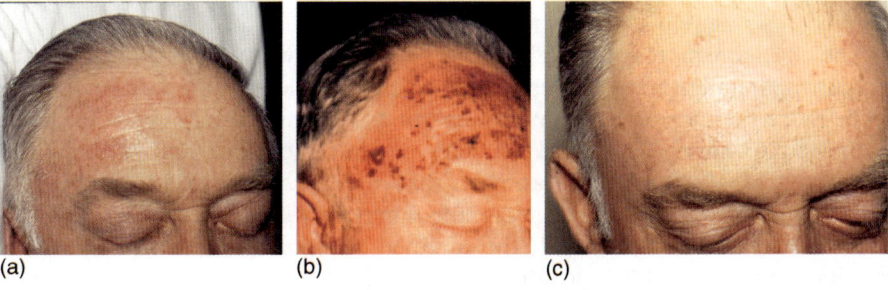

(a)                              (b)                              (c)

**Figure 5.** (a) Actinic keratosis prior to treatment. (b) Actinic keratosis 3 days after ALA-PDT treatment with red light. (c) Actinic keratosis 30 days after ALA-PDT. Note remaining nodule.

   (vi) Parasitic infections – cutaneous and mucocutaneous leishmaniasis.
  (vii) Rapid quantification of response to treatment – leukemia, T-cell
       lymphoma of skin, recurrent carcinoma of breast, malaria
 (viii) Cosmetic procedures – rejuvenation of skin, removal of excess hair

## 1.4. Interaction between ALA-PDT and the Immune System

By itself, ALA-PDT is able to kill malignant cells by either necrosis or apoptosis. However, there is strong suggestive evidence that, in at least some clinical situations, the immune system acts in cooperation with ALA-PDT to destroy those malignant cells.

### 1.4.1. Squamous Cell Carcinoma of the Skin

In the presence of a normal immune system, squamous cell carcinoma of the skin usually can be destroyed quite readily by ALA-PDT. However, if the immune system is suppressed (either by uncontrolled AIDS or by the strong immunosuppressive therapy that is given routinely to patients with kidney or heart transplants), the destruction of squamous cell carcinomas of the skin by ALA-PDT becomes very much more difficult, and may in fact be impossible. In contrast, patients whose AIDS is under control and transplant patients whose maintenance dose of immunosuppressive agents has been significantly reduced respond to ALA-PDT like patients who have a normal immune system.

### 1.4.2. Kaposi's Sarcoma

Patients whose immune system has been suppressed by uncontrolled AIDS often develop a form of Kaposi's sarcoma that fails to show any clinically significant response to ALA-PDT. In contrast, patients who develop the classical slowly progressive age-onset type of Kaposi's sarcoma respond to ALA-PDT very well. In some cases, the treatment of lesions at one anatomical sight results in the apparent spontaneous eradication of previously untreated lesions at distant sites.

    The mechanism responsible for the synergy between ALA-PDT and the immune system has not yet been identified, but in at least some cases it appears to be more than a simple "mopping up" of a few residual malignant cells that happened to survive the PDT. It may be that the oxidation of biological material generates strong new antigens that can cross-react with the malignant cells. Another possibility is that the inflammatory response induced by ALA-PDT in close association with the targeted malignant cells stimulates non-specific immune and phagocytic cells, which then attack and destroy any residual malignant cells. Alternatively, products of the inflammatory response may stimulate a tumor-specific cellular immune reaction that has the capacity to destroy malignant cells throughout the body.

## 1.5. Conclusions

ALA-PDT can be very helpful in clinical situations in which standard forms of therapy are not appropriate, such as patients with basal cell nevus syndrome who cannot be treated with ionizing radiation and yet have far too many lesions to treat by surgery. It can be used also to salvage certain lesions that have failed maximum safe dose radiation therapy, and it is being used routinely as the primary form of treatment for superficial basal cell carcinoma, Bowen's disease, and actinic keratosis. The following chapters discuss certain other indications.

*Chapter 2*

# Basic Principles

## B. Krammer, Z. Malik, R. Pottier and H. Stepp

**Table of Contents**

## Abstract

This chapter outlines the basic scientific principles involved in photodynamic therapy, both in general and specifically for ALA-PDT. Thus this chapter is of a scientific nature and covers details ranging from historical aspects to current research. It includes up-to-date data from various fields (physics, chemistry, biology, *etc.*) that shed light on possible mechanisms of action. While clinicians will certainly find here more in-depth knowledge on the scientific aspects of ALA-PDT, those who are solely interested in the clinical aspects of ALA-PDT can opt to skip this chapter.

## 2.1. Introduction

Photodynamic therapy (PDT) has recently been accepted as a new tool for the selective destruction of pathological tissue.[1] In principle, PDT is a simple adaptation of chemotherapy. A photosensitizer, light and oxygen work together to cause death (by necrosis or apoptosis) of the diseased tissue. The photosensitizer itself is non-toxic, but it can transfer the energy of absorbed light to molecular oxygen and thus produce chemically aggressive oxygen. The activating light by itself (visible or near infrared (IR) radiation) is harmless too. The intriguing potential of PDT lies in its ability to act *selectively* on the target tissue without the need for precise light targeting. In such a situation, it is not necessary to precisely identify the location or borders of the lesion to be treated. Such an approach opens the possibility of treating very early lesions, the existence of which may be known but not their exact location within the organ (*e.g.* whole bladder wall PDT[2,3]). In addition, PDT can be performed repetitively without the accumulation of serious side effects, and it can be combined with most other treatment modalities. Its minimally invasive nature and scar-free wound-healing have made it a good option for the treatment of skin cancers and other skin disorders.[4,5] For palliative treatment of invasive cancers, PDT causes less trauma than most alternative treatments, and it is sometimes the only remaining treatment option (*e.g.* for recurrent brain tumour[6]). In addition, PDT has proven to be effective in treating conditions where an increased risk of cancer is present (*e.g.* Barrett oesophagus[7]).

A variety of other photosensitizers have already obtained regulatory approval (Photofrin, Visudyne, Foscan, MetVix) or are currently in clinical studies (Lutetium-Texaphyrin, tin-Etiopurpurin, Mono-L-Aspartyl Chlorin e6, Pheophorbide a, WST09). However, aminolevulinic acid (ALA) is unique in several ways. It is not a photosensitizer as such, but it can induce the intracellular synthesis of a photosensitizer (protoporphyrin IX, PpIX) by entering into the biochemical pathway for the synthesis of haem. Exogenous ALA can produce photosensitizing amounts of PpIX in a variety of tissues, and especially in those that are abnormal. The metabolic conversion of exogenous ALA into a photosensitizing concentration of PpIX is relatively fast, requiring only 1–3 h. Furthermore, since the PpIX is part of a biosynthetic pathway, it has a built-in clearance mechanism that restricts possible systemic patient photosensitivity to less than 2 days. The fluorescence properties of PpIX can be exploited also to locate occult cancer.[8]

## 2.2. A Brief History of PDT

Around 1900, malaria was still endemic in Europe. In 1880, the French military doctor Charles L. A. Laveran (1845–1922) had discovered the parasite responsible for the disease. In 1898, the British military doctor Ronald Ross finally identified the Anopheles mosquito as the vector. Much effort was then directed toward the development of chemical agents (drugs) to combat the disease. Hermann von Tappeiner, who had become the director of the Pharmacological Institute of the University of Munich in 1887, accidentally got involved with key experiments that led to the earliest scientific investigations on PDT. He assigned the task of studying a series of potential anti-malarial agents to his young medical student Oscar Raab, who started his thesis in 1897. While studying the toxicity of acridine in cultures of infusoria in hanging drops, he made the astonishing discovery that the toxicity of acridine was greatly enhanced in the presence of light.[9,10] The discovery stimulated research into the new phenomenon. Within a few years, several substances that showed light-enhanced toxicity were identified, among which tetrabromo-fluorescein (eosin) exhibited the strongest differential effect and appeared to be clinically suitable, as it had previously been used in patients for the treatment of epilepsy.

The first clinical application of eosin-PDT was performed in 1903, when Hermann von Tappeiner and Albert Jesionek treated malignant skin lesions[11] (Figure 1). In the same year, Georges Dreyer from the Finsen Institute in Copenhagen used a different drug, erythrosine, to cause light-induced tissue necrosis.[12] In 1904, von Tappeiner coined the expression 'photodynamic' for this 'light effective' phenomenon.[13] He believed that the mechanism causing cellular destruction was closely connected with the fluorescence of photodynamic substances, and therefore was different from the mechanism underlying the sensitization of photographic plates that also was being extensively studied at that time. This assumption of a fluorescence-mediated mechanism was shared neither by Dreyer in Copenhagen nor by Albert Neisser in Breslau. Vigorous arguments were exchanged among the researchers.[14,15] However, all were in agreement that oxygen was required.[16,17]

The first series of clinical applications of PDT did not fully meet expectations, and although further research was performed and haematoporphyrin (Hp) was identified as a new and potent sensitizer by Walter Hausmann[18] and modified by Samuel Schwartz and Richard Lipson (Hp derivative, HpD[19,20]), it was not until 1973 that Tom Dougherty of Buffalo, NY founded a research group for PDT. Photofrin, an enriched and standardized formulation of HpD, was the photosensitizer promoted by Dougherty *et al.*, and it became the first PDT drug approved for clinical oncological treatment. As HpD suffers from some restrictions in therapeutic effectiveness and produces long-term skin photosensitization, other drugs have also been developed and investigated.

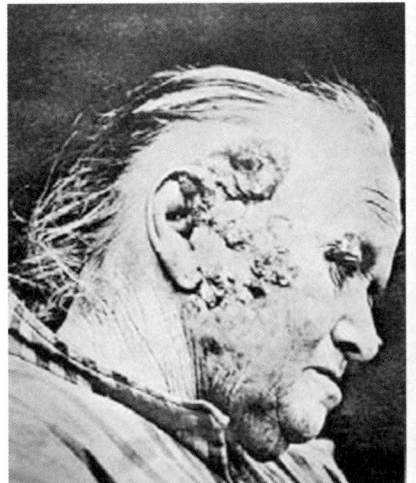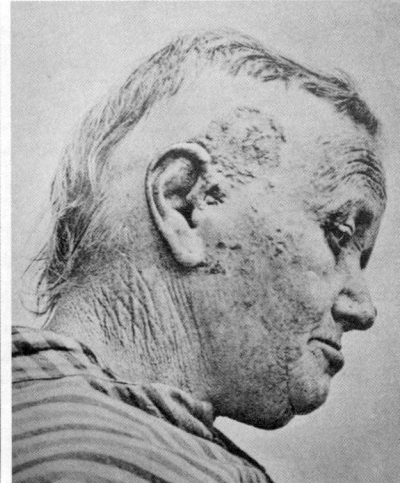

**Figure 1.** 'Photograms' of the first-documented patient treated with eosin and sunlight for 'multiple carcinomas' of the skin (70-year-old 'daytaller's widow'). Left picture was taken on Sept. 10, 1903 and right picture on November 14, 1903.

Among these are Foscan or mTHPC, the most potent photosensitizer, and 5-ALA, which probably causes the least side effects.[21–23] The light-induced fluorescence of ALA in particular has been studied for its potential diagnostic value.[24–26]

PDT and fluorescence diagnosis (FD) using ALA offer several clinically significant advantages, and they have been studied extensively in preclinical and clinical investigations since 1987.[27,28] The first indications that ALA might be a promising drug for PDT came from five independent sources:

In 1987, Zvi Malik and H. Lugaci[29] reported the use of 5-ALA to induce endogenous porphyrin synthesis in Friend erythroleukaemic cells. Photoirradiation of the cells with 'black light'-induced deformations and cell disintegration in more than 95% of the cells when examined by scanning electron microscopy (SEM) (Figure 2). The dependence of the process on the dose of light showed a relationship between the photodynamic effect and porphyrin accumulation. Both necrotic and apoptotic features were expressed, including disintegration of the plasma membrane (shown in Figure 2d and e), mitochondrial damage (Figure 3b), chromatin condensation and blebbing of the nuclear envelope (Figure 2e and f).

Mohammed El-Far had discussed in meetings[30] the possibility of using ALA, but had not published his results.

Also in 1987, Johan Moan reported an evaluation of 5-ALA as a photosensitizer in mice. However, he failed to find photosensitizing concentrations of porphyrins in either tumours or normal tissues.[31]

Having investigated ALA-stimulated porphyrin synthesis in plant tissues as early as 1975,[32] Alcira del Batlle in Buenos Aires reported on tumor selective build-up of porphyrins in ALA-incubated tissue-explants in 1988.[33]

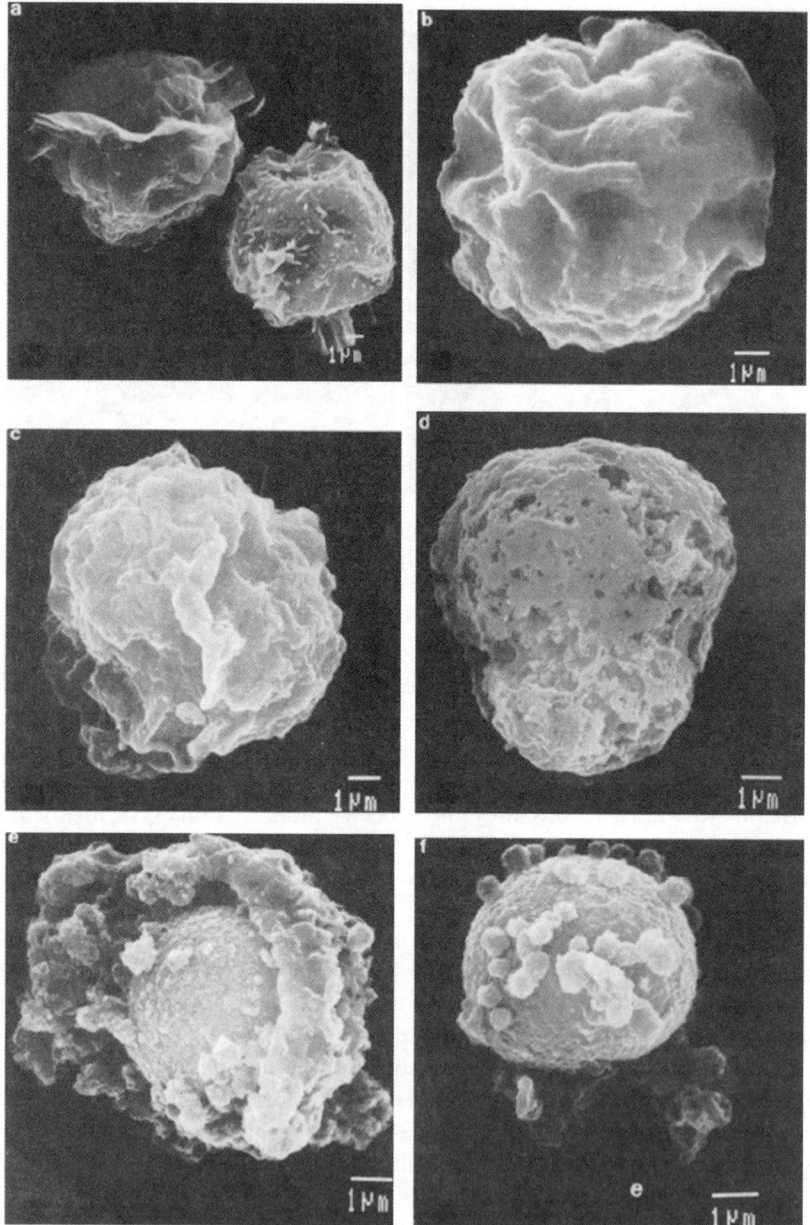

**Figure 2.** ALA-PDT of erythroleukaemic cells treated with ALA and exposed to light as revealed by scanning electron microscopy. a and b, Control cells; c–f, photoactivated cells.

James Kennedy had been investigating the porphyrin fluorescence produced in mice by intravenous or intraperitoneal injections of ALA since 1981,[34] and from time to time he had attempted, without success, to use ALA-induced porphyrins to destroy various types of malignant tissues by photodynamic

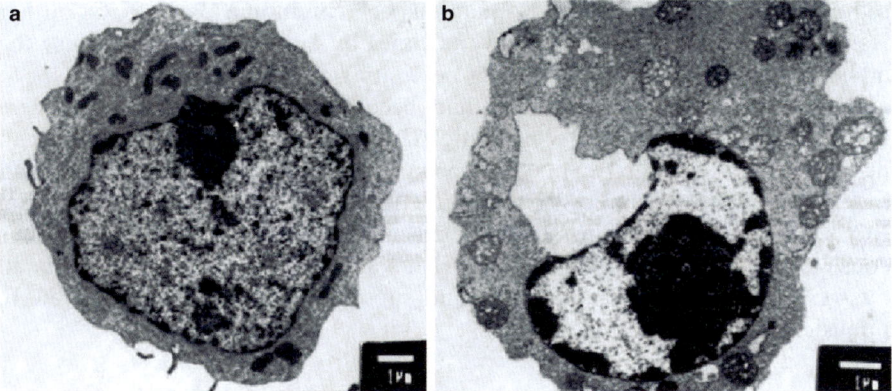

**Figure 3.** Transmission electron microscopy of the cells in Figure 2, depicting (a) control cells, (b) subcellular damage in mitochondria, nucleus, endoplasmic reticulum and nuclear envelope.[29]

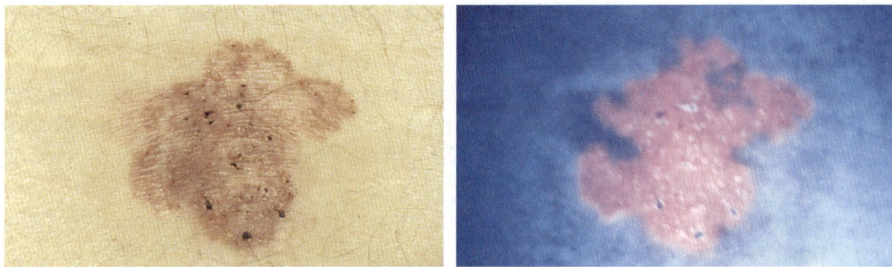

**Figure 4.** Superficial basal cell carcinoma prior to treatment by ALA-PDT. Note the lack of an exact correlation between the white light and the fluorescent photos.

action. These failures were rather discouraging, but he persisted because he did not understand how ALA-induced porphyrins could be present at high concentration in some of the tissues yet be so obviously ineffective at photosensitizing them. A major advance in technology occurred in the spring of 1985, when Kennedy and Pottier spent the entire night in their laboratory measuring the fluorescence emission spectrum of mice that had been injected with various doses of ALA. They already knew from previous experiments in which skin fluorescence had been estimated visually (rather than quantified by spectrophotofluorometer) that the ALA-induced fluorescence rose to a peak 4–6 h after the injection of ALA (Figure 4), and then decreased exponentially to near background levels within 36–48 h. However, they now had identified the main component responsible for the fluorescence by its spectrum, and understood its pharmacokinetic behaviour more clearly.[35]

It was in 1987 that the first patients were treated with ALA-PDT. The very first patient was a woman with extensive areas of actinic keratosis (AK), basal cell carcinoma (BCC) and baso-squamous cell carcinoma (baso-SCC) on her forehead. She had been treated several times by HpD-PDT, but because the intervals between treatments were of necessity rather short, she had begun to accumulate HpD in her skin and had become quite photosensitive. There seemed to be nothing to lose by trying ALA-PDT, and 'ethics approval' had just been obtained. The treatment turned out to be quite effective. The second patient was an elderly man with a mixture of nodular, ulcerated and superficial BCCs, besides other more serious medical problems. Since his physicians considered him far too ill for surgical removal of the tumours, he was treated by ALA-PDT. The results of these and subsequent clinical studies[36,37] were published in 1990 and 1992.

## 2.3. Principles of ALA-Based PDT

### 2.3.1. Phototoxicity

Three components are required in order to cause phototoxicity in targets such as tumours. These components are

- A photosensitizer: Its role is to capture (absorb) the energy of a photon and then transfer it to another molecule (the substrate). The photosensitizer used for ALA-PDT is PpIX. This is synthesized from its precursor ALA in a five-step biosynthetic process.
- Light: This must be of a wavelength that can be absorbed by the photosensitizer.
- Substrate: This is a molecule that accepts energy from the photosensitizer. The substrate in ALA-PDT is molecular oxygen, $O_2$. The excited substrate (singlet oxygen) initiates chemical destruction of the tumour by a process that involves oxidation.

An ideal photosensitizer for PDT should satisfy the following requirements:

- non-toxic in the absence of light;
- highly efficient at absorbing light energy, and in transferring it to the substrate;
- able to absorb long-wavelength light (above 600 nm) for deep penetration of tissue;
- accumulates preferentially within the target tissue; and
- clears rapidly from normal tissues.

Figure 5 illustrates the essential steps that occur during ALA-based PDT: When ALA is administered either systemically or topically, the photosensitizer PpIX is synthesized by the target tissue more rapidly than it can be converted into haem, excreted or otherwise lost (see following Section 2.3.2). PpIX therefore accumulates in the target tissue. Light is delivered to the target either directly or *via* fibre-based applicators. The resulting cell destruction is either caused by or triggered by an oxidative process.

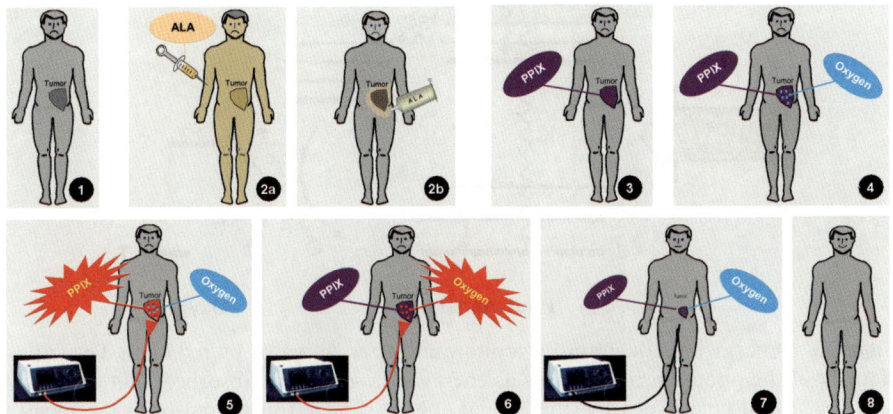

**Figure 5.** (1) Patient with tumour, (2) delivery of ALA (a) systemic (b) topical, (3) ALA has selectively produced PpIX in the tumour, (4) tissue oxygen is already present, (5) light energy is absorbed by and stored in the PpIX molecules (excited states), (6) the excitation energy is transferred from PpIX to oxygen (the oxygen excited by photosensitization is a potent oxidizing agent that is capable of rupturing double bonds found, for example, in cell membrane constituents), (7) following irradiation, the toxic processes that were initiated during irradiation cause tumour cells to die within hours or days. Part of the PpIX has been 'used up' by self-destruction (photobleaching) and (8) successful PDT.

As a photosensitizer, PpIX molecules are able to store absorbed light energy in an excited energy level. This stored energy can be transferred to molecular oxygen, as schematically illustrated in Figure 6. However, once a PpIX molecule has been electronically excited by the absorption of a photon of suitable wavelength, it has several paths by which it can return to the ground state. One path leads only to heat, molecular vibrations that are finally transformed into kinetic energy of neighbouring molecules. Another path involves emission of a photon whose wavelength reflects the energy difference between the upper and the lower electronic states. The process of emitting excitation energy between an excited- and a ground-singlet state is called fluorescence. Both the production of heat (non-radiative relaxation) and fluorescence are very fast (nanosecond) processes.

The third path is the one involved in PDT. This path is called 'intersystem crossing', which requires a spin change of the excited electron. A basic physicochemical principle states that molecules in a 'singlet' manifold (an ensemble of states all having their external electrons in anti-parallel spins) cannot cross over to states in a 'triplet' manifold (an ensemble of states all having their external electrons in parallel spins). However, in higher excited states, the close proximity of states allows a certain 'looseness' of this forbidden spin rule, and there can be a certain degree of 'cross-over' from the first excited singlet state of the photosensitizer to its lowest triplet state. The energy difference between the triplet state of a molecule and its fundamental singlet ground state is much larger than that found in intersystem crossing, and the 'spin forbiddeness' rule

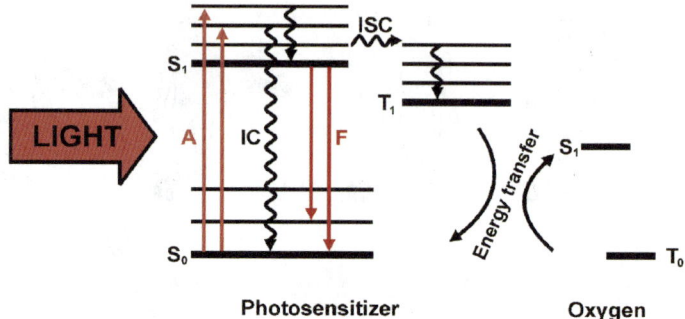

**Figure 6.** Energy level diagram for a photosensitizer activated by red light. The triplet state $T_1$ of the photosensitizer serves as energy storage for the absorbed light energy. If an oxygen molecule is nearby, this stored energy can be transferred onto the oxygen molecule, producing aggressive singlet oxygen. $S_0$, ground state; $S_1$, excited singlet state; A, absorption; IC, internal conversion (loss of some of the excited energy *via* molecular vibrations without the emission of light); F, fluorescence emission; ISC, intersystem crossing between 'singlet' system and 'triplet' system; $T_1$, excited triplet state of the photosensitizer.

is more closely followed. Thus, the excited triplet state stays 'in limbo' (retains its excitation energy) for a much longer time (micro- or milliseconds), in comparison to pico- or nanoseconds for transitions between similar spin states.

Oxygen, in contrast to most molecules, has a triplet state as its fundamental ground state. Thus, the photosensitization (energy transfer) process occurs between the photosensitizer triplet state and an oxygen excited triplet state. This is a transition between two triplet states – an allowed transition. Its efficiency is therefore high. Now, the excited oxygen (in its excited triplet state) quickly crosses over to a slightly lower singlet state (the reactive state), and again trapped in limbo for a relatively long time (a few microseconds). This allows plenty of time for the excited oxygen to collide and chemically react with the double bonds present in many cellular constituents (see Section 2.4).[38,39] Note that for simplicity of presentation, the excited triplet state of oxygen is not illustrated in Figure 6.

The transfer of energy from a molecule of PpIX to a molecule of oxygen returns the PpIX molecule back to its ground state, at which point it is again ready to absorb another photon. The excited oxygen molecule produced by the photosensitization process could not have been excited by a photon directly due to quantum mechanical rules. Thus, PpIX acts as a catalyst, an energy trap that catches the energy of a photon of light, transfers it to an oxygen molecule and then returns to its original, unexcited (ground) state. Activation and deactivation can thus be repeated many times, producing a high concentration of singlet-(activated) oxygen molecules.

However, many molecules of PpIX are destroyed by reacting with the excited singlet oxygen. This destruction of photosensitizer molecules by the excited oxygen is called photobleaching, and is clinically a very important property of ALA-induced PpIX.

### 2.3.2. ALA Metabolism

ALA-hydrochloride is a white powder that dissolves easily in water, giving a clear, acidic solution (pH=2.2 at 1%) with absorption bands in the ultraviolet spectral range. The molecular weight is 167.59 g mol$^{-1}$ (35.45 g mol$^{-1}$ for the chloride ion). The chemical structures of ALA and PpIX are given in Figure 7.

Cells synthesize ALA, and from ALA haem is synthesized, which is the active centre in a variety of proteins such as cytochromes and haemoglobin and enzymes of the respiratory chain. Eight molecules of ALA are used to build four pyrrol rings that are enzymatically connected to form a larger macro-ring, PpIX. Haem is produced by insertion of a ferrous ion.

The following aspects of the biochemical conversion of ALA into PpIX (Figure 8) are relevant for either PDT or FD:

- The synthesis of ALA normally controls the synthesis of haem. ALA is the first committed precursor in the biosynthetic pathway for haem, while haem is the last. The synthesis of ALA is controlled by a feedback mechanism that reflects the intracellular concentration of free (unused) haem. An increase in the concentration of free haem causes a decrease in the synthesis of ALA from glycine and succinyl coenzyme A, and therefore a decrease in the synthesis of haem. However, in the presence of a large amount of exogenous ALA, this feedback mechanism is bypassed, and every step in the biosynthetic pathway then tries to operate at its maximum capacity. Any 'check point' in the pathway may then lead to the accumulation of the intermediate that is synthesized immediately upstream and is the normal substrate for the check point.

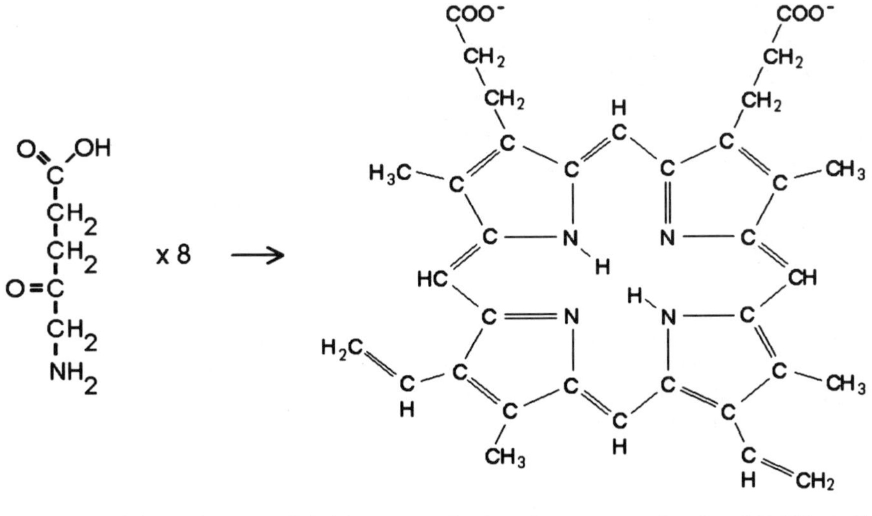

**Figure 7.**   Eight molecules of ALA are required to form one molecule of PpIX *via* the haem biosynthesis pathway.

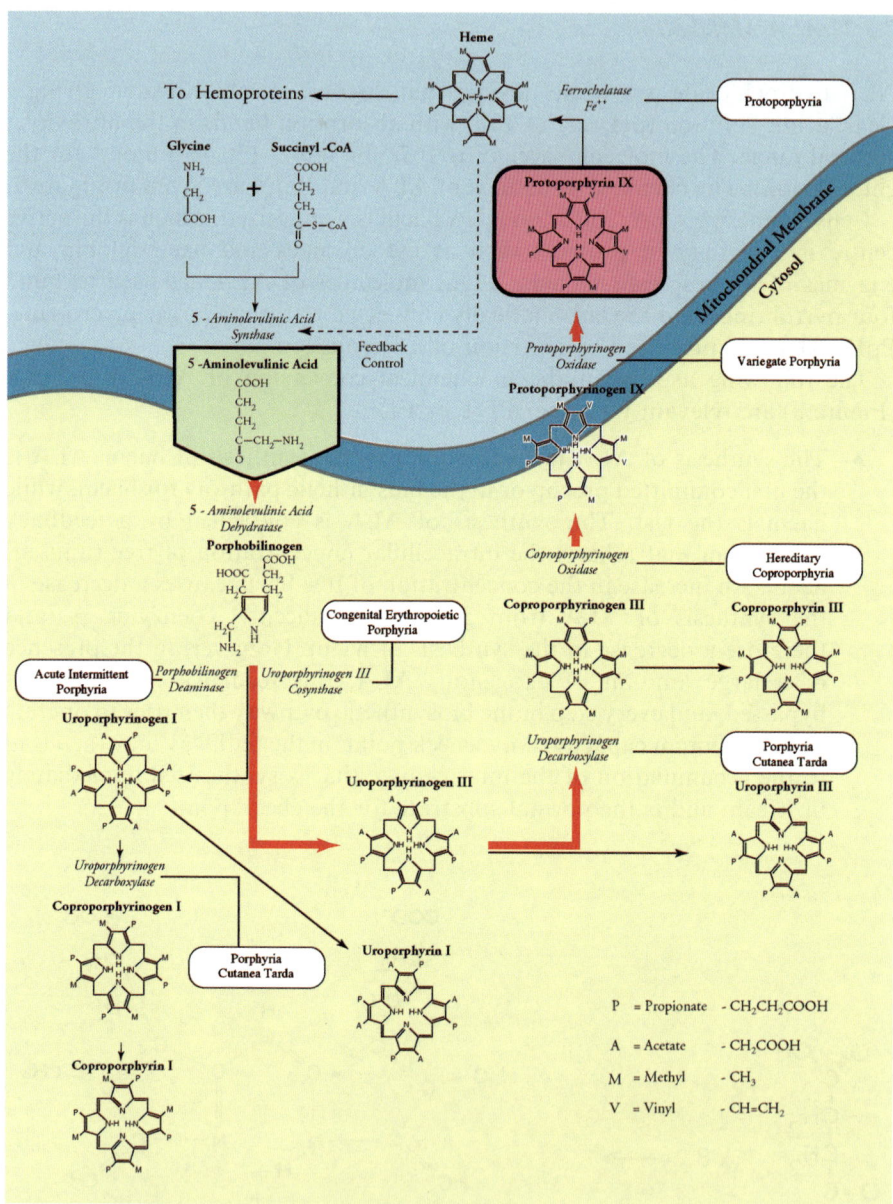

**Figure 8.** Protoporphyrin IX is produced during haem biosynthesis. It accumulates selectively within cells when the normal rate-controlling feedback mechanism is bypassed by the addition of exogenous 5-aminolevulinic acid. (Drawing reproduced with permission from Journal of Clinical Laser Medicine and Surgery, 14/5, 1996, pp. 289–304, Mary Ann Liebert, Inc., publishers).

- Under normal conditions, the check point is the synthesis of ALA, and PpIX does not accumulate. However, in the presence of an excess amount of exogenous ALA, there are two check points – porphobilinogen deaminase (PBGD), which processes porphobilinogen to uroporphyrinogen III, and ferrochelatase, which is responsible for turning PpIX into haem by inserting a molecule of ferrous iron into the centre of the tetrapyrrole ring.

Deviations from normal conditions occur in porphyric patients who suffer from inherited deficiencies of one of the enzymes in the haem biosynthetic pathway. Even in the absence of exogenous ALA, these check points may cause various intermediates in the pathway to accumulate and produce a variety of symptoms.[40] The enzymes that are involved in the different forms of phorphyria are shown in connection with the white 'ovals' in Figure 8. Patients suffering from erythropoietic protoporphyria, for instance, have a deficiency in ferrochelatase (FC), accumulate PpIX, and develop skin photosensitivity.[41] Acute intermittent porphyria is associated with an increased excretion of ALA and porphobilinogen and is characterized by severe neurological abnormalities, often in the absence of skin photosensitization.

The rate at which ALA is converted into PpIX (and eventually into haem) is related to cellular requirement for haem. A reduction in the activity of FC and an increase in the activity of the enzyme PBGD have been found in cancer cells, as compared to their normal counterparts. Decreased FC activity may be related to (either responsible for, or caused by) a shift in tumour cell energy metabolism from oxidative phosphorylation to glycolysis, while the increased activity of PBGD might be connected with an increase in the demand for energy that is now being produced relatively inefficiently by anaerobic processes.

All else being equal, an increase in PBGD would allow more PpIX to be produced, while at the same time the reduction in FC would limit the conversion of PpIX into haem and would therefore tend to cause the accumulation of PpIX. Spontaneous accumulation of PpIX in tumour tissue has indeed been detected,[42,43] although this is a rather occasional observation and may sometimes be due to bacteria that are growing in necrotic tissue. Although PpIX accumulation in tumours in the absence of exogenous ALA is metabolically mediated, it is not determined only by the changes in the PBGD and FC activities.[44]

However, when a suitable amount of exogenous ALA is added, all tumours develop PpIX fluorescence, although different types of tumours vary in the intensity of their fluorescence. Exogenous ALA bypasses the normal feedback control point (the synthesis of ALA) and pushes the haem biosynthetic pathway to operate at its maximum possible capacity. Since tumours have a decreased capacity to convert PpIX into haem, under such conditions PpIX will accumulate.

This is the phenomenon that is exploited by ALA-PDT. In the presence of exogenous ALA, certain types of cells accumulate relatively large amounts of PpIX, while other cells accumulate very much less. Techniques for inducing such a selective accumulation of PpIX in specific types of cells are described in detail in the various clinical chapters of this book.

### 2.3.3. Underlying Biological Concepts

Genetic alterations in cancer cells lead to self-sufficiency in growth signals, insensitivity to growth-inhibitory signals, evasion of programmed cell death (apoptosis), unlimited replicative potential, sustained angiogenesis and tissue invasion and metastasis. Because the malignant phenotype is manifested by unrestricted cell growth both in the tumour itself and in the surrounding tissue (*e.g.* angiogenesis), it is not surprising that there is also an increased requirement for energy by these tissues. The increased energy requirement for neoplastic growth is a general phenomenon of malignancy that should be re-examined in light of recent data on the regulation of haem biosynthesis.

Let us suppose that the decrease in FC activity in malignant cells is the primary defect. Such a decrease would result in a decrease in the production of haem, and therefore in decreased production of the haem-containing enzymes of the tricarboxylic acid cycle. But haem is required by every nucleated cell for the production of energy. Normally, the aerobic metabolic pathway is used to generate high-energy bonds such as ATP, which the cell needs for reproduction and numerous other functions (Figure 9). Since the malignant cell does not have enough haem, it is unable to meet its energy requirements *via* this route. However, there is another pathway that it can use. Anaerobic respiration is less efficient than aerobic, and therefore generates much less ATP per unit of substrate. However, by working overtime at low efficiency ('spinning its wheels'), the cancer cell can produce the required amount of ATP, although in so doing it uses up large amounts of substrate.

Malignant cells show a strong tendency to replace a good part of their aerobic reactions with anaerobic reactions, and cancers tend to 'suck energy'

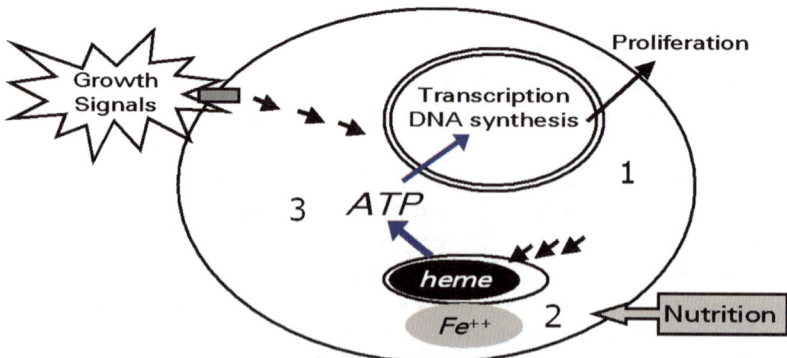

**Figure 9.** Haem synthesis is central for cytochrome activity and ATP synthesis. Anaerobic (1) and aerobic biochemical pathways (for which haem biosynthesis (2) is essential) supply ATP (3) for DNA synthesis and cell proliferation.

from patients (weight loss, unusual fatigue, *etc.*). Partially blocked at the level of FC but with the ALA feedback mechanism still intact, cancer cells would try to synthesize more ALA in an attempt to produce more haem. An increase in PBGD might be another way of attempting to produce more haem. In any case, these attempts at compensation would fail because of the partial block at the FC level, and PpIX might accumulate to some extent even in the absence of exogenous ALA. In its presence, the haem biosynthetic pathway would be primed to generate large amount of PpIX.

Possible reasons for the tumour-selective production/accumulation of PpIX following exposure to 5-ALA have been suggested in several reviews.[28,45,46] One of the suggested mechanisms for this phenomenon is linked to porphobilinogen, since this intermediate in haem biosynthesis has been observed elevated in various human tumours.

PBGD, one of the rate-limiting enzymes in haem biosynthesis, is expressed as two isoforms, a 'housekeeping' isoform present in all cell types and a second isoform expressed only in erythroid cells. Recently, Greenbaum *et al.*[47] have studied the expression and cellular localization of PBGD in cancer cells by the independent techniques of fluorescence immunostaining and the expression of PBGD fused to green fluorescent protein. Much to their surprise, it was discovered that in rapidly proliferating glioma and other cells, a major fraction of the PBGD was found in the nucleus despite the fact that haem biosynthesis occurs in the cytoplasm. Figure 10 reveals nuclear and cytoplasmic localization of PBGD in SH human melanoma cells using immunostaining with anti-human PBGD. During mitosis, chromatids were intensely stained for PBGD in comparison to the interphase chromatin.

These findings suggest a possible dual role for housekeeping PBGD in fast dividing cancer cells, one related to the haem biosynthesis pathway and another coupled to nuclear function, which might be linked to tumourigenesis.

In additional testing, RanBPM, a nuclear Ran-binding protein, was identified as an interacting partner of PBGD in the nucleus. Thus, it was suggested that PBGD possibly has a nuclear role related to transformation and differentiation, in addition to its cytosolic enzymatic activity required for haem

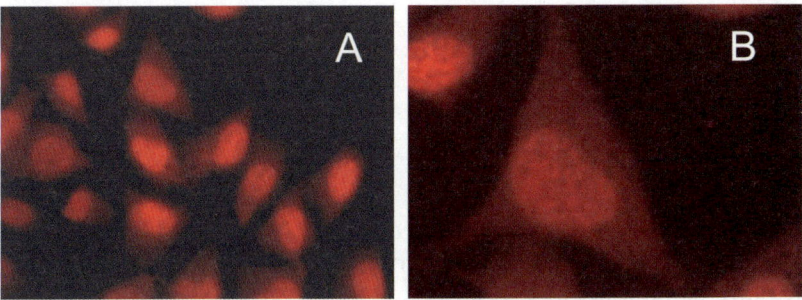

**Figure 10.** Nuclear and cytosolic localization of PBGD in B16 human melanoma cells detected by fluorescence immuno-labelling. Control cells, in which anti-PBGD antibody was withheld from the samples, show no signal. Original magnification (A) ×40, (B) ×100.

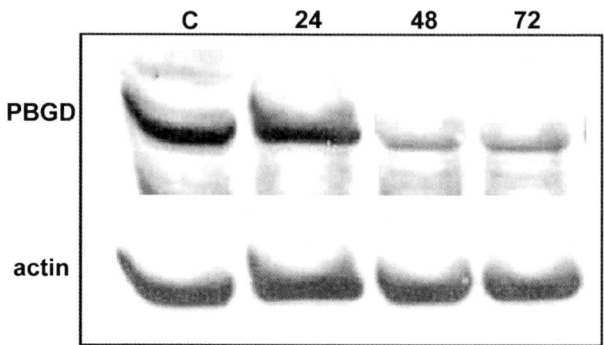

**Figure 11.** PBGD and actin protein levels during butyrate-induced differentiation of B16 human melanoma cells. Total PBGD and actin were detected by Western blotting with anti-PBGD and anti-actin following 24, 48 and 72 h of 2.5 mmol butyrate treatment. C: control.

synthesis. Furthermore, substantially decreased nuclear and cytoplasmic PBGD levels were observed in differentiating cells whose proliferation had been arrested.[48]

Human melanoma cells treated with sodium butyrate, which is a well-known agent of differentiation, are arrested in the $G_1$ phase and show increased melanin accumulation and morphological changeover to mature melanocyte-like cells. Figure 11 depicts the dependence of PBGD levels on the differentiation stage of human melanoma cells. Terminal differentiation achieved after 48–72 h is accompanied by a major reduction in PBGD and therefore in porphyrin biosynthesis capacity.

In conclusion, it has been proposed that PBGD is likely to have a regulatory function that is related to neoplastic transformation, in addition to its enzymatic function in haem biosynthesis. It appears that PBGD is transported in and out of the nucleus and that this trafficking is related to the state of differentiation of the cell. Therefore, enhanced protoporphyrin synthesis by cancer cells is largely dependent upon PBGD enzymatic activity and, in addition, on PBGD's second and as yet unknown function within the nucleus.


## 2.4. Biological Mechanisms

### 2.4.1. Basic Cellular Mechanisms

#### 2.4.1.1. ALA Delivery and Cellular Uptake
As a prodrug, ALA is able to diffuse through skin and other structures to target tissues such as tumours or dysplastic lesions, an effect largely dependent upon the composition of the cellular and extracellular structures. ALA can be given either systemically or topically, which implies the possibility that a broad range of concentrations of the prodrug may be effective.

Attempts to enhance ALA drug delivery when topical applications are used have included debulking tumours with curettage prior to the application of ALA, repeated PDT treatment, iontophoresis, and addition of DMSO or EDTA.[46] By using ALA derivatives such as ALA esters with increased lipophilicity, drug delivery can be enhanced after topical application, and deeper tissue penetration may be achieved.[46]

ALA is taken up intracellularly by an active, energy-consuming mechanism *via* β-amino acid and GABA carriers.[49] Cells of renal or intestinal origin incorporate ALA *via* peptide transporters (PEPT-1 and PEPT-2).[50] Uptake of the charged ALA molecule into the cell can be somewhat improved by the use of lipophilic ALA esters such as methyl- and hexyl-esters, which are not taken up by membrane transporters such as the GABA carrier. Instead, passive diffusion seems to be the method by which some ALA esters enter the cell.[51] The methylesters entered partly *via* transporters of non-polar amino acids and partly by carriers for glycine.[52] Once the ester is in the cell, free ALA is regenerated by intracellular esterases, and the following processes are the same as for non-esterified ALA.

In human cell lines, esterification of ALA with long-chain alcohols has been found to reduce 30–150-fold the amount of ALA needed to reach the same level of PpIX accumulation as with non-esterified ALA.[53,54] However, it must be remembered that experiments in tissue culture and in a living animal system may be quite different, since both serum and extracellular fluid have strong non-specific esterase activities.

### 2.4.1.2. PpIX Accumulation

After uptake/diffusion of ALA (esters) into the cell, ALA enters the haem biosynthesis pathway and PpIX synthesis is stimulated. At this point, it does not matter whether the ALA present in the cell was obtained from ALA esters or was the simple ALA hydrochloride.

Among the many parameters influencing PpIX generation and accumulation, changes in the activity of enzymes in the haem biosynthetic pathway such as FC (conversion of PpIX into haem) and PBGD (synthesis of uroporphyrinogen, a non-fluorescing and non-photoactive precursor with eight carboxylic acid groups) are most important.

Besides ALA (ester) concentration and incubation time, other biochemical parameters such as the rate of PpIX degradation and excretion downstream, intracellular localization of the PpIX,[55,56] intracellular and extracellular pH,[57] state of differentiation,[58] cell type, culture density[51] and the presence or absence of serum in the culture[59] have been shown to influence the efficiency of PpIX formation.

ALA induces elevated PpIX levels on most (but not all) tissues that line body surfaces or body cavities. It can be shown that PpIX preferentially accumulates in the mucosa of the rat colon, but to a much lesser extent in the submucosa and muscle layers.[60] Mucosal selectivity occurs for other organs (stomach, bladder) *in vivo* when ALA is administered either intravenously or orally.[61] Tissues that show no ALA-induced PpIX include muscle (striated, smooth and cardiac), dermis, blood vessels, mature erythrocytes and most (but not all) other tissues of mesodermal origin.[62] However, since many of these mesodermal tissues have multiple components, a more detailed examination might reveal that a minority component in some does in fact show a measurable degree of ALA-induced PpIX.

Malignant tissues usually accumulate more PpIX than do the corresponding normal tissues from which they were derived. Increased PpIX accumulation can be found in many malignant cell lines, as compared to their non-malignant counterparts.[63] There are also certain non-malignant but altered tissues that show enhanced PpIX production in comparison to cells in their normal state. Examples of these tissues are psoriatic, actinic keratotic and pre-keratotic lesions. Thus, while an enhancement of ALA-induced PpIX accumulation is not specific for malignant tissues, in general it does indicate the presence of some abnormality.

Since PpIX is synthesized intracellularly, the relative contribution of microcirculatory damage in comparison with direct cytotoxicity seems to be less significant than with preformed exogenous porphyrin-like photosensitizers.[64]

### 2.4.1.3. PpIX Localization

PpIX is produced in the mitochondria (Figure 12), usually with a higher rate in neoplastic than in untransformed cells.

After longer incubation times, PpIX is redistributed from mitochondria into the cytosol, to the perinuclear region in membrane-rich organelles such as the endoplasmic reticulum and to the nuclear envelope excluding the nucleus.[29,65,66] Localization in the nuclear membrane could be a reason for the moderate dark toxicity detected.[67]

PpIX can be found to some extent in the plasma membrane and in lysosomes, and thus may be responsible for subcellular sites of damage following light irradiation.[68]

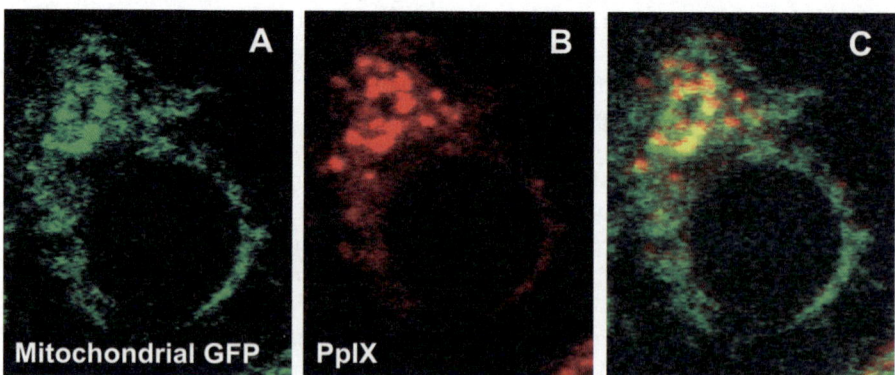

**Figure 12.** PpIX synthesis and localization in C6 glioma cells stably transfected with mitochondrial-GFP and treated for 4 h with ALA. A, mitochondrial localization revealed by mitochondrial GFP; B, PpIX fluorescence; C, the merged image of A + B showing co-localization of GFP and PpIX.

### 2.4.1.4. Photodynamic Damage

The site of localization of PpIX, its concentration and the irradiation protocol largely determine the kind and amount of damage caused upon photoactivation of the photosensitizer.

The localization of PpIX molecules clearly determines the site of action in the cell. As do other porphyrin-like sensitizers, PpIX acts primarily *via* an energy transfer process that generates singlet oxygen. This highly oxidizing reactive oxygen has an action radius of about 10–20 nm and cannot diffuse far enough to initiate damage elsewhere in the cell.[69] Oxidation processes lead to inactivation of proteins and lipid peroxidation. As a result of the damage, and with some variation depending upon the metabolic state of the cell, either important functions and structures of the cell are irreversibly damaged and the cell dies *via* necrosis, or processes for repair and/or apoptosis are induced.

At the molecular level, genes and their products can be induced, up-regulated or down-regulated, depending on the various protocols and cell lines used. Cell protection can be enhanced by induction of the gene for the heat shock protein HSP70, by the up-regulation of the gene for haem oxygenase-1[70] and by increasing intracellular free calcium.[71,72] Elevation of c-Jun *N*-terminal kinase activity, important for cell survival, has been observed, as well as changes in the expression of genes for proliferation such as c-myc.[73] Cells respond to oxidative stress by induction of early response genes, by changes in the expression of genes relevant for apoptosis/survival, such as bcl-2,[70,74] or by phosphorylation of p38 mitogen-activated protein kinase (MAPKs), a stress-responsive protein involved in apoptosis.[73]

Since mitochondria are the main targets of ALA-PDT, at least after short incubation times, a decrease in ATP concentration can generally be found. Mitochondrial benzodiazepine receptors may affect PpIX phototoxicity. Possibly a ligand of benzodiazepine receptor with PpIX has the potential to rescue cells from porphyrin-induced photolysis.[75]

Results to date reflect a variety of possible reactions of different cell lines to different degrees of damage. Current knowledge on signal transduction pathways induced by ALA-PDT is somewhat inconsistent and incomplete, except for apoptosis, which is fairly well understood. The repair processes of PpIX-induced sublethal photodamage take very specific courses for each cell line and treatment condition, and therefore cannot be reduced to a common pathway. If the photodamage cannot be repaired, it leads finally to cell death. Depending on the conditions for each cell line, apoptosis or necrosis can be induced.[76]

The photodynamic effects of ALA-PDT could be further enhanced *in vitro*, when combined with other anti-cancer modalities such as ionizing radiation[77] or hyperthermia.

### 2.4.1.5. Pain During and After ALA-PDT

The occurrence of localized pain during and shortly after exposure to photo-activating light can be prevented by either local or general anaesthesia, but not by opiates, non-steroidal anti-inflammatory drugs or anti-histamines. There appear to be two types of pain – a sunburn-like pain, which responds in the usual way to analgesia, and a neurogenic pain that does not. If the treated tissue lies directly above a major nerve, the neurogenic pain may be experienced throughout the length of that nerve.

The sunburn-like pain responds to the application of cool water, to topical anaesthesia and to analgesia. As with sunburn, the pain may be significant for about the first 24 h after treatment, and then gradually fades away. In contrast, the neurogenic pain responds somewhat to physical stimulation of the treated surface by touch (gentle pressure or gentle rubbing), by a current of cool air or by a spray of cool water. This is the pain that makes the patient want to scratch. It fades away after perhaps 30–60 min, but it can be most annoying and uncomfortable while it is present. It could be caused by the degradation of PpIX by haem oxygenase 2, which localizes to neurons,[78] or by the neuro-toxicity of ALA itself.[79]

Large nerves generally lie deep to the surface, and neurogenic pain may be reduced by using blue or green (superficial) rather than red (deeply penetrating) light to minimize phototoxic damage to the nerves.[80] Reducing the time between the application of ALA and exposure of the target area to the light may be helpful also, since this would reduce the depth of tissue penetration by ALA. It has been reported that pain is reduced by using ALA esters.[81] This could be due to the different uptake routes to the cell: ALA but not ALA-ME uses – on tape-stripped normal skin – the GABA receptors to enter the cell, which transport ALA also to the peripheral nerve endings.[82]

### 2.4.2. Mechanisms of Cell Death

#### 2.4.2.1. ALA-PDT Induces Apoptotic and Necrotic Cell Death of Tumours

ALA-based PDT results in a sequence of photochemical and photobiological processes that cause irreversible injuries to tumour cells, with a major influence on mitochondria. A variety of subcellular targets are affected during light activation of endogenous PpIX, the specific target being determined largely by ALA incubation and irradiation regimes as well as the histological origin of the cancer cells.

Two cellular mechanisms are responsible for cell death: necrosis and apoptosis. Both are observed at lethal and/or supra-lethal doses of PDT. Necrosis is passive cell death or 'cell killing', mainly by plasma membrane destruction, while apoptosis is a physiological cell death, a 'suicide' in which the cell breaks up into membrane-enclosed particles that are engulfed and recycled by neighbouring cells. Necrosis induces an inflammatory reaction of the immune system, while apoptosis does not. Apoptosis and necrosis are both post-transcriptional events that occur within a few hours after the treatment. The balance between apoptosis and necrosis appears to be dependent not only on the intensity of the PDT, but also on many other features such as the genetic and metabolic potential, the site of localization in subcellular compartments and the local concentrations of PpIX. Necrosis is characterized by swelling, distortion and degradation of subcellular organelles, including the mito-chondria, endoplasmic reticulum and nucleus, and finally disintegration of the plasma membrane. Necrotic cell death occurs following a wide variety of subcellular injuries that are induced by singlet oxygen and other activated oxygen species. If oxidation processes in the cell cause too much damage and/or the intracellular energy level required for apoptosis or even repair processes decreases under the threshold specific for that cell, necrosis takes place. While the genetic and biochemical pathways of necrosis that occur downstream of the individual insults are not fully resolved, some signalling pathways that involve Jak1 and Tyk2 protein kinases have been described as inducing necrosis.[76,83]

The apoptotic pathways activated by photosensitization of cancer cells using exogenous and endogenous sensitizers have recently been reviewed by Oleinick *et al.*[84] and Piette *et al.*[85] Several steps in the induction and execution of apoptosis require high levels of ATP, which might be affected by rapid mitochondrial damage. Whenever a rapid drop in the cellular ATP pool was observed, caspase activation and nuclear fragmentation appeared as clear indications of apoptosis.[84-86] Since most of the apoptosis signalling pathways converge in mitochondria, which play a key role in apoptosis execution, a limited photodamage to the mitochondria themselves may be a strong apoptosis inducer. ALA-PDT is a good candidate in this respect, since PpIX is generated in the mitochondria and after relatively short incubation times can cause phototoxic damage there.

In general, the damage produced by *in vivo* PDT may activate endonucleolysis and chromatin condensation. Apoptosis is therefore an early event in tumour-shrinkage following PDT. Several photosensitizers currently used in clinical or pre-clinical studies localize in or have a major influence on mitochondria, which are directly involved in triggering the apoptotic pathway. Several signal transduction apoptotic pathways are activated during PDT both *in vitro* and *in vivo*. Some of these signalling processes initiated by mitochondrial damage are stress responses aimed at cell protection, while others are likely to contribute to cell death.[87]

The apoptosis rate in cell killing by PDT varies considerably according to cell type and the structure of the endogenous or exogenous sensitizer molecules. A typical *in vitro* study on an HL60 cell model using ALA-based photodynamic treatment revealed the induction of signalling processes that lead to two apoptotic processes in parallel: one representing the mitochondrial pathway and the other involving disruption of calcium homeostasis and activation of the endoplasmic reticulum stress-mediated pathway. The disruption of mitochondrial membrane potential was paralleled by a decrease in ATP level, unmasking of the mitochondrial antigen 7A6, release of cytochrome c into the cytoplasm, activation of caspases 9 and 3 and cleavage of poly(ADP-ribose) polymerase, which was followed by DNA fragmentation. In another case (chronic myelogenous leukaemia derived cell line K562), early apoptotic features resulted in the down-regulation of Bcl-2 expression, decrease of the mitochondrial and plasma membrane potential and the release of cytochrome c into the cytoplasm, while late apoptotic events were lacking. Neither caspase-3 activation nor DNA fragmentation occurred, but rapidly progressing cell necrosis resulting from plasma membrane damage was observed instead.

Induction or up-regulation of anti-apoptotic heat shock proteins could be responsible for the inhibition of the apoptotic process upstream of caspase activation. Moreover, signal transduction pathways involving cyclin-dependent kinase(s) play an important role in arresting cells and delaying apoptosis.[88,89]

Similarly, light-activated PpIX was found to induce apoptosis in breast cancer cells when the PpIX was mainly localized in mitochondria, but led to necrosis when PpIX diffused to other sites, including the plasma membrane.[90]

*In vivo* ALA-PDT of AK revealed that one day after PDT, all layers of the epidermis exhibited slightly degenerative necrosis, with chromatin condensation in the lower layer of the epidermis. TUNEL staining revealed that apoptosis was involved in tumour cell death after PDT in patients with AK. Necrosis in all layers of the epidermis and lymphocyte infiltration in the dermis were found 3 days after PDT.[91] Similarly, ALA-PDT of nasopharyngeal carcinoma cells *in vitro* induced 80% cell death *via* apoptosis within the first 12 h following irradiation, with necrosis accountable for less than 20% of the cell death.[92]

In summary, the current data concerning photodynamic cell death following ALA treatment of tumours show that both apoptosis and necrosis are seen side by side in the treated tumours, which reflects physical parameters such as local singlet-oxygen production during the treatment interval in addition to biological parameters that depend on tumour origin, PpIX synthesis capacity, subcellular accumulation and cell metabolism.

The amount of apoptosis/necrosis can be shifted, at least at the cellular level, by changing the protocol and/or the cellular energetic status. Since moderate necrosis stimulates the immune system and even induces anti-tumour immune reactions, this mode of cell death should not be completely avoided by selecting only apoptotic conditions. On the other hand, massive necrosis may result in generalized shock. The advantage of apoptosis induction is that the patient can be treated with lower doses than required for necrosis. However, with our current state of knowledge, no general recommendation can be made.

### 2.4.3. Interaction with Tissue

#### 2.4.3.1. Tumour Differentiation and ALA-PDT
The relationship between tumour cell differentiation and PpIX accumulation during ALA-PDT is poorly defined. In an *in vitro* model of colon carcinoma CT26 with cell differentiation induced by sodium butyrate, differentiation caused a reduced accumulation of endogenous PpIX.[86] On the other hand, murine 3T3 L1 pre-adipocytes increased their capacity to accumulate PpIX as they became more differentiated.[93] There does not seem to be any general rule about the effect of differentiation on ALA-PDT.

#### 2.4.3.2. Hyperthermia, Tissue Oxygenation and ALA-PDT
Hypoxia, often present in the centre of tumours, significantly reduces ALA-induced PpIX synthesis and PDT efficiency in comparison to well-oxygenated conditions. The magnitude of the effect of hypoxia on PpIX synthesis is

dependent on cell density and proliferation rate. A dramatic decrease in PpIX fluorescence has been observed for high-density unfed-plateau cells under hypoxic conditions.[94]

On the other hand, during ALA-PDT, hyperthermic conditions are produced. This heating of skin tumours possibly affects blood supply, tissue oxygenation and development of hypoxic conditions. Tissue oxygenation, one of the key dosimetric factors involved in clinical PDT, is dependent upon tissue temperature during irradiation. Surface temperature differences between normal and tumour tissues after exposure to 170 J cm$^{-2}$ light dose have been reported to be 3.3 + /−0.5 °C in the forehead areas, 2.5 + /−0.4 °C in the nose areas and 0.8 + /−0.3 °C in the ear areas of patients with solar keratoses (SK) and BCC.[95]

A large increase in tumour oxygenation (pO$_2$ from 3 to 9.5 mm Hg) was monitored during ALA-PDT treatment, with a small residual increase that returned back to baseline levels by 48 h after treatment. The increased tumour pO$_2$ was attributed primarily to altered local blood flow, but decreased local metabolic oxygen consumption owing to cellular damage may contribute to the overall effect.[96]

It has been suggested that IR-induced hyperthermia in combination with ALA-PDT would enhance the photodamage of the treated tumour.[97] A significant increase in tumour damage has indeed been observed after simultaneous application of ALA-PDT and IR-induced hyperthermia, which caused a mild heating of the tumour to 39–43 °C at a 3-mm depth. An enhanced tumour growth delay, based on tumour volume measurement, was observed, indicating a significant improvement in the anti-tumour effect after application of ALA-PDT+IR conditions. It should be noted that photothermic treatment of pigmented B16 mouse melanoma tumours without PDT induces pronounced damage to endothelial cells, with swelling of mitochondria, melanosomal disruption, and changes of the entire structure of tumour microvasculature.[98]

In summary, hyperthermia develops during ALA-PDT treatment as a direct consequence of the non-coherent red-light illumination, which normally delivers some IR radiation also. The mild hyperthermia increases blood flow in microvasculature and oxygen supply to the target tissue, and might thus have a beneficial outcome to the PDT treatment. When using broadband red light, one might consider not filtering out the IR component in the light source.

*2.4.3.3. Photobleaching of PpIX, Photoproducts and Light Fractionation Effects*
Photobleaching, the loss of PpIX fluorescence during irradiance and the formation of hydroxyaldehyde-chlorin photoproduct, is seen during ALA-PDT, especially when using low-irradiance doses. Fluorescence monitoring during clinical ALA-PDT of patients with superficial and nodular BCCs, SCCs and Kaposi's sarcomas indicate that photoirradiation reduces the 635 nm PpIX fluorescence signal in all tumours. The rate of PpIX photobleaching in superficial BCC and SCC tumours is significantly higher than in large nodular BCC tumours. These differences in the kinetics of fluorescence reduction have been attributed to tumour thickness.

After a short post-irradiation interval, recovery of PpIX fluorescence has been observed in large and deeply penetrating BCC and Kaposi's sarcoma lesions.[99] A similar process occurs also in normal skin. ALA is not photoactive, and any ALA that had not yet been converted into PpIX was therefore not destroyed when exposed to the photoactivating light. The keratin layer and the stratum spinosum that lie above the basal layer can function as a slow-release trap for the ALA, so that it is quite routine to wipe out the PpIX fluorescence by photobleaching only to see more PpIX fluorescence appear a few hours later. However, this PpIX fluorescence generally is restricted to the normal skin surrounding the tumour, presumably because the normal skin still remains capable of generating PpIX while the tumour does not.

The process of hydroxyaldehyde-chlorin photoproduct generation during photobleaching can be monitored by an increase in fluorescence emission centred at about 672 nm. In addition, the appearance of a fluorescence emission peak near 620 nm is consistent with accumulation of uroporphyrin/coproporphyrin in response to mitochondrial damage. The kinetics of the fluorescent photoproduct exhibits irradiance dependence, with greater peak accumulation at lower irradiance. These findings are consistent with an *in vivo* predominantly oxygen-dependent photobleaching reaction mechanism, which provides spectroscopic evidence that PDT delivered at low irradiance deposits a greater photodynamic dose for given light fluences than at higher irradiance.[100,101]

Light fractionation has been shown to enhance the photodynamic effect of ALA-PDT in some *in vivo* cases. For instance, a single short interruption of 150 s in the light irradiation may dramatically reduce the net light dose required to achieve extensive necrosis as reported for an experimental colon tumour model.[102] Similarly, in a small number of patients, light fractionation during ALA-PDT improved the effectiveness of the treatment for oesophageal cancer, but increased side effects such as the occurrence of mild oesophageal stenosis.[103] On the other hand, in an experimental animal model of amelanotic melanoma and ALA-PDT, fractionating the PDT or reducing the light intensity appears not to have been successful with regard to the therapeutic effect 28 days after PDT. It was concluded that there was no significant effect of fractionation in that particular model.[104]

In conclusion, the available clinical data show some improvement on the effectiveness of ALA-PDT from fractionated light doses. Thus, when patient convenience allows prolonged treatment sessions, one might consider fractionating the dose of light.

*2.4.3.4. Alteration of the Immune Response Following ALA-PDT*
The involvement of immune responses in the clearance of tumours after ALA-PDT is a challenging question. It is conceivable that adding exogenous ALA to a target tissue will induce PpIX synthesis in the entire repertoire of cells, including cells of the immune system. Both tissue macrophages and dendritic cells produce and accumulate PpIX immediately, without preceding activation, in contrast to lymphocytes that accumulate PpIX only when activated by

specific antigens.[105] ALA-PDT treatment of lymphocytes exposed to an antigen induced selective eradication of these antigen-primed cells without harming resting cells.[106] Dendritic cells that accumulate the photosensitizer following ALA treatment are functionally altered by the cytosolic PpIX. Antigen-presenting cells do appear to have a decreased potential to present antigens to lymphocytes following ALA–PDT. It has been proposed that specific immune responses can be modulated by PDT to selectively eliminate activated T-cell subsets mediating graft- *versus* host-disease without harming resting T-cells.[107] Thus, potentially, ALA-PDT could also be used in transplantation or in treating autoimmune diseases, and information should be obtained on how this photosensitizer affects all immune cells.

Recently some effort has been made to investigate whether and how PDT can be used as anti-tumour vaccine, since in many cases immunostimulation could be observed following PDT. Even induction of a cytotoxic T-cell response by PDT-generated murine tumour cell lysates could be found *in vivo*, together with maturation of dendritic cells and IL-12 expression.[108] This PDT vaccine could be used to efficiently prevent growth of tumour cells. The mechanisms for this effect are not yet clear, but it can be suggested that heat shock proteins play a role.

Eradication of SCC of the skin by ALA-PDT depends upon an intact immune system. Kidney transplant patients on full immunosuppressive therapy and patients with poorly controlled AIDS have defective immune systems, and also a defective response to treatment of their cancers by ALA-PDT. On the other hand, kidney transplant patients whose immunosuppressive treatment has been reduced to very low levels show a response that is indistinguishable from normal, as do well-controlled patients with AIDS. Patients who have epidemic Kaposis sarcoma (AIDS-related) respond very poorly when treated with ALA-PDT, while patients with the age-onset variety do very well, sometimes clearing distant lesions that were not being treated as well as the lesions that were. There is a definite link between ALA-PDT and the immune system, and this link has clear clinical relevance.

### 2.4.3.5. PpIX Pharmacokinetics

After a bolus injection of ALA, its concentration in plasma decreases rapidly with a half life of less than 1 h[109–111] owing to an efficient excretion into the urine and metabolization in the liver and some other organs.[112,113] From the liver, synthesized porphyrins are released into the blood circulation, PpIX is also excreted into the intestines *via* bile.[51] After oral administration, plasma levels of ALA peak at approximately 30 min with excretion half lives only slightly longer. Interestingly, ALA bioavailability is in a high range of 40–60% compared to intravenous application,[109,111] rendering oral uptake of ALA the preferred form of systemic application.[61]

The build-up of PpIX upon a bolus-like delivery of ALA varies broadly for different types of tissue, both in peak intensities and in times to peak. As far as normal tissues are concerned, mucosal epithelia show a much higher accumulation of PpIX at moderate doses of ALA than underlying stroma and muscle.[51,61] Differential accumulation of PpIX in normal tissues can be clinically used to identify the small parathyroid gland by intraoperative fluorescence detection.[114,115] Normal brain synthesizes very little PpIX,[113,116,117] human cortex only minute amounts, providing good contrast to malignant glioma.[118,119] However, considerably more PpIX is found in brain tissue with the hexyl derivative of ALA.[120] Liver metabolises ALA efficiently and PpIX is detected in comparably high amounts.[112,113] The times needed to reach maximum PpIX concentration vary typically from 30 min to 3 h for tissues in mice and rats following *i.v.* application, between 2 and 4 h after oral delivery.[112,113]

Clinical measurements of PpIX pharmacokinetics have been made in blood plasma and from some easily accessible sites by fluorescence spectroscopy after different routes of ALA delivery.[111,121–124] Perhaps the most important result is that PpIX has returned to baseline levels within 48 h, wherever measured and irrespective of the route of administration.[121,123,125] Topical forms of ALA delivery (skin ointment, bladder instillation, lung inhalation) resulted in considerably less systemic PpIX compared to oral or intravenous uptake. For instance, Rick *et al.*[123] compared oral *versus* inhalation and intravesical delivery of ALA (40 mg kg$^{-1}$ bw, 500 mg (10%) and 2% solution, respectively). Oral ALA resulted in a more than 50-fold PpIX concentration in plasma compared to inhalation. Instillation gave the earliest PpIX plasma peak and a further reduced concentration. Their results are summarized in Table 1. Times to PpIX peak for oral ALA delivery were shorter in plasma (6.7 h) compared to skin (6.5–9.8 h) measurements. Interestingly, the strongest skin fluorescence was encountered on the lips. The topical routes of ALA application lead to PpIX plasma levels that are within or only slightly above the range of normal subjects (0–8 μg L$^{-1}$).[126] Side effects owing to circulating PpIX are therefore highly improbable for topical application. Systemic delivery, however, leads to a significant increase of both, ALA and PpIX levels and side effects are observable: nausea and vomiting, hypotension and tachycardia are the most frequently reported side effects after oral uptake of at least 30 mg/kg body weight of ALA.[123,127,128] A transient elevation in liver function tests can also be detected at lower doses (20 mg kg$^{-1}$). Patients with porphyria are usually excluded from ALA-FD or PDT to avoid the risk of inducing an acute phase or unpredictable PpIX levels. A general skin photosensitivity has to be considered for up to 48 h after oral uptake of ALA, including precautions to prevent exposure to bright day or excess artificial room light during this time period.

**Table 1.** Averaged results of curve fitted PpIX plasma kinetics (corresponding standard deviations in brackets). Mean maximum concentration $c_{max}$ (μg L$^{-1}$) was reached at the mean peak time $t_{max}$ and decreased to values lower than 5% of the maximum at $t_{0.05}$ (n.d.= not detected)

| Application | $t_{max}$ (SD) (h) | $t_{0.05}$ (SD) (h) | $c_{max}$ (SD) (μg L$^{-1}$) |
|---|---|---|---|
| Oral ($n = 10$) | 6.7 (0.8) | 35 (4.6) | 742 (87) |
| Inhalation ($n = 5$) | 4.1 (0.8) | 21 (7.8) | 12 (2.9) |
| Instillation ($n = 5$) | 2.9 (0.5) | n.d. | 1 (0.8) |

If the treatment is performed without anaesthesia, local pain may be a limiting factor for the applicable light intensity. This issue mostly concerns dermatological ALA-PDT and is described in more detail in the corresponding clinical chapter.

There was some discussion, whether the PpIX measured in a specific tissue is synthesized locally in that tissue or is accumulated from circulating PpIX that has been synthesized e.g. in the liver.[37,129] Since injected PpIX and 5-ALA-induced PpIX show very different tissue specificity, one may assume that the main part of PpIX in the tissue is indeed synthesized in situ.

When ALA is applied topically to patients with AK, BCC or psoriasis, and the PpIX is monitored via surface-detected fluorescence, a three-component pharmacokinetic profile is observed.[130] During the first 3–5 h, there is a rapid increase in the ALA-induced PpIX fluorescence. This initial phase is predominantly due to the overall rate of bioconversion of ALA into PpIX, including the rate at which ALA penetrates the skin barrier. After approximately 3 h of ALA application, there is normally a smaller rate of increase of PpIX fluorescence intensity. Often, at this time, there appears a second phase in the pharmacokinetic curve where the observed fluorescence intensity is either quite constant or fluctuates. Such a phase is also observed in surface-detected ALA-induced PpIX fluorescence from normal human skin,[131] and probably reflects the balance between the rate of formation of PpIX (including the rate of delivery of ALA to the reaction site, the mitochondria) and the rate of clearance of PpIX from the measurement site. This second phase was often observed to be longer for thick psoriatic plaques. There may be several reasons for this: thick plaques may have a different rate of delivery of ALA to the site of bioconversion than thin psoriatic lesions; they may provide a large reservoir of ALA in the stratum corneum; or their clearance rate may be lower due to less-efficient clearance mechanisms from the stratum corneum. PpIX may contribute most strongly to the observed fluorescence signal as it is closer to the surface. The last phase that can be observed in most of the pharmacokinetic curves is the clearing component. At this stage, a quasi-first order clearance can be observed, with characteristics similar to the clearance of PpIX from animal skin.[132] At this time, the ALA reservoir in the skin has been depleted, and the observed clearance rate is a reasonably good indication of how fast the body is cleared of the extra PpIX that was induced by the exogenous ALA.

### 2.4.4. Advantages of ALA Versus Pre-Formed Photosensitizer for PDT and FD

#### 2.4.4.1. Compatibility

When pre-formed photosensitizers are introduced into the body, it usually treats them as foreign material that must be detoxified and eliminated. However, there is no pathway dedicated to this process, which therefore may take a significant period of time.

While PpIX has no natural function as a photosensitizer, it is the only porphyrin that is natural to the body. Consequently, when one temporarily induces abnormally high levels of PpIX in tissue, natural control mechanisms ensure that the extra porphyrin will be eliminated quickly by established pathways until normal equilibrium concentrations are re-established. Thus PpIX is truly unique in terms of its compatibility to the human body.

### 2.4.4.2. No Overdosing of the Photosensitizer

Since the PpIX photosensitizer is produced *via* a biosynthetic route, there is a natural ceiling or threshold concentration above which the body will not convert any more ALA into PpIX. Thus one only needs to be concerned with the toxicity level of ALA itself. It produces no detectable neurotoxicity even at a concentration of 1.5 mM, which is a much higher concentration than that typically used in ALA-PDT.[133] *In vivo* and *in vitro* studies have found that near lethal intraperitoneal injection of ALA only results in a transient depression of motor nerve conduction.[133] ALA also failed to inhibit neurite outgrowths in chick embryo neuroblasts even at high doses.[134]

Thus it is not critically important to control the administered concentration of ALA in comparison to pre-formed photosensitizers, since the ALA is relatively non-toxic and the biochemically induced PpIX concentration is self-limiting.

### 2.4.4.3. Photobleaching

PpIX is one of the most photolabile porphyrins known. During the course of ALA-PDT, PpIX molecules react with some of the singlet oxygen that was produced *via* the photosensitization process, and can be converted to a product (or products) that can no longer photosensitize and has a quite different absorption spectrum. This photobleaching of PpIX has several clinically important consequences:

(i) PpIX functions as an optical filter, screening out all wavelengths that can activate (be absorbed by) PpIX. The presence of PpIX in the upper layers of a thick tumour will prevent some of those wavelengths from reaching the deeper layers. Photobleaching destroys the optical filter effect and thus permits deeper penetration by the light of the target tissue.

(ii) PpIX that has been photobleached is no longer a photosensitizer. During treatment, the relatively small amount of PpIX that accumulates within normal cells is inactivated by the light long before it can cause more than transient damage to the normal tissues. This makes it possible to give a lethal dose of light to the target tissue without destroying adjacent or overlying normal tissues. Since light intensity decreases with increasing thickness of tissue, and since the total dose of light is the product of intensity × duration, a photosensitizer that does not photobleach readily could cause lethal phototoxic damage to the overlying normal tissues if an attempt is made to destroy deeper tissues by increasing the duration of exposure.

This photobleaching phenomenon has some very practical applications. One can select a dose of ALA and a duration of ALA exposure that will produce only mild and transient damage to normal tissues within the treatment field. Then if one overdoses with light, the additional light will not cause additional damage to the normal tissue (since all of the PpIX in the tissue has already been inactivated), but it will add to the effective depth of cell killing. A similar procedure greatly simplifies light dosimetry calculations involving complex curved surfaces such as the nose and adjacent cheeks. It is almost impossible to give every part of such an area the same dose of photoactivating light. One normally chooses the most deeply shadowed area of the field and then calculates the required light dose as though that area were the standard. Less shaded parts of the treatment field will receive an excess of light, but that causes no problems because the photosensitizer in the normal tissues in the less shaded areas will have been inactivated by being photobleached. It is necessary only to arrange conditions such that the concentration of ALA in the normal tissues cannot cause more than transient damage when exposed to a large excess of light. This is done by modifying the concentration of ALA that is applied and/ or the interval between its application and the exposure to light.

### 2.4.4.4. Versatility of Administration
No other photosensitizer can be applied both topically and systemically. ALA readily penetrates the epidermal layer of human skin, a point that is especially important for dermatological applications. It can be administered by topical application, by mouth, as a suppository, by aerosol, by injection into a body cavity, or by intradermal, subcutaneous or intravenous injection.

### 2.4.4.5. Photodetection
ALA can be used to detect tiny areas of abnormal tissue by ALA-induced fluorescence. These will not necessarily be malignant, but they certainly will be abnormal. If the skin surface is involved, the detection procedure is rapid, simple and safe. Examination of surfaces such as the bladder wall or the oesophagus is more complicated and requires a special scope.

### 2.4.4.6. Rapid Evaluation of Effectiveness of Other Therapies
Since the synthesis of ALA-induced PpIX is dependent upon intact mitochondrial function, the failure to synthesize PpIX is an indicator of decreased metabolic activity. This can be used to evaluate the effectiveness of radiation therapy or chemotherapy. For example, 2% ALA applied to the skin of a patient with cutaneous T-cell lymphoma (mycosis fungoides) may allow visualization of the fluorescence produced by masses of tumour cells lying within or under the skin. A decrease in the intensity or area of fluorescence may reflect a positive response to radiation therapy or chemotherapy. Similarly, the response to chemotherapy or hormone therapy of cutaneous or subcutaneous nodules of breast cancer that are too small to detect by palpation may allow rapid evaluation of its sensitivity or resistance to a specific agent. The response of leukaemia to therapy can be followed by inducing PpIX *in vitro* and by measuring its intensity *via* flow cytometry. In all cases it is essential to use fluorescence intensity standards.

*2.4.5. Limitations of ALA for PDT and FD*

At present, there are two main limitations to ALA-PDT. One limitation is related to its spectral region of absorptivity. PpIX has a long-wavelength absorption maximum at 635 nm, which is not the optimal wavelength for light penetration through human tissue. The deepest 'optical window' in human tissue would be around 800 nm. Thus the depth of light penetration is limited, which limits the treatment depth. A similar situation exists for the depth of drug penetration in tissue. ALA does not penetrate as deep in tumour tissue as one would like in order to eradicate large tumours. For example, small superficial lesions of AK are well treated by ALA-PDT. However, larger lesions such as nodular BCCs often show only partial response when treated with ALA-PDT. One approach that seems to show promise for this last limitation is to use an ester derivative of ALA. The different polarity of ester derivatives seems to facilitate deeper penetration into tumour tissue. Once inside the tissue, the ester derivatives are enzymatically cleaved to produce ALA *in situ*. Thus ester derivatives of ALA may permit destruction of larger tumour volumes.

Finally, the two steps – ALA uptake into the cell and PpIX formation – are more difficult to control than drug uptake only.

## 2.5. Light Sources for PDT

*2.5.1. Basic Requirements*

For PDT, the tissue must be irradiated with light of appropriate wavelengths (within the absorption spectrum of PpIX) and dose (to produce enough singlet oxygen to fully eradicate the neoplastic tissue). When endoscopic applications are necessary, the activating light has to be delivered through thin optical fibres, and laser light sources best meet such requirements. Their main disadvantage is the high price tag associated with some lasers. Therefore, incoherent light sources (lamps, LED-based light sources) have also been developed and hold much promise, especially for dermatological applications. A comprehensive review of light sources for PDT has been published by Brancaleon and Moseley.[135] A list of suppliers is provided in the appendix.

*2.5.1.1. Lasers*
Light Amplification by Stimulated Emission of Radiation produces light beams with low divergence (easily focused), with coherent light (synchronized waves) and with a single colour (monochromatic). Laser light can be produced from gases, liquids and solids. Some types produce a fixed wavelength only, while others are tuneable over a broad range.

For PDT with fibre-based applicators, laser light offers the advantage of being efficiently injected into fibres. The most cost-effective, robust and long-lasting laser systems suitable for ALA-based PDT are certainly diode laser

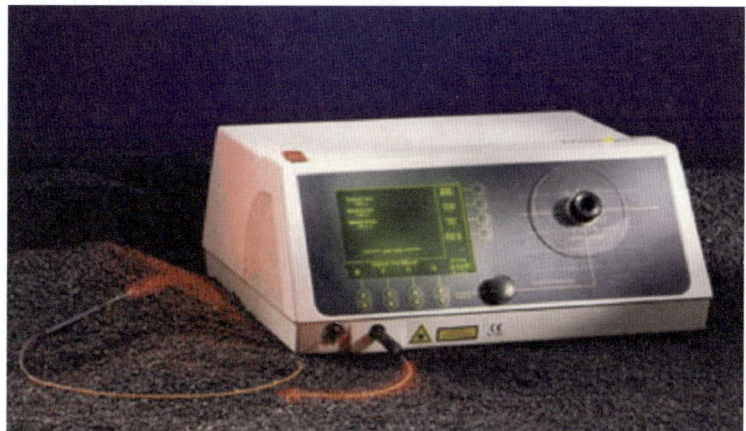

**Figure 13.** Diode laser system for ALA-PDT with fibre-based applicators. Ceralas PDT (www.biolitec-us.com). Image with permission of BioLitec AG.

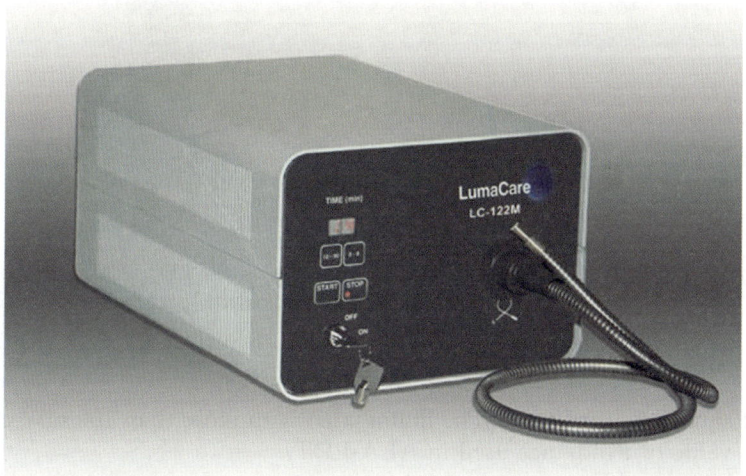

**Figure 14.** Broadband light source that can be used for fluorescence diagnosis and PDT. Light is delivered through special fibre bundles. LumaCare™ LC-122M (www.lumacare.com). Image with permission of LumaCare.

systems (Figure 13). Also, many laser diodes can be successfully coupled to provide large surface irradiation.

### 2.5.1.2. Lamp Systems

For non-endoscopic PDT applications, lamp systems provide filtered light that can match the absorption band of PpIX (Figures 14 and 15). They are more cost-effective and easier to handle than laser systems. For PDT in dermatology, suitable lamp systems including light dosimetry are available, where several

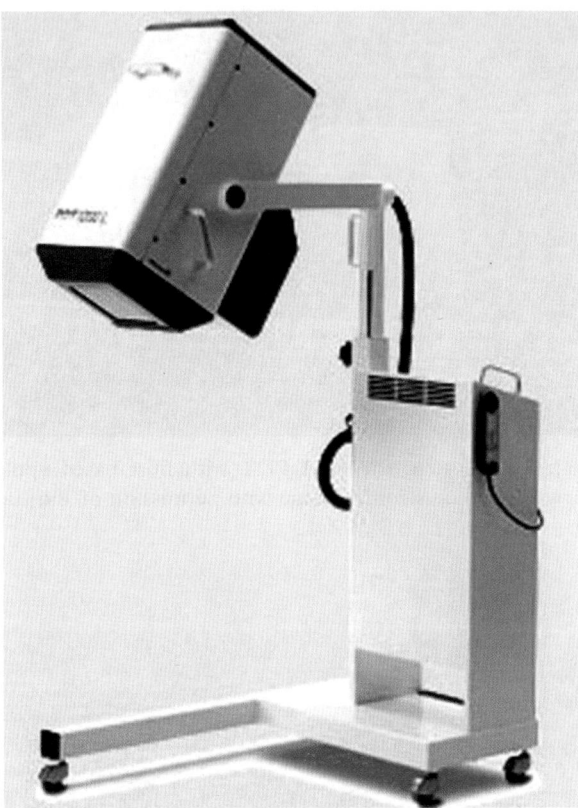

**Figure 15.** Incoherent red light source, frequently used for PDT of skin lesions. PDT
1200 L (www.waldmann-medizintechnik.com). Image with permission of Waldmann
Medizintechnik.

tens of square centimetres can be illuminated with more than 100 mW cm$^{-2}$ (a
typical value of irradiance used in ALA-PDT).

Highly focused lamp systems enable transmission of several watts of broadband light through fibre bundles or fluid light guides.

### 2.5.1.3. LED Sources

Light Emitting Diodes (LEDs) are semi-conductor light sources emitting
quasi-narrow bandwidth light (*ca.* 30 nm) with high efficiency. A single LED
does not emit much power, but LEDs are very cheap and small, and several
of them can be stacked in arrays that are very well suited for superficial
PDT. Recently, LED arrays of several tens of diodes in a single package
have been developed, emitting in excess of 100 mW (Figure 16). Currently, light
from LEDs cannot easily be fed into thin fibres, but further development
of semi-conductor light sources may well be able to replace lasers for PDT in
the future.

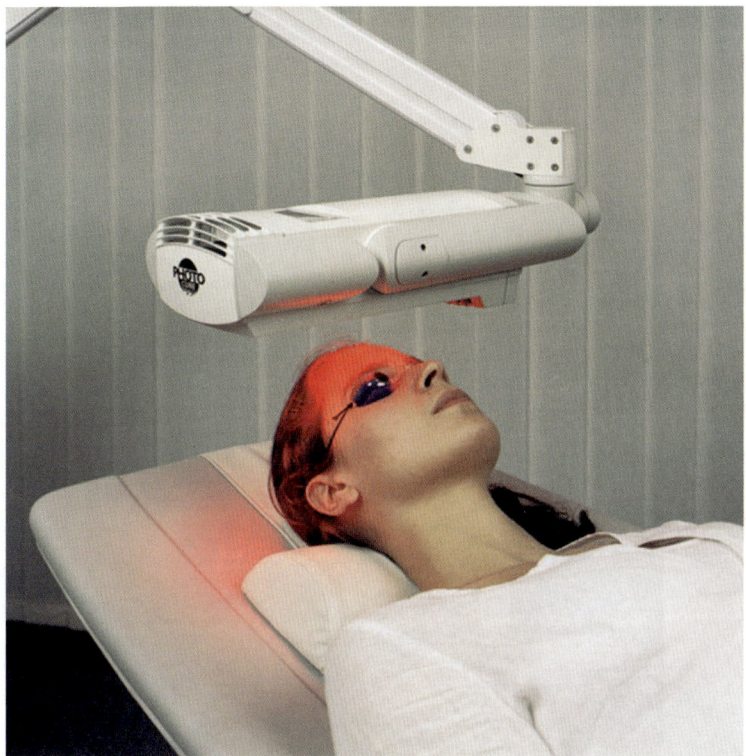

**Figure 16.** LED light source for PDT of skin lesions. Aktilite CL 128 (www.photo-cure.com). Image with permission of PhotoCure ASA.

### 2.5.1.4. Fluorescence Sources

For very shallow lesions, *e.g.* early AK, it is possible to use fluorescent lights that have been specifically designed to produce blue light that corresponds to the Soret band of PpIX. One example of such a device is shown in Figure 17.

### 2.5.2. Applicators

The production of light energy with suitable wavelengths and intensity for PDT purposes is quite straightforward. However, in order to deliver this light to the tissue in the most efficient way, applicators at the distal end of the light source are often needed. Successful PDT requires light application systems allowing for a homogenous irradiation of the target tissue. For surface irradiation, quasi-flat (*e.g.* skin), cylindrical (*e.g.* oesophagus, bronchi) or spherical (*e.g.* bladder, resection cavities) geometries are distinguished. For bulk tumour tissue (*e.g.* brain, prostate, pancreas), light application is performed interstitially by insertion of multiple fibres into the tumour (Figure 18). Market availability of light diffusers certified for human use is still rather limited at this stage (see appendix).

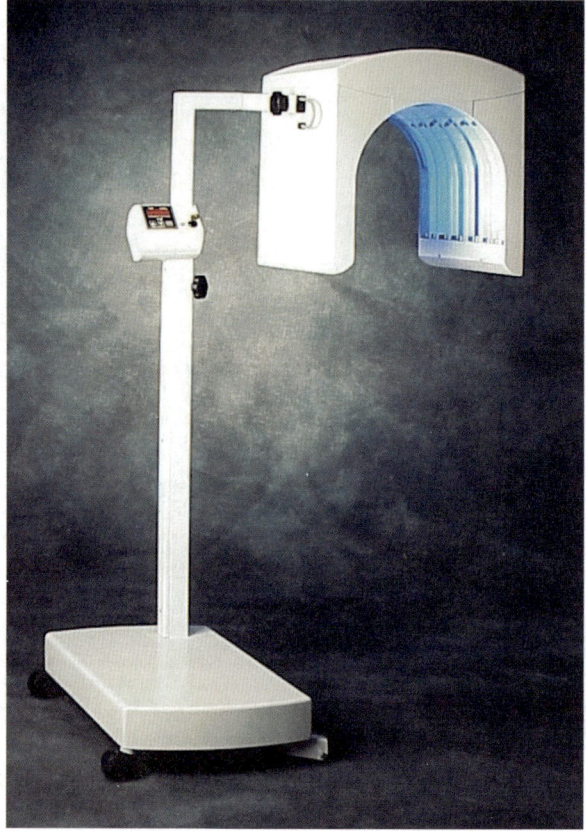

**Figure 17.** The BLU-U contains a curved set of fluorescent tubes whose emission spectrum matches the Soret absorption band of PpIX. This commercial lamp is used to treat actinic keratosis *via* ALA-PDT. Picture from DUSA Pharmaceuticals Inc., with permission.

The irradiation of *flat surfaces* is achieved either by large surface emitters (lamp systems or LED arrays) or by fibre-based systems with micro-lens tips. The micro-lens tips produce an enlarged image of the polished plane end of the fibre on the tissue surface. As the light exits the fibre end uniformly, the intensity distribution over the image of the fibre end on the tissue is also uniform with sharp boundaries at the outer edge of the irradiation zone. The same principle can be used with fluid light guides when utilized with an appropriate incoherent lamp system.

The irradiation of *cylindrical surfaces* is a little bit more challenging. Applicators developed for such purposes are fibre-based and equipped with radially diffusing fibre tips. Some of these applicators can achieve homogenous radiation from a cylindrical surface of several centimetres in length and a few millimetres in diameter by using multiple backscattering foils.[136] The adaptation of applicator diameter to the organ is important in organs that contain folds, unless being appropriately distended.[137] Small-diameter (*ca.* 1 mm) radial

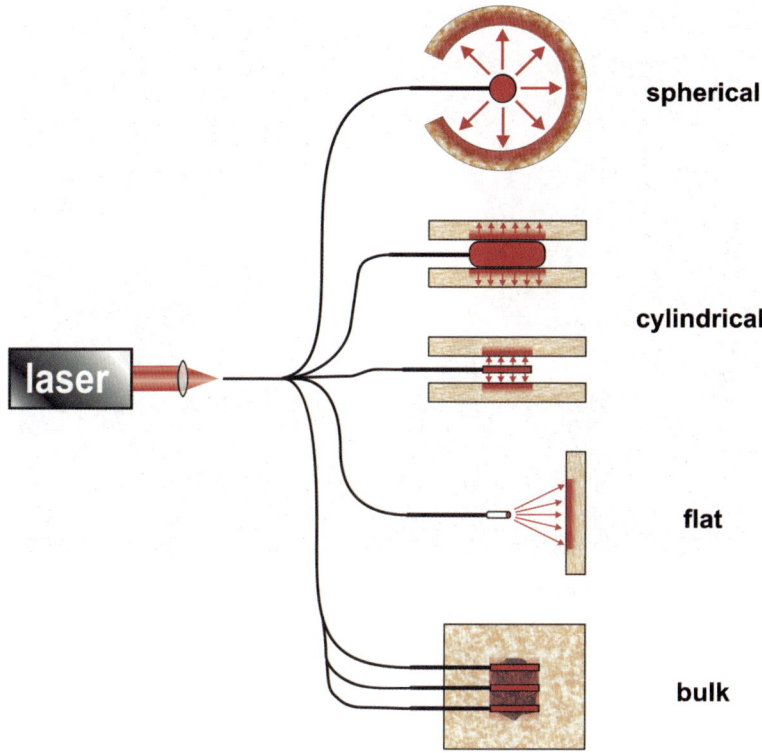

**Figure 18.**   Schematic drawing of different types of fibre-based light diffusers. The fibre
is usually a flexible quartz fibre with a core diameter of 400–600 μm.

diffusers can be placed directly into small-diameter organs or inside transparent
balloon catheters for central placement. They are available in lengths up to
several centimetres.

   *Spherical organs* can be irradiated with fibre applicators equipped with a
bulb-shaped tip, dispersing the light uniformly in all directions. It is, however,
not always easy to ensure a central positioning of the spherical diffuser during
PDT. Therefore, different types of spherical light application systems have been
developed, partly relying on multiple backscatter to enhance homogeneity.[138]

   In order to irradiate bulk tumour tissue, *interstitial* placement of one or more
fibre(s) is indicated. In general, the use of small-diameter radial diffuser tips
is recommended rather than bare fibres, in order to avoid excess heating at
the fibre tip. Irradiation delivered is typically less than 200 mW cm$^{-1}$ length of
the radial diffuser, in order to avoid thermal effects. For the placement of the
fibres, modified instrumentation for brachytherapy or stereotaxis can be used.
For proper light dosimetry, information on the optical tissue parameters as
detailed in refs. 139 and 140 is required in order to obtain a good estimate for
the light penetration depth and thus for the appropriate inter-fibre distance.[141]

   Future applications of PDT may require specially adapted light applicators.
Irradiation of portio and cervix, for instance, can either be done separately with

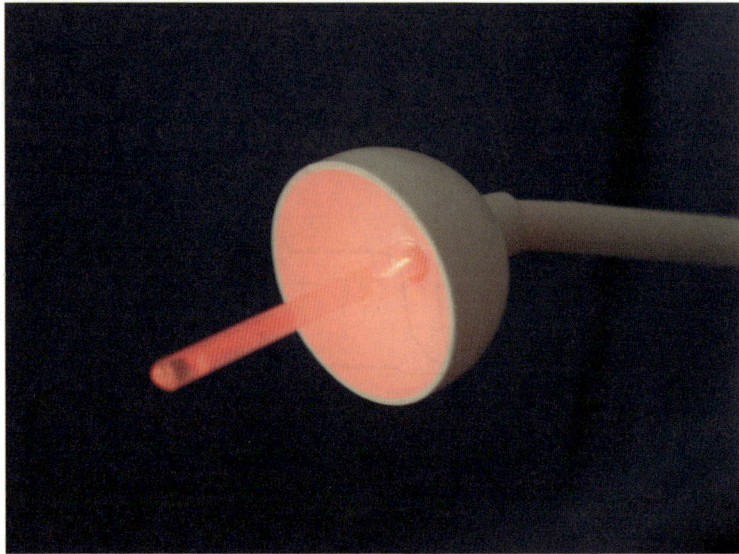

**Figure 19.**   Photograph of a light applicator for PDT on the portio, irradiating endo-
and ecto-cervix simultaneously. The applicator is designed to emit the same intensity
over the whole radiant surface. The portio surface is irradiated with a transparent semi-
sphere covered by a backscattering foil, the cervical canal is irradiated with a radial
diffuser of variable diameter (BioLitec AG, Jena, Germany).

a micro-lens fibre for the portio and a radial diffuser inside the cervical canal, or
with a more complex but also more convenient applicator (Figure 19).

'Intelligent' applicators include a means for measuring the applied irradiance
or a signal providing information on tissue optical properties to individualize
light dosimetry.[136,138,142] Measurements of fluorescence bleaching during PDT
may also be able to predict the individual light dose necessary to achieve the
desired effect.[143,144]

### 2.5.3.  Dosimetry

In order to trigger cell death, a minimum number of aggressive singlet-oxygen
molecules has to be produced in each cell within a limited period of time. This
number of singlet-oxygen molecules (per tissue volume) is the threshold 'photodynamic dose' required for tissue destruction. The photodynamic dose depends
upon the sensitizer concentration, the oxygen concentration and the fluence
(effective light dose), and of course also on the intracellular localization of the
photosensitizer.

For a given clinical application there usually exists a recommended 'incident
light dose' or 'energy density' or 'fluence', given either in Joules per square
centimetre ($J\ cm^{-2}$) of irradiated surface or in Joules per centimetre of fibre
length ($J\ cm^{-1}$) of a radial diffuser. The irradiation parameter determines the
'amount of light' that has been successfully delivered to the tissue and has

produced a certain number of singlet-oxygen molecules within that volume of tissue. A given irradiation can be applied over a shorter or longer time interval, depending on the intensity of light impinging on the tissue. This parameter is called irradiance ('incident intensity' or 'power density'), usually given in milliwatts per square centimetre (mW cm$^{-2}$) or mW cm$^{-1}$ of fibre length.

A simple relationship holds between irradiation $H$, irradiance $E$ and time $t$:

$$H = E \cdot t$$

*Example:*

A circular spot of diameter 4 cm is to be treated with an irradiation of 100 J cm$^{-2}$ and an irradiance of 50 mW cm$^{-2}$. How long is the irradiation time and what (laser) power has to be emitted from the applicator during that time?

Answer:

(i) $H = E \cdot t$, $t = H / E = (100 \text{ J}) / (50 \cdot 0.001 \text{ W}) = 2000$ s, (1 J $= 1$ W $\cdot$ s), (ii) Area A is $A = r^2\pi = 2 \cdot 2 \cdot \pi \cdot \text{cm}^2 = 12.6 \text{ cm}^2$, applied at 50 mW $\cdot$ cm$^{-2}$ gives laser power $P = A \cdot E = 12.6 \cdot 50$ mW $= 630$ mW $= 0.63$ W. Increasing the irradiance to 100 mW $\cdot$ cm$^{-2}$ reduces the irradiation time by half, but requires delivery of 1.26 W total (laser) light power emitted from the applicator.

If there is a clear recommendation for the irradiation to be applied, the only care that has to be taken is the proper selection of the lesion geometry. One has to take into account that the light fluence is highest on the surface of the tissue but decreases exponentially underneath the surface. A sufficient therapeutic effect may therefore be obtained only within a certain depth of the tissue. The treatment of thicker lesions thus requires a more careful planning of the irradiation parameters.

It should be mentioned at this point that the bleaching (self-destruction) of PpIX during irradiation has certain implications on light dosimetry. At first, one might think that this effect would be negative, but such is not the case. Bleaching prevents the possible overdosing of light. Once all of the ALA-induced PpIX is used up (by self-destruction), any further irradiation has no additional effect, especially on adjacent normal tissue. Of course, if the initial concentration of PpIX is too low, there will not be enough singlet-oxygen molecules produced to kill the tissue. A minimum concentration of PpIX is therefore required in the target tissue. The surrounding normal tissue can, however, be perfectly protected from lethal effects if the concentration of PpIX is below the threshold to produce lethal damage on prolonged irradiation. Even very high light doses would then not be harmful. The details of this consideration are expanded in Section 2.7.

## 2.6. Principles of ALA-Based FD

### 2.6.1. Physics behind Fluorescence

The physical origin of fluorescence lies in the interaction of light photons with the outer electrons of molecules: Once a molecule has been electronically

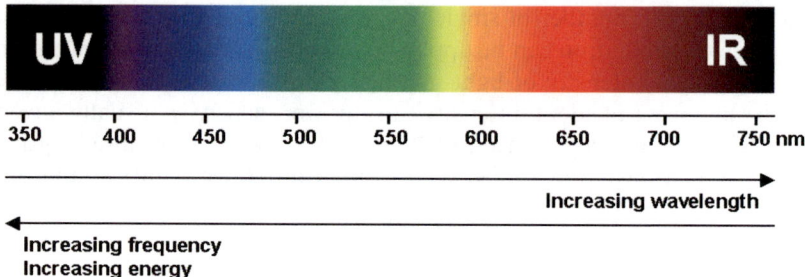

**Figure 20.** Representation of the visible spectrum of electromagnetic radiation. *Note*: to the 'red end' are longer wavelengths and lower photon energies, to the 'blue end' are shorter wavelengths and higher photon energies. UV: ultraviolet, IR: infrared.

excited by absorption of a photon of appropriate wavelength, it has several possibilities to return to its ground state. One of these possibilities involves emission of a secondary photon that carries the energy difference of the transition from the upper to the lower electronic state. The higher this difference, the higher the frequency or the shorter the wavelength of the emitted photon. High-energy photons can excite the 'blue' cones in our retina, and low-energy photons only the 'red' ones (Figure 20). Owing to the law of energy conservation, the emitted photon can carry energy that is only lower or equal to the energy of the exciting photon. So, fluorescence light is usually red shifted (to longer wavelengths) with respect to the light used for excitation. This loss of energy is due to some vibrations in the excited molecule, resulting in a small loss of energy as heat. The corresponding wavelength shift can be exploited for very sensitive techniques to detect fluorescent substances, as the excitation light can be blocked by appropriate filters and the absence of background light provides excellent contrast.

One can measure the fluorescence intensity, which is proportional to the fluorochrome concentration, although dependent on a few other parameters. Another feature that can be measured is the time delay between the absorption and the emission processes. This fluorescence lifetime usually is in the nanosecond time-scale, requiring fast electronics to measure and resolve small differences in fluorescence lifetimes. The signals can be generated in the time domain by using short-pulsed lasers or in the frequency domain using high-frequency modulated lasers. Time-gated imaging or spectroscopy can be used to enhance the relative fluorescence intensity of a specific fluorochrome, thus reducing background.[145] Real lifetime images code the local lifetimes in pseudo colours, and are no longer dependent upon the fluorochrome concentration but can provide additional information such as chromophore local binding site or tissue pH.

### 2.6.2. Autofluorescence for FD

For medical diagnostic purposes, one may ask whether endogenous fluorochromes (autofluorescence) are able to provide contrast between diseased and normal tissues. In general, connective tissue shows a brighter fluorescence than epithelial tissue. This is due to the presence of various fluorescent structures in collagen and elastin fibrils. Apart from these, enzymes of the respiratory chain such as the reduced form of nicotinamide adenine dinucleotide (NADH) or the flavins (FAD) show significant fluorescence also.

The longest wavelength that can excite an amino acid (tryptophan) is 310 nm (UV). Clinically important diagnostic systems that rely on autofluorescence operate with excitation wavelengths from 337 nm (nitrogen laser) to approximately 450 nm. The contrast obtained with these systems is largely due to structural features of the tissue. The early lung cancer detection systems exploit mainly the fact of epithelial thickening during the early growth of bronchial cancer. The strongly fluorescent connective tissue is covered by an increasingly thick layer of weakly fluorescent (malignant) epithelium as the cancer grows. Thus, a malignant lesion will reveal a reduced fluorescence intensity when compared to the surrounding normal tissue. As this effect is more pronounced for green fluorescence than for red (which experiences lower absorption), there is also a shift from green to red (from lower to higher wavelengths). Thus, small malignant lesions usually appear as dark red-brown spots within a brightly green fluorescent background.[146] Systems that extract more spectral information from tissue autofluorescence[147,148] or that exploit fluorescence lifetime differences between diseased and normal tissue autofluorescence[149] are currently under clinical assessment.

### 2.6.3. ALA-Induced PpIX for FD

In comparison to autofluorescence, a substantial increase in contrast may be obtained if a fluorochrome is available that selectively labels the target tissue. The majority of photosensitizers used for PDT show fluorescence emission, and many photosensitizers have been investigated for their FD potential. However, the tolerable level of potential side effects of a drug is considerably lower for diagnostic purposes as compared to therapeutic intervention. Therefore, Photofrin, a potent photosensitizer that is approved for treatment in several countries, did not emerge as a drug that can be used routinely for diagnostic purposes because it causes a prolonged and generalized photosensitization. Many other photosensitizers suffer from a similar problem too. However, ALA is an exception. Its advantages are its rapid clearance rate that limits generalized phototoxicity to about 24 h, its suitability for topical applications, and the fact that it is an endogenous substance that is converted to the fluorochrome intracellularly. Today, it is the most widely used drug for FD. Despite its lack of full official approval (to date, it is approved only for treating AK, while its methyl ester is approved for the treatment of AK and BCC), the detection of bladder carcinoma using ALA is quite popular in Germany. The STORZ, Olympus and Wolf companies have developed 'PDD' sets for cystoscopy, and several hundreds of them have already been sold. Multi-centre approval studies are currently being conducted by the Medac Company, Hamburg, for the topical use of 5-ALA for the detection of bladder cancer, and for its systemic (oral) use for the detection of malignant glioma. PhotoCure has obtained European approval for Hexvix, a hexyl derivative of ALA, for the detection of bladder cancer.

PpIX is a fluorochrome with an extremely strong shift between its main excitation wavelength and its wavelength of fluorescence emission. There is a strong absorption in the Soret band at the far blue end of the visible spectrum at 400 nm (Figure 21 left, blue curve) that induces a characteristic fluorescence spectrum with two prominent peaks in the red at 630 and 700 nm, when evaluated in solution (Figure 21 left, red curve). In tissue, blood absorption causes changes in the relative peak intensities and a different microenvironment causes a slight shift in peak positions (Figure 21 right).

There is clear evidence for preferential PpIX accumulation in epithelial tissue compared to connective tissue or muscle, with a further differential increase in malignant tissue.[51] Thus a higher PpIX fluorescence intensity is normally observed in cancer tissue. Since ALA can be applied topically and needs no 'leaky vessel' accumulation mechanism, we can even expect the staining of very early lesions that have not yet initiated tumour angiogenesis.

As we know from fluorescence microscopy, good contrast requires proper filtering of the fluorescence excitation and fluorescence emission light paths. We can detect the intensity of the fluorescent light within a specified wavelength band, or we can detect the whole spectrum of the fluorochrome, the fluorescence lifetime, polarization or energy transfer features. If we first focus on the simple detection of fluorescence intensity over the whole emission wavelength, we are faced with some unfavourable features of optical tissue diagnostics: multiple light scattering, tissue layering, autofluorescence background, inhomogenous composition, irregular observation geometry and a limited specificity in fluorochrome accumulation. To handle as many as possible of these potential drawbacks, it is advantageous to combine the emission of several fluorochromes (PpIX fluorescence and autofluorescence) with the remitted light (in a narrow wavelength band) of the same image but in different colours.

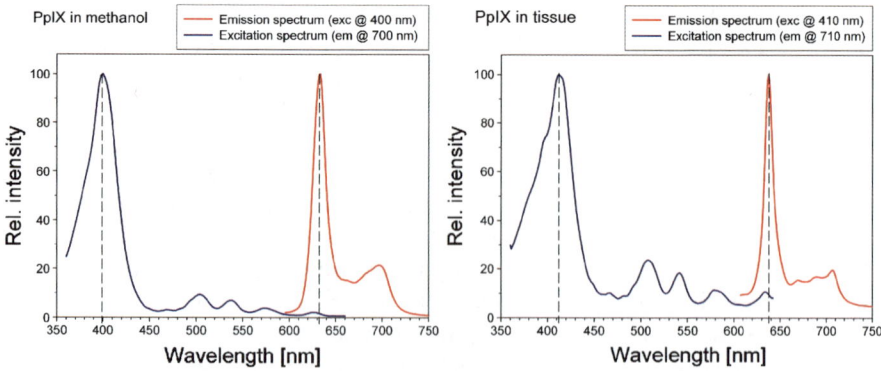

**Figure 21.** Left: Spectra of fluorescence excitation (blue) and fluorescence emission (red) of PpIX in methanol. Right: The same spectra from tissue (malignant glioma, measured *ex situ*).

The wavelength of the re-emitted light should be selected such that the dependence of the remitted intensity on absorption approximates that of the overall fluorescence intensity. Monte Carlo simulations of light transport in tissue show that wavelengths of 450–470 nm, close to the local absorption minimum of blood in this spectral region, are best suited to serve as a remission standard.[150] A small amount of light at these wavelengths makes normal tissue appear blue, although tissue autofluorescence and some non-specific PpIX accumulation may be present. A higher PpIX content turns the colour to red. This change in colour occurs irrespective of the observation distance and is largely independent on blood absorption, whereas a simple measurement of fluorescence intensity is strongly affected by these non-specific factors. Recognition of suspicious tissue (higher intracellular PpIX content) is thus more reliably achieved by observation of the colour contrast than by identifying bright or faint red PpIX fluorescence (intensity contrast).

The colour contrast approach is illustrated in Figure 22. The RGB (red–green–blue) image in the bottom part of the figure shows the colour contrast fluorescence image of a urinary bladder wall area bearing two small exophytic tumours as it is generated by the D-light system. A straightforward fluorescence imaging approach would have focused on the PpIX fluorescence intensity. A bandpass filter transmitting only the red spectral band where PpIX fluorescence occurs would have been used. The resulting image is displayed in the upper left of Figure 22. It contains all the artefacts discussed above: inhomogenous

**Figure 22.** Demonstration of how colour contrast fluorescence imaging provides easy recognition of suspicious lesions by reducing artefacts from inhomogenous illumination, non-specific PpIX fluorescence, and autofluorescence background, as well as tissue absorption. Explanation: refer to text.

illumination, unspecific background fluorescence, some autofluorescence and most significantly, the smaller tumour within the circle gives no contrast at all. The reason for that becomes evident in the blue remission image (at $450 \pm 10$ nm). There is a selectively increased absorption in the tumour, leading to a reduced remission intensity. The increased absorption also interferes with the increased PpIX accumulation and prevents a positive contrast in the red fluorescence image. The combination of fluorescence and remission images, however, clearly eliminates the absorption artefact and displays in red colour what has to be considered suspicious. In this case, the reduced autofluorescence of the cancer epithelium of the exophytic tumours (green image) further increases the contrast.

### 2.6.4. Technical Implementation

The fluorescence of PpIX can be excited by any light source that can provide a sufficiently narrow band of light that covers one of the absorption peaks of PpIX. This might be a Wood's lamp, similar to those used at discotheques to produce a 'futuristic' atmosphere by exciting the fluorescence of various substances (mainly the 'whiteners' in washing powder). But they can also be used to excite PpIX fluorescence as is shown in Figure 23.

The excitation wavelength of Wood's lamps is largely invisible to the human eye. Induced fluorescence therefore can be observed without any additional filtering.

Endoscopic fluorescence excitation requires, however, light sources that can be focused to a small spot, so that the illumination fibres of the endoscope can transmit enough light to efficiently excite the PpIX in tissue. As lasers provide the smallest focus of all light sources and as the Kr-ion-laser has a line at 406 nm, perfectly matching the absorption maximum of PpIX, Kr-ion-lasers were the first endoscopically used light sources for PpIX fluorescence detection.[151] Again, as with Wood's light, no filtering was necessary for fluorescence observation, as the sensitivity of the human eye is low enough for the 406-nm line to provide a good balance between red PpIX fluorescence and violet backscatter.

Unfortunately, Kr-lasers are very expensive, quite inefficient and short-lived light sources. Therefore, commercially available incoherent lamps with high-intensity bulbs (short-arc Xenon lamp) (K. Storz, R. Wolf, Olympus, see appendix) are often used for the purpose of PpIX excitation. Such lamps are often equipped with filter wheels, allowing for switching between regular white-light mode and fluorescence mode. As they operate with excitation wavelengths further in the visible, the backscattered excitation light has to be filtered away in the detection light path. This is done with filters in the endoscopes' eyepieces. They are designed such that they block excitation light completely up to about 440 nm, but transmit a little backscattered light around 450 nm in order to produce the colour contrast described in Section 2.6.3. This filter characteristic is displayed schematically in Figure 24.

With the systems described thus far, PpIX fluorescence can be observed with the naked eye. Documentation with video cameras is not straightforward, as

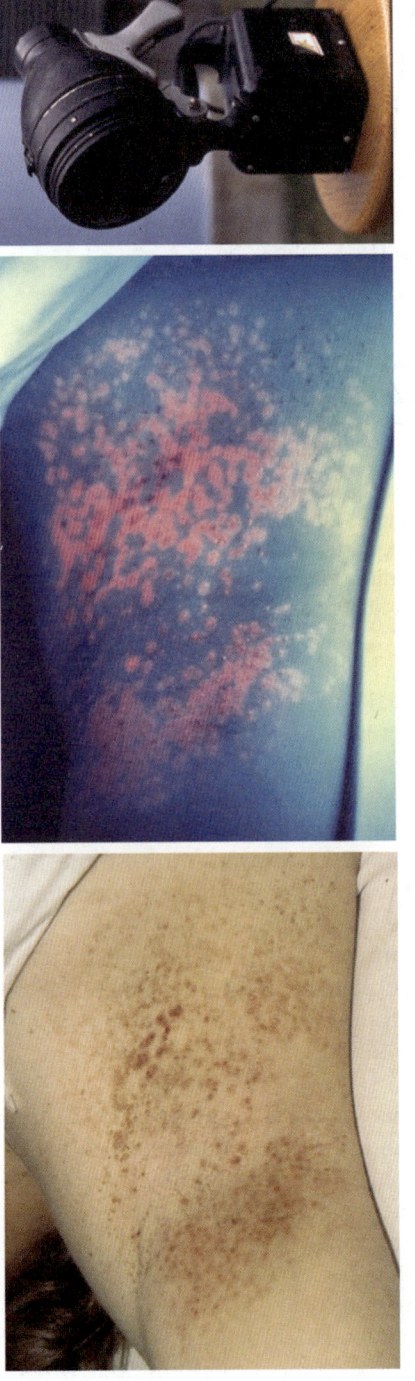

**Figure 23.** Basal cell nevus syndrome induced by ionization radiation (left). ALA-induced PpIX fluorescence induced in basal cell nevus syndrome (middle), excited with a Wood's lamp (right).

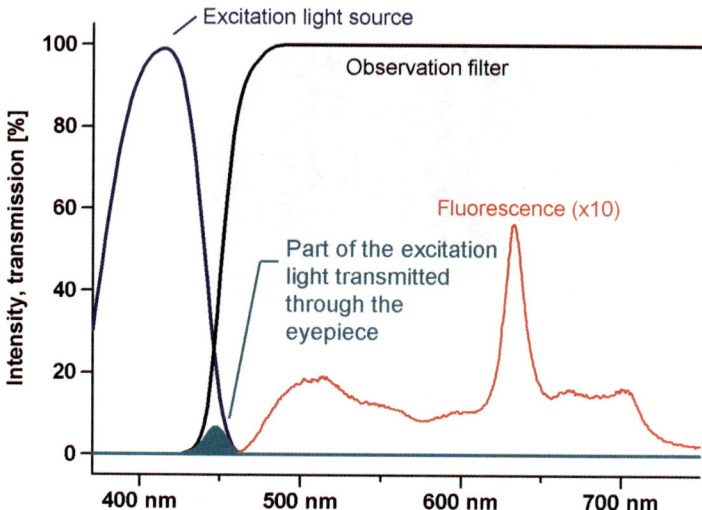

**Figure 24.** Characteristics of the excitation filter in the light source and the observation filter in the endoscope's eyepiece. Due to a small overlap of both transmission curves at about 450 nm, part of backscattered light of this wavelength is transmitted to the observer (blue). The observation filter is completely transparent in the green (autofluorescence) and red (PpIX fluorescence) parts of the spectrum.

the fluorescence is too weak for most standard cameras. For use with endoscopes, cameras with increased sensitivity had to be developed. This was achieved by either integrating some image frames to obtain high-quality fluorescence images but with reduced frame repetition rate or implementing very low noise/high gain amplifiers in order to obtain real time-frame repetition rate but reduced image quality. A connection of the light system with the camera controller is used to switch both units simultaneously between the two 'white light' and 'fluorescence' modes.

The equipment that employs such enhanced sensitivity cameras is used for most clinical studies performed in order to evaluate the additional merits of ALA-induced PpIX FD compared to white-light tissue inspection. The same equipment is used also to investigate ALA derivatives or hypericin in clinical settings.[152]

For the future, numerous technical advances may be anticipated: LEDs may replace lamp systems, and flexible video endoscopes may provide high-resolution fluorescence images. As with LEDs, a very fast switching between white light and fluorescence is possible, and an overlay of fluorescence information onto the white-light image may be realized. For the FD of skin lesions, such an overlay imaging has been realized with flash-lamp fluorescence excitation (BioCam, see the appendix).

Tissue fluorescence includes much more information than just fluorescence intensity of a single fluorochrome (regarding the other fluorescent tissue information as undesired background). Attempts to extract most of this information are being performed and may lead to enhanced fluorescence detection systems. First of all, calculating quantitative information from the images may objectify the fluorescence finding.[153] Further, spectroscopy provides a possibility of detecting very subtle changes in relative contributions of different fluorochromes or peak wavelengths. Computing algorithms ranging from multi-variate linear regression to neural networks and principal component analysis have been proposed to extract the most relevant features from spectral measurements. Spectral imaging, as shown by the contribution of Malik *et al.*, provides images of computed spectral features.[66] Next, fluorescence lifetime imaging can be used to eliminate background fluorescence or to provide information on the microenvironment of selected fluorochromes.[154–156]

There is also a justified hope that molecular biology will identify clinically applicable markers for the fluorescence staining of diseased tissue with an increased specificity or even sensitivity when compared to ALA-induced PpIX. One such development may be the detection of the (over) expression of matrix proteinases by 'proteinase activated fluorescence': Self-quenching fluorochromes are bound densely to a substrate that is specifically digested by certain matrix proteinases that are secreted by cancer cells. As long as the substrate is intact, there is no fluorescence, but after digestion, the fluorochromes are separated, the quenching mode is removed, and fluorescence intensity increases.[157]

Finally, fluorescence detection cannot be regarded as a stand-alone diagnostic tool. Properly analysed remission signals, OCT, confocal techniques and endoscopic ultrasound, navigational techniques based on CT, MRI or optical tomography will have to be combined reasonably. This will certainly present a great step forward. It may well be expected that highly sensitive endoscopic and non-endoscopic tumour detection can thus successfully be complemented by online staging and grading, which could finally reduce the requirement for multiple biopsies and might even render instantaneous sectioning partly dispensable.

### 2.6.5. Clinical Relevance

X-rays, discovered by Wilhelm Conrad Röntgen in late 1895, were immediately recognized as a major breakthrough for medical diagnostics. Within a single year, hospitals had founded radiological units to exploit the fact that one could look inside a patient. During the 20th century, X-ray imaging has been further developed to today's multi-slice spiral CT-instrumentation allowing virtual 3-D displays of the surface or interior of any organ of the human body. Nuclear Magnetic Resonance Imaging (NMRI), Scintigraphy and Positron Emission Tomography (PET) have joined in the imaging armamentarium of modern medical diagnostics. We must not forget ultrasound scanners, a cheap and powerful source of online diagnostics even for general practitioners, and endoscopy, an indispensable prerequisite for keyhole surgery.

Hence, one might think that there is no need for additional diagnostic tools. But, imagine for instance that you were a urologist, looking through a cystoscope that is inserted into a patients' urinary bladder. You can easily see the papillary tumour protruding into the bladder's lumen that had been discovered pre-operatively in a urogram and a CT-scan. An electrical resection loop allows you to remove the tumour. Still, from statistics you know that it is rather likely that you will not have cured the patient – there is a recurrence probability of more than 50% due to 'invisible' flat malignant extensions of the tumour or additional flat lesions elsewhere in the bladder.[158] None of the imaging technologies discussed so far will be able to localize such early stage tumours and guide you during endoscopic surgery. What you need is a switch on your endoscope to change illumination and highlight all malignant lesions with a bright contrast, as the safety marks of banknotes will fluoresce under UV illumination.

Figure 25 illustrates such a situation: With the standard white-light endoscopic view (left) you might be satisfied after locating the small exophytic tumour on the sidewall of the urinary bladder. However, you would have missed an equally malignant flat lesion a little off from the first. The fluorescence of selectively accumulated PpIX upon instillation of an ALA solution into the bladder 2 h prior to endoscopy reveals even very flat malignant lesions, and provides an online guide for more complete endoscopic tumour surgery.[152,159]

Another clinically important example is brain tumour surgery. Malignant gliomas are the most common primary brain tumours, and have a dismal prognosis. Average survival is as short as approximately 12 months despite surgery and radiotherapy or chemotherapy. It is obviously not possible to perform surgery with an extensive safety margin. Still, the more radically the tumour has been removed, the longer is the survival expectancy. Again, PpIX fluorescence provides a powerful intra-operative means to discriminate malignant from normal tissue. Of course, MRI, CT or PET are invaluable tools for

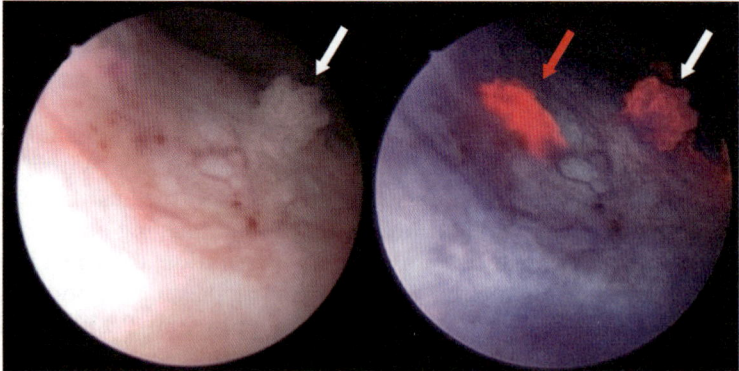

**Figure 25.** Left: papillary malignant tumour of the urinary bladder by standard white-light endoscopy (white arrow), right: same area in fluorescence mode after instillation of ALA. The papillary tumour shows a selective red fluorescence. Distant from this tumour, an additional circumscribed area, not conspicuous in white light, shows red fluorescence (red arrow): biopsy proved early stage malignant tumour (pTaG1). (Photos: courtesy of D. Zaak).

pre-operative planning. But intra-operative decision of where exactly to continue or cease resection is not sufficiently supported by these imaging methods.

### 2.6.6. Caveats

As with all diagnostic procedures in medicine, ALA-induced PpIX fluorescence is not 100% sensitive (some tumours may not be stained, meaning there will be 'false negatives') and not 100% specific (there may be staining of benign tissue, meaning there will be 'false positives'). What that means clinically and whether there is still a benefit from FD or not is discussed in the clinical chapters. But apart from imperfect fluorescence staining, there are some caveats to bear in mind in order to avoid misinterpretations during fluorescence endoscopy.

The major drawback of PpIX fluorescence is that it fades away during prolonged and intense excitation. This process is called photobleaching, and it limits the length of a fluorescence inspection. Depending on the intensity setting, photobleaching also occurs during white-light inspection, so that one should generally minimize light exposure and try to obtain the fluorescence information as quickly as possible. Photobleaching occurs, however, only in tissues exposed to light.

One should bear in mind the distance dependence of fluorescence intensity observed through endoscopes. It decreases with the square of the observation distance. As the dynamic range of a video camera is quite limited, a fluorescence positive tissue area may easily be missed if observed from too long a distance. If the lesion is also very small, its area also decreases with the square of the distance, and the number of photons observed from this lesion decreases with the 4th power of the distance. Thus, one should always keep as close as possible to the tissue so as to ensure sufficient brightness of the fluorescence image.

A camera artefact, more pronounced with single-chip cameras than with three-chip cameras, may suggest fluorescence where there is none. Overexposure leads to saturation of the blue channel, whereas the red channel still increases its signal, leading to a bright purple colour in such overexposed areas. Such purple areas will move over the tissue when the optical instrument is moved. One should reduce the excitation intensity in such cases (e.g. by increasing the distance a little) and try to achieve a more homogeneous excitation of the observed tissue surface.

One should be aware that ALA, if topically applied, will penetrate no further than approximately 1 mm into the tissue. Accordingly, the fluorescence of a thicker tumour will be limited to a depth of 1 mm, and the entire volume of the tumour will not be stained.

Especially for bladder cancer detection, there are the following important precautions: First, urine shows a strong green fluorescence (which one might even use to identify the ureteral orifices) that can disturb the observation like a dense fog. So, a careful voiding of the bladder prior to insertion of the cystoscope is indispensable and repetitive rinsing with the irrigation liquid is recommended in order to retain a clear view of the bladder wall. A clear view is also impeded by blood in the irrigation liquid. Fluorescence cystoscopy requires careful haemostasis and thorough irrigation. Finally, erroneous positive fluorescence may occur, as when the tissue is observed at a sharp angle (change to side-view optics in such cases) or when the bladder wall is not properly unfolded (increase the filling volume).

In case of an air–tissue interface, reflections of the blue light at 450 nm will occur, and this may either screen real fluorescence or suggest positive fluorescence where there is none. To find out, one can either move around with the instrument or dry the tissue surface.

In summary, some caveats have to be considered, and there is certainly a learning curve until the limitations of FD as described can be handled routinely. But the learning curve is short and the advantages of fluorescence detection are quickly evident.

## 2.7. Consideration of PpIX Photobleaching in Light Dosimetry

In a simplified model for irradiation of photosensitized tissue, we may assume a constant oxygen concentration (valid for low-light fluence rates and non-compromised blood flow) and a reproducible intracellular localization of PpIX. The photodynamic dose $D$ then is simply proportional to the product of the sensitizer concentration '$c$' and the light dose '$\Phi t$', where $\Phi$ is the space irradiance or light fluence rate and $t$ is the irradiation time.

$$D = \alpha \cdot c \cdot \Phi \cdot t \tag{1}$$

where

$D$ = photodynamic dose;

$\alpha$ = proportionality constant;

$c$ = concentration of photosensitizer;

$\Phi$ = light fluence rate; and

$t$ = time of light application.

Of all the parameters in Equation 1, the most difficult parameter to assess is the light fluence rate, $\Phi$. While it is relatively easy to determine the fluence rate at the surface of skin, the determination of the active light fluence rate inside the tissue at the site of photosensitization is not trivial. Light scattering and absorption within the tissue have to be taken into account. Light scattering is quantitatively considered by the two parameters $\mu_s$ (mean number of scattering events per distance travelled, given in [mm$^{-1}$]) and $g$ (isotropy parameter, a number between 0 for completely isotropic and 1 for completely forward scatter). Absorption events are described with the mean number of such events per distance travelled, $\mu_a$, also given in [mm$^{-1}$]. Figure 26 illustrates the light fluence rate dependence on the depth inside the tissue for a perpendicularly incident light beam of fluence rate '100%' and representative absorption and scattering parameters for red light around 635 nm. First of all, it is apparent

Fluence rate relative to incident beam [%]

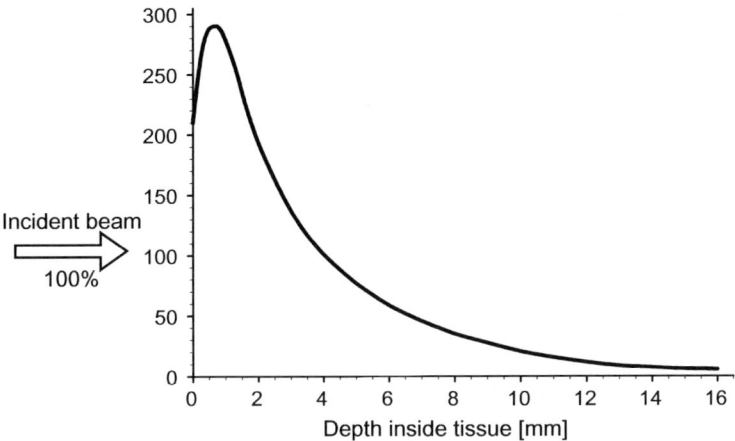

**Figure 26.** Result of a Monte Carlo simulation of light transport in turbid media. At an absorption of $\mu_a = 0.4$ mm$^{-1}$ and a scattering of $\mu_s = 4$ mm$^{-1}$ and forward scatter of $g = 0.8$, the fluence rate underneath the tissue surface is significantly higher compared to the incident plane wave (set at 100%). An exponential decrease toward deeper parts of the tissue limits the PDT effects to several millimetres.

that within the superficial tissue layers, multiple light scattering produces an increase in fluence rate compared to the incident beam. Deeper inside the tissue, however, the fluence rate decreases exponentially, and at a certain depth the light dose in a given irradiation time will fall below the threshold for cell death. The depth of tumour destruction achievable is thus always limited by the penetration depth of the light. With prolonged irradiation, the depth of destruction may be increased, but unfortunately only to a rather limited extent for feasible irradiation times and when thermal effects are avoided.[160]

Photosensitizer concentration is the second factor that determines the photodynamic dose. This may also vary with the depth of the tissue, or even with the irradiation time due to self-destruction (photobleaching). In this case, the concentration '$c$' is a function of time, and the concentration decreases exponentially over time:

$$c = c_0 \cdot e^{-b \cdot \Phi \cdot t} \tag{2}$$

where

$c = $ concentration of the photosensitizer;

$c_0 = $ initial sensitizer concentration;

$b = $ bleaching constant;

$\Phi = $ light fluence rate; and

$t = $ time of light application.

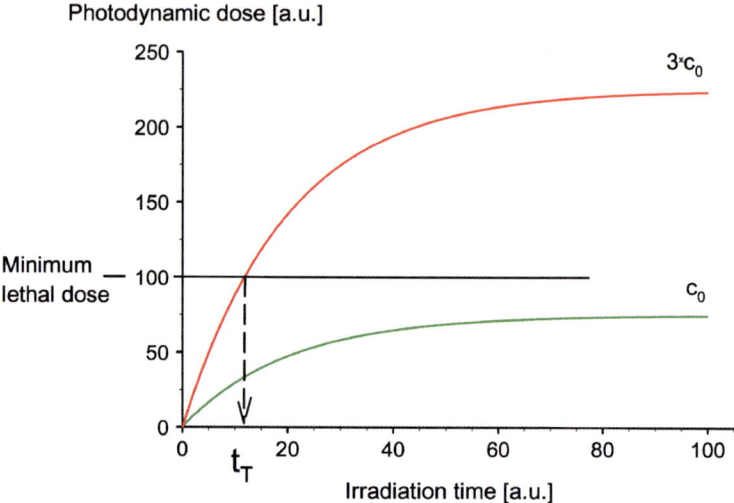

**Figure 27.** Plot of Equation 3 for two different initial PpIX concentrations differing by a factor of three. The arbitrarily chosen threshold for tissue destruction (minimum lethal photodynamic dose, set at 100) is reached only with the higher initial concentration of photosensitizer after time $t_T$, whereas photobleaching causes the photodynamic effect to saturate below this level for tissue with a sufficiently low initial PpIX concentration.

Inserted into Equation 1 and integrated over time, the photodynamic dose ($D$) shows time dependence as depicted in Figure 27 and expressed by the following equation:

$$D = \frac{\alpha \cdot c_0}{b}\left(1 - e^{-b \cdot \Phi \cdot t}\right) \tag{3}$$

It is obvious that for a photosensitizer that shows significant photobleaching – which is specifically true for ALA-induced PpIX – the maximum achievable photodynamic effect saturates at a level that is primarily determined by the initial sensitizer concentration '$c_0$'. At this level, all photosensitizer is used up and further irradiation has no additional effect. This limitation is, however, not necessarily unfavourable. Consider a situation where the tumour has accumulated a significantly higher photosensitizer concentration $c_{0T}$ compared to the surrounding or deeper normal tissue $c_{0N}$:

$$c_{0T} > c_{0N} \tag{4}$$

As schematically shown in Figure 27, the photodynamic dose for tumour tissue exceeds the threshold dose for irreversible tissue damage very soon after time $t_T$ (or light dose $\Phi t_T$), whereas for normal tissue, the initial photosensitizer concentration may be low enough to be unable to trigger sufficient damage to produce significant cell death at all, even for arbitrarily long irradiation times. Photobleaching can thus simplify light dosimetry, as a tumour-selective destruction can be achieved with any light dose above $\Phi \cdot t_T$ without the risk of adverse effects on normal tissue.

The considerations above are simplified, and a more profound description of the mathematics around drug and light dosimetry in PDT have been reported by Jacques[161,162] and Star.[163]

Although PpIX has its highest absorption in the violet spectral range (around 410 nm), PDT is usually performed with red light. There are several reasons for this. First, the absorption spectrum of haem has a broad absorption peak around 420 nm, but two to three orders of magnitude less absorption above 620 nm. Thus, a reasonable light penetration of more than 1 mm into the tissue is only achieved with red light. The question of homogeneity and reproducibility of the PDT effect also favours red light. Shorter wavelength light is always subject to strong local intensity variations within the tissue that are caused by the presence of small or large blood vessels. Tumour cells close to a vessel might thus be shielded from PDT light and survive irradiation. This is much less critical when red light is used. Still, even with light around 800 nm in the absorption minimum of blood and tissue, the penetration depth is limited to few millimetres, mainly due to light scattering. In practice, the action radius of PDT is usually less than 1 cm, and thicker tumours require multiple irradiations, *e.g.* by inserting several fibres into the tumour volume.

It has been pointed out previously that proper light dosimetry can be quite difficult. It is already difficult to measure the incident light intensity in a sterile environment, and it is even more difficult to individually take the absorption and scattering of the tissue being irradiated into account. For this reason, dosimetry is rarely done. The instructions for light dosimetry are purely empirical values obtained from animal experiments or clinical studies, and always refer to the incident intensity.

Light dose recommendations for different organs must be compared very cautiously. Not only differences in optical tissue parameters are to be considered, but also the organ and irradiation geometry. In whole bladder wall irradiation, for instance, diffuse backscatter from one side of the bladder wall may re-enter on the opposite side and add to the primary incident light. For red light, the backscatter intensity is about 50% of the primary incident intensity, and considering multiple backscatter, the effective light dose that can induce PDT effects is composed of only 40% primary incident light but 60% multiple backscatter (Figure 28). Thus, 50 J cm$^{-2}$ applied in the bladder compares to 125 J cm$^{-2}$ applied to surface skin.

Although several of different effects impede a precise light dosimetry, the practical clinical application turns out to be less critical than one might expect. One of the main reasons is the preservation of connective tissue in most PDT situations. Thus the risk of inducing perforation or breakdown of the integrity of an organ wall is rather low.[164] It is only when sensitizers are used that show significant accumulation in muscle and connective tissue layers that muscle constriction or stenosis formation may result from excess irradiation,[164,165] and

**Influence of diffuse backscatter (R) in the bladder**

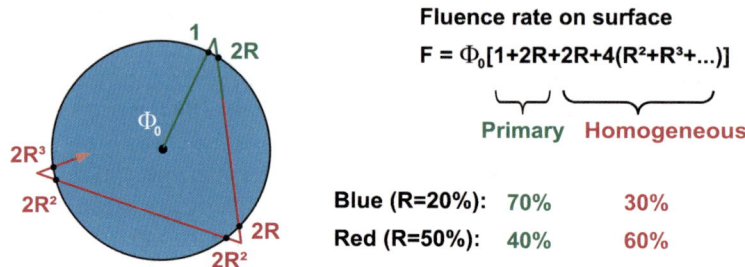

**Figure 28.** Schematic drawing of a bladder with an incident photon emitted from a small central spherical diffuser. Compared to a flat geometry (e.g. skin surface), in hollow organs the primary irradiation accounts only for approximately 40% (red wavelength range). Backscattered light re-enters the tissue and contributes to the incident irradiance to a degree that is dependent on the reflectivity $R$ of the tissue (for blue light, the reflectivity of the bladder wall is approximately 20%, and multiple backscatter accounts for only 30% of the overall irradiance).

therefore more care has to be taken in calculating the dose of light. Overall, ALA-PDT is very safe owing to the very low accumulation of PpIX in connective tissue and muscle layers, and owing to the photobleaching of PpIX as described above.

## 2.8. Conclusions

### 2.8.1. ALA Versus Other Photosensitizers

While the pros and cons of ALA-PDT have already been discussed (Sections 2.4.4 and 2.4.5), it is interesting to examine some other differences that exist between ALA-induced PpIX as a photosensitizer for PDT and the use of other preformed photosensitizers. The main difference is that while preformed photosensitizers need a mechanism of selective accumulation/delivery to tumour and normal tissues, PpIX is selectively *synthesized* in tumour tissue. As such, the site of action is often different for PpIX than for other photosensitizers. While most preformed photosensitizers tend to induce the primary oxidative reactions on cell membranes of the microvasculature, ALA-PDT's primary site of damage is located in the mitochondria. A consequence of this is the induction of a high degree of apoptosis as a mechanism of cell death, in comparison to the necrosis often found with preformed photosensitizers.

Once the PpIX molecule has been synthesized, one could reason that it would undergo a process of distribution within the body similar to that of any other preformed porphyrin-based photosensitizer. There is, however, a major difference. The bioconversion of ALA into PpIX places the photosensitizer molecule at a site that cannot easily be reached by the introduction of pre-formed PpIX.

Thus, even the same photosensitizer does not distribute in the same way if introduced as a pre-formed photosensitizer or as a precursor drug. This being said, once a photosensitizing amount of ALA-induced PpIX has been synthesized, it will in most cases leave the mitochondria and redistribute to different parts of the body in the same way that other drugs do, *albeit* from a different starting point. Thus ALA-PDT has two phases in which one may wish to perform the irradiation. The first phase is at approximately 3 h after the administration of ALA, when a large fraction of PpIX will be found in the vicinity of the mitochondria (thus producing a high degree of apoptosis). The second phase would be at about 16–20 h post-ALA administration, when some of the PpIX would have redistributed somewhat and the degree of apoptosis would be less (more necrosis).

From the point of view of the efficacy of production of reactive oxygen species, PpIX is not particularly more effective than other approved photosensitizers such as Photofrin. The main difference remains in the way in which PpIX is produced, *i.e.*, *via* a route that is natural to the body. It remains the closest 'natural' photosensitizer to date, and at this stage seems to be the easiest and safest one to use.

### 2.8.2. Clinical Significance of ALA-PDT

ALA-PDT is easy to use and has a large safety margin for both the dose of light and the dose of drug. This makes it particularly attractive to use instead of preformed photosensitizers. At this stage, its limited depth of activation probably means that ALA-PDT may well find its most useful role in non-oncological, dermatological applications. It remains to be seen to what extent this 'unique' approach will find medical applications.

# References

1. T. Patrice, Comprehensive series in photochemistry and photobiology, *Photodynamic Therapy*, vol 2, Royal Society of Chemistry, UK, 2003.
2. A.P. Berger, H. Steiner, A. Stenzl, T. Akkad, G. Bartsch and L. Holtl, Photodynamic therapy with intravesical instillation of 5-aminolevulinic acid for patients with recurrent superficial bladder cancer a: single-center study, *Urology*, 2003, **61**, 338–341.
3. R. Waidelich, W. Beyer, R. Knuechel, H. Stepp, R. Baumgartner, J. Schroeder, A. Hofstetter and M. Kriegmair, Whole bladder photodynamic therapy with 5-aminolevulinic acid using a white light source, *Urology*, 2003, **61**, 332–337.
4. R.M. Szeimies and M. Landthaler, Photodynamic therapy and fluorescence diagnosis of skin cancers, *Recent Results Cancer Res.*, 2002, **160**, 240–245.
5. R.M. Szeimies, S. Karrer, A. Sauerwald and M. Landthaler, Photodynamic therapy with topical application of 5-aminolevulinic acid in the treatment of actinic keratoses: an initial clinical study, *Dermatology*, 1996, **92**(3), 246–251.

6. S.A. Friesen, G.O. Hjortland, S.J. Madsen, H. Hirschberg, O. Engebraten, J.M. Nesland and Q. Peng, 5-Aminolevulinic acid-based photodynamic detection and therapy of brain tumors (Review), *Int. J. Oncol.*, 2002, **21**, 577–582.

7. C.J. Kelty, S.L. Marcus and R. Ackroyd, Photodynamic therapy for Barrett's esophagus: a review, *Dis. Esophagus*, 2002, **15**, 137–144.

8. M. Kriegmair and R. Baumgartner, Intravesikuläre Instillation von Delta-Aminolävulinsäure (ALA) – Eine neue Methode zur photodynamischen Diagnostik und Therapie, *Laser-Medizin*, 1992, **8**, 83.

9. O. Raab, Ueber die Wirkung fluoreszierender Stoffe auf Paramaecien, *Z. Biol.*, 1898, **524**.

10. O. Raab, Über die Wirkung fluoreszierender Stoffe auf Infusorien, *Z. Biol.*, 1900, **39**, 524–526.

11. H.v. Tappeiner and A. Jesionek, Therapeutische Versuche mit fluoreszierenden Stoffen, *Muench. Med. Wochenschr.*, 1903, **47**, 2042–2044.

12. G. Dreyer, Lichtbehandlung nach Sensibilisierung, *Dermatol. Z*, 1903, **10**, 6.

13. H.v. Tappeiner, Zur Kenntnis der lichtwirkenden Stoffe, *Dtsch. Med. Wochenschr.*, 1904, **16**, 579–580.

14. H.v. Tappeiner, Beruht die Wirkung der fluoreszierenden Stoffe auf Sensibilisierung?, *Muench. Med. Wochenschr.*, 1904, **16**, 714–715.

15. L. Halberstaedter and A. Neisser, Zur Kenntnis der Sensibilisierung, *Dtsch. Med. Wochenschr.*, 2003, **18**, 805–807.

16. C. Ledoux-Lebards, Action de la lumiere sur la toxicite de l'eosine et de quelques autres substances, *Ann. l'Institut. Pasteur*, 1902, **16**, 587–593.

17. A. Jodlbauer and H. von Tappeiner, Die Beteiligung des Sauerstoffs bei der Wirkung fluorescierender Stoffe, *Dtsch. Arch. Klin. Med.*, 1905, **82**, 520–546.

18. W. Hausmann, Über die sensibilisierende Wirkung tierischer Farbstoffe und ihre physiologische Bedeutung, *Biochem. Z.*, 1908, **14**, 275.

19. S. Schwartz, K. Absolon and H. Vermund, Some relationships of porphyrins, *x-rays and tumors*, *Med. Bull.*, 1955, **27**, 7–13.

20. R.L. Lipson and E.J. Baldes, The photodynamic properties of a particular hematoporphyrin derivate, *Arch. Dermatol.*, 1960, **82**, 508–516.

21. D.E.J.G. Dolmans, D. Fukumura and R.K. Jain, Photodynamic therapy for cancer, *Nat. Rev. Cancer*, 2003, **3**, 380–387.

22. T. Hasan, A. Moor and B. Ortel, *Photodynamic therapy of cancer*, in *Cancer Medicine*, J.F. Holland and E. Frei (eds), B.C. Decker, Hamilton, Ontario, 2000, 489–502.

23. C.H. Sibata, V.C. Colussi, N.L. Oleinick and T.J. Kinsella, Photodynamic therapy: a new concept in medical treatment, *Braz. J. Med. Biol. Res.*, 2000, **33**, 869–880.

24. H. Messmann, E. Endlicher, C.M. Gelbmann and J. Scholmerich, Fluorescence endoscopy and photodynamic therapy, *Dig. Liver Dis.*, 2002, **34**, 754–761.

25. T. Filbeck, U. Pichlmeier, R. Knuechel, W.F. Wieland and W. Roessler, Clinically relevant improvement of recurrence-free survival with 5-aminolevulinic acid induced fluorescence diagnosis in patients with superficial bladder tumors, *J. Urol.*, 2002, **168**, 67–71.

26. D. Zaak, M. Kriegmair, H. Stepp, H. Stepp, R. Baumgartner, R. Oberneder, P. Schneede, S. Corvin, D. Frimberger, R. Knuchel and A. Hofstetter, Endoscopic detection of transitional cell carcinoma with 5-aminolevulinic acid results of 1012 fluorescence endoscopies, *Urology*, 2001, **57**, 690–694.

27. J.C. Kennedy, S.L. Marcus and R.H. Pottier, Photodynamic therapy (PDT) and photodiagnosis (PD) using endogenous photosensitization induced by

5-aminolevulinic acid (ALA): mechanisms and clinical results, *J. Clin. Laser Med. Surg.*, 1996, **14**, 289–304.

28. C.J. Kelty, N.J. Brown, M.W.R. Reed and R. Ackroyd, The use of 5-amino-laevulinic acid as a photosensitiser in photodynamic therapy and photodiagnosis, *Photochem. Photobiol. Sci.*, 2002, **1**, 158–168.

29. Z. Malik and H. Lugaci, Destruction of erythroleukaemic cells by photoactivation of endogenous porphyrins, *Br. J. Cancer*, 1987, **56**, 589–595.

30. M. El-Far, A. Setate and M. El-Maadawy, Photodynamic therapy with amino-levulinic acid and meso-tetrahydroxyphenylchlorin: first initial clinical experience, in *SPIE proceedings*, SPIE, Bellingham, Washington, 1995, **2371**, 236–242.

31. J. Moan, Q. Peng, J.F. Evensen, K. Berg, A. Western and C. Rimington, Photo-sensitizing efficiencies, tumor, and cellular uptake of different photosensitizing drugs relevant for photodynamic therapy of cancer, *Photochem. Photobiol.*, 1987, **46**, 713–721.

32. A. Batlle, E.B.C. Llambias, E.W. Xifra and H.A. Tigier, Porphyrin biosynthesis in the soybean callus tissue system-XV. The effect of growth conditions, *Int. J. Biochem.*, 1975, **6**, 591–606.

33. N.M. Navone, A.L Frisardi, E.R. Resnik, A. Batlle and C.F. Polo, Porphyrin synthesis in human breast cancer. Preliminary mimetic in vitro studies, *Med. Sci. Res.*, 1988, **16**, 61–62.

34. D.X. Divaris, J.C. Kennedy and R.H. Pottier, Phototoxic damage to sebaceous glands and hair follicles of mice after systemic administration of 5-aminolevulinic acid correlates with localized protoporphyrin IX fluorescence, *Am. J. Pathol.*, 1990, **136**, 891–897.

35. R.H. Pottier, Y.F. Chow, J.P. LaPlante, T.G. Truscott, J.C. Kennedy and L.A. Beiner, Non-invasive technique for obtaining fluorescence excitation and emission spectra *in vivo*, *Photochem. Photobiol.*, 1986, **44**, 679–687.

36. J.C. Kennedy, R.H. Pottier and D.C. Pross, Photodynamic therapy with endog-enous protoporphyrin IX: basic principles and present clinical experience, *J. Photochem. Photobiol. B*, 1990, **6**, 143–148.

37. J.C. Kennedy and R.H. Pottier, Endogenous protoporphyrin IX, a clinically useful photosensitizer for photodynamic therapy, *J. Photochem. Photobiol. B*, 1992, **14**, 275–292.

38. C.J. Gomer, Preclinical examination of first and second generation photosensi-tizers used in photodynamic therapy, *Photochem. Photobiol.*, 1991, **54**, 1093–1107.

39. C.J. Gomer, A. Ferrario, N. Hayashi, N. Rucker, B.C. Szirth and A.L. Murphree, Molecular, cellular, and tissue responses following photodynamic therapy, *Lasers Surg. Med.*, 1988, **8**(5), 450–463.

40. A.M. Batlle, Porphyrins, porphyries, cancer and photodynamic therapy – a model for carcinogenesis, *J. Photochem. Photobiol. B*, 1993, **20**, 5–22.

41. G.M. Murphy, Diagnosis and management of the erythropoietic porphyrias, *Dermatol. Ther.*, 2003, **16**, 57–64.

42. R. Sroka, R. Baumgartner, A. Buser, C. Ell, D. Jocham and E. Unsöld, Laser assisted detection of endogenous porphyrins in malignant diseases, *SPIE Proceed-ings of Physiological Monitoring and Early Detection Diagnostic Methods*, SPIE, Bellingham, Washington, 1992, **1641**, 99–106.

43. T. Bocher, J. Beuthan, M. Scheller, J.U. Hopf, M. Linnarz, R.D. Naber, O. Minet, W. Becker and G.J. Mueller, Combined quantitative and qualitative two-channel optical biopsy technique for discrimination of tumor borders, *SPIE Proceedings of Optical Biopsy*, SPIE, Bellingham, Washington, 1995, **2627**, 118–124.

44. R.C. Krieg, H. Messmann, J. Rauch, S. Seeger and R. Knuechel, Metabolic characterization of tumor cell-specific protoporphyrin IX accumulation after exposure to 5-aminolevulinic acid in human colonic cells, *Photochem. Photobiol.*, 2002, **76**, 518–525.

45. S.B. Brown, The role of light in the treatment of non-melanoma skin cancer using methyl aminolevulinate, *J. Dermatolog. Treat.*, 2003, **14**(3), 11–14.

46. N.C. Zeitouni, A.R. Oseroff and S. Shieh, Photodynamic therapy for non-melanoma skin cancers. Current review and update, *Mol. Immunol.*, 2003, **17–18**, 1133–1136.

47. J.F. Greenbaum, Y. Gozlan, D. Schwartz, D.J. Katcoff and Z. Malik, Nuclear distribution of porphobilinogen deaminase (PBGD) in glioma cells: a regulatory role in cancer transformation, *Br. J. Cancer*, 2002, **18/86**, 1006–1011.

48. L. Greenbaum, D.J. Katcoff, H. Dou, Y. Gozlan and Z. Malik, A porphobilinogen deaminase (PBGD) Ran-binding protein interaction is implicated in nuclear trafficking of PBGD in differentiating glioma cells, *Oncogene*, 2003, **22**, 5221–5228.

49. E. Rud and K. Berg, *Characterisation of the cellular uptake of 5-aminolevulinic acid*, in *Photochemotherapy of Cancer and Other Diseases, Proceedings of the SPIE*, B. Ehrenberg and K. Berg (eds), SPIE, Bellingham, Washington, 1999, **3563**, 28–37.

50. F. Doring, J. Walter, J. Will, M. Focking, M. Boll, S. Amasheh, W. Clauss and H. Daniel, Delta-aminolevulinic acid transport by intestinal and renal peptide transporters and its physiological and clinical implications, *J. Clin. Invest.*, 1998, **101**, 2761–2767.

51. Q. Peng, K. Berg, J. Moan, M. Kongshaug and J.M. Nesland, 5-Aminolevulinic acid-based photodynamic therapy: principles and experimental research, *Photochem. Photobiol.*, 1997, **65**, 235–251.

52. E. Rud, O. Gederaas, A. Hogset and K. Berg, 5-aminolevulinic acid, but not 5-aminolevulinic acid, esters is transported into adenocarcinoma cells by system BETA transporters, *Photochem. Photobiol.*, 2000, **71**, 640–647.

53. P. Uehlinger, M. Zellweger, G. Wagnieres, L. Juillerat-Jeanneret, B.H. van den and N. Lange, 5-Aminolevulinic acid and its derivatives: physical chemical properties and protoporphyrin IX formation in cultured cells, *J. Photochem. Photobiol. B*, 2000, **54**, 72–80.

54. J.M. Gaullier, K. Berg, Q. Peng, H. Anholt, P.K. Selbo, L.W. Ma and J. Moan, Use of 5-aminolevulinic acid esters to improve photodynamic therapy on cells in culture, *Cancer Res.*, 1997, **57**, 1481–1486.

55. K.P. Uberriegler, E. Banieghbal and B. Krammer, Subcellular damage kinetics within co-cultivated WI38 and VA13-transformed WI38 human fibroblasts following 5-aminolevulinic acid-induced protoporphyrin IX formation, *Photochem. Photobiol.*, 1995, **62**, 1052–1057.

56. H. Liang, D.S. Shin, Y.E. Lee, D.C. Nguyen, T.C. Trang, A.H. Pan, S.L. Huang, D.H. Chong and M.W. Berns, Subcellular phototoxicity of 5-aminolaevulinic acid (ALA), *Lasers Surg. Med.*, 1998, **22**, 14–24.

57. C. Fuchs, R. Riesenberg, J. Siegert and R. Baumgartner, H-dependent formation of 5-aminolaevulinic acid-induced protoporphyrin IX in fibrosarcoma cells, *J. Photochem. Photobiol. B*, 1997, **40**, 49–54.

58. B. Ortel, N. Chen, J. Brissette, G.P. Dotto, E. Maytin and T. Hasan, Differentiation-specific increase in ALA-induced protoporphyrin IX accumulation in primary mouse keratinocytes, *Br. J. Cancer*, 1998, **77**, 1744–1751.

59. M. Grubinger, P. Hammerl, E. Banieghbal and B. Krammer, *Accumulation of aminolevulinic acid-induced protoporphyrin IX as a photosensitizer in L-929 cells*, in *Research Advances in Photochemistry and Photobiology*, R.M. Mohan (ed), vol. 1, Global Research Network, Kerala India, 2000, 137–145.

60. J. Bedwell, A.J. MacRobert, D. Phillips and S.G. Bown, Fluorescence distribution and photodynamic effect of ALA-induced PpIX in the DMH rat colonic tumour model, *Br. J. Cancer*, 1992, **65**, 818–824.

61. C.S. Loh, A.J. MacRobert, J. Bedwell, J. Regula, N. Krasner and S.G. Bown, Oral versus intravenous administration of 5-aminolaevulinic acid for photodynamic therapy, *Br. J. Cancer*, 1993, **68**, 41–51.

62. R. Pottier and J.C. Kennedy, *Photodynamic therapy with 5-aminolevulinic acid, basic principals, and applications*, in *Photochemotherapy Photodynamic Therapy and Other Modalities, Proceedings of the SPIE*, B. Ehrenberg, G. Jori and J. Moan (eds), SPIE, Bellingham, Washington, 1996, **2625**, 2–10.

63. S. Collaud, A. Juzeniene, J. Moan and N. Lange, On the selectivity of 5-amino-levulinic acid-induced protoporphyrin IX formation, *Curr. Med. Chem. Anti. Canc. Agents*, 2004, **4**, 301–316.

64. I. Wang, S. Andersson-Engels, G.E. Nilsson, K. Wardell and K. Svanberg, Superficial blood flow following photodynamic therapy of malignant non-melanoma skin tumours measured by laser Doppler perfusion imaging, *Br. J. Dermatol.*, 1997, **136**, 184–189.

65. B. Krammer and K. Uberriegler, In-vitro investigation of ALA-induced protoporphyrin IX, *J. Photochem. Photobiol. B*, 1996, **36**, 121–126.

66. Z. Malik, M. Dishi and Y. Garini, Fourier transform multipixel spectroscopy and spectral imaging of protoporphyrin in single melanoma cells, *Photochem. Photobiol.*, 1996, **63**, 608–614.

67. D.M. Fiedler, P.M. Eckl and B. Krammer, Does delta-aminolaevulinic acid induce genotoxic effects?, *J. Photochem. Photobiol. B*, 1996, **33**, 39–44.

68. J.M. Gaullier, M. Geze, R. Santus, M.T. Sa, J.C. Maziere, M. Bazin, P. Morliere and L. Dubertret, Subcellular localization of and photosensitization by protoporphyrin IX human keratinocytes and fibroblasts cultivated with 5-aminolevulinic acid, *Photochem. Photobiol.*, 1995, **62**, 114–122.

69. J. Moan and K. Berg, The photodegradation of porphyrins in cells can be used to estimate the lifetime of singlet oxygen, *Photochem. Photobiol.*, 1991, **53**, 549–553.

70. T. Verwanger, R. Sanovic, F. Aberger, A.M. Frischauf and B. Krammer, Gene expression pattern following photodynamic treatment of the carcinoma cell line A-431 analysed by cDNA arrays, *Int. J. Oncol.*, 2002, **21**, 1353–1359.

71. O.A. Gederaas, K. Thorstensen and I. Romslo, The effect of brief illumination on intracellular free calcium concentration in cells with 5-aminolevulinic acid-induced protoporphyrin IX synthesis, *Scand. J. Clin. Lab. Invest.*, 1996, **56**, 583–589.

72. D. Grebenova, P. Halada, J. Stulik, V. Havlicek and Z. Hrkal, Protein changes in HL60 leukemia cells associated with 5-aminolevulinic acid-based photodynamic therapy. Early effects on endoplasmic reticulum chaperones, *Photochem. Photobiol.*, 2000, **72**, 16–22.

73. L.O. Klotz, C. Fritsch, K. Briviba, N. Tsacmacidis, F. Schliess and H. Sies, Activation of JNK and p38 but not ERK MAP kinases in human skin cells by 5-aminolevulinate-photodynamic therapy, *Cancer Res.*, 1998, **58**, 4297–4300.

74. T. Verwanger, G. Schnitzhofer and B. Krammer, Expression kinetics of the (proto) oncogenes c-myc and bcl-2 following photodynamic treatment of normal and

transformed human fibroblasts with 5-aminolaevulinic acid-stimulated endogenous protoporphyrin IX, *J. Photochem. Photobiol. B*, 1998, **45**, 131–135.

75. S.L. Ratcliffe and E.K. Matthews, Modification of the photodynamic action of delta-aminolaevulinic acid (ALA) on rat pancreatoma cells by mitochondrial benzodiazepine receptor ligands, *Br. J. Cancer*, 1995, **71**, 300–305.

76. D. Schwartz, Y. Gozlan, L. Greenbaum, T. Babuschkina, D.J. Katcoff and Z. Malik, Differentiation dependent photodynamic therapy regulated by porphobilinogen deaminase in B16 melanoma, *Brit J. Cancer*, 2004, **90**(91), 1833–1841.

77. K. Berg, Z. Luksiene, J. Moan and L. Ma, Combined treatment of ionizing radiation and photosensitization by 5-aminolevulinic acid-induced protoporphyrin IX, *Radiat. Res.*, 1995, **142**, 340–346.

78. X. Li and J.D. Clark, The role of heme oxygenase in neuropathic and incisional pain, *Anesth. Analg.*, 2000, **90**, 677–682.

79. R.J. Riopelle and J.C. Kennedy, Some aspects of porphyrin neurotoxicity *in vitro*, *Can. J. Physiol. Pharmacol.*, 1982, **60**, 707–714.

80. C. Fritsch, H. Stege, G. Saalmann, G. Goerz, T. Ruzicka and J. Krutmann, Green light is effective and less painful than red light in photodynamic therapy of facial solar keratoses, *Photodermatol. Photoimmunol. Photomed.*, 1997, **13**, 181–185.

81. S.R. Wiegell, I.M. Stender, R. Na and H.C. Wulf, Pain associated with photodynamic therapy using 5-aminolevulinic acid or 5-aminolevulinic acid methylester on tape-stripped normal skin, *Arch. Dermatol.*, 2003, **139**, 1173–1177.

82. S.R. Wiegell, I.M. Stender, R. Na and H.C. Wulf, Pain associated with photodynamic therapy using 5-aminolevulinic acid or 5-aminolevulinic acid methylester on tape-stripped normal skin, *Arch. Dermatol.*, 2003, **139**, 1173–1177.

83. N.N. Danial and S.J. Korsmeyer, Cell death: critical control points, *Cell*, 2004, **116**, 205–219.

84. N.L. Oleinick, R.L. Morris and I. Belichenko, The role of apoptosis in response to photodynamic therapy what: where, why, and how, *Photochem. Photobiol. Sci.*, 2002, **1**, 1–21.

85. J. Piette, C. Volanti, A. Vantieghem, J.Y. Matroule, Y. Habraken and P. Agostinis, Cell death and growth arrest in response to photodynamic therapy with membrane-bound photosensitizers, *Biochem. Pharmacol.*, 2003, **66**, 1651–1659.

86. C.B. Oberdanner, T. Kiesslich, B. Krammer and K. Plaetzer, Glucose is required to maintain high ATP-levels for the energy-utilizing steps during PDT-induced apoptosis, *Photochem. Photobiol.*, 2002, **76**, 695–703.

87. L. Bourre, N. Rousset, S. Thibaut, S. Eleouet, Y. Lajat and T. Patrice, PDT effects of m-THPC and ALA, phototoxicity and apoptosis, *Apoptosis*, 2002, **7**, 221–230.

88. K. Kuzelova, D. Grebenova, M. Pluskalova, I. Marinov and Z. Hrkal, Early apoptotic features of K562 cell death induced by 5-aminolaevulinic acid-based photodynamic therapy, *J. Photochem. Photobiol. B*, 2004, **73**, 67–78.

89. D. Grebenova, K. Kuzelova, K. Smetana, M. Pluskalova, H. Cajthamlova, I. Marinov, O. Fuchs, J. Soucek, P. Jarolim and Z. Hrkal, Mitochondrial and endoplasmic reticulum stress-induced apoptotic pathways are activated by 5-aminolevulinic acid-based photodynamic therapy in HL60 leukemia cells, *J. Photochem. Photobiol. B*, 2003, **69**, 71–85.

90. T. Kriska, W. Korytowski and A.W. Girotti, Hyperresistance to photosensitized lipid peroxidation and apoptotic killing in 5-aminolevulinate-treated tumor cells overexpressing mitochondrial GPX4, *Free Radic. Biol. Med.*, 2002, **33/10**, 1389–1402.

91. H. Nakaseko, M. Kobayashi, Y. Akita, Y. Tamada and Y. Matsumoto, Histological changes and involvement of apoptosis after photodynamic therapy for actinic keratoses, *Br. J. Dermatol.*, 2003, **148/1**, 122–127.

92. C.S. Betz, J.P. Lai, W. Xiang, P. Janda, P. Heinrich, H. Stepp, R. Baumgartner and A. Leunig, *In vitro* photodynamic therapy of nasopharyngeal carcinoma using 5-aminolevulinic acid, *Photochem. Photobiol. Sci.*, 2002, **1**, 315–319.

93. G. Li, M.R. Szewczuk, R.H. Pottier and J.C. Kennedy, Effect of mammalian cell differentiation on response to exogenous 5-aminolevulinic acid, *Photochem. Photobiol.*, 1999, **69**, 231–235.

94. I. Georgakoudi, P.C. Keng and T.H. Foster, Hypoxia significantly reduces aminolaevulinic acid-induced protoporphyrin IX synthesis in EMT6 cells, *Br. J. Cancer*, 1999, **79**, 1372–1377.

95. A. Orenstein, G. Kostenich, H. Tsur, L. Kogan and Z. Malik, Temperature monitoring during photodynamic therapy of skin tumors with topical 5-aminolevulinic acid application, *Cancer Lett.*, 1995, **93**, 227–232.

96. B.W. Pogue, J.A. O'Hara, I.A. Goodwin, C.J. Wilmot, G.P. Fournier, A.R. Akay and H. Swartz, Tumor PO(2) changes during photodynamic therapy depend upon photosensitizer type and time after injection, *Comp. Biochem. Physiol. A Mol. Integr. Physiol.*, 2002, **132**, 177–184.

97. A. Orenstein, G. Kostenich, Y. Kopolovic, T. Babushkina and Z. Malik, Enhancement of ALA-PDT damage by IR-induced hyperthermia on a colon carcinoma model, *Photochem. Photobiol.*, 1999, **69**, 703–707.

98. G. Kostenich, T. Babushkina, Z. Malik and A. Orenstein, Photothermic treatment of pigmented B16 melanoma using a broadband pulsed light delivery system, *Cancer Lett.*, 2000, **157**, 161–168.

99. A. Orenstein, G. Kostenich and Z. Malik, The kinetics of protoporphyrin fluorescence during ALA-PDT in human malignant skin tumors, *Cancer Lett.*, 1997, **120**, 229–234.

100. C.D. Sudworth, M.R. Stringer, J.E. Cruse-Sawyer and S.B. Brown, Fluorescence microspectroscopy technique for the study of intracellular protoporphyrin IX dynamics, *Appl. Spectrosc.*, 2003, **57**, 682–688.

101. J.C. Finlay, D.L. Conover, E.L. Hull and T.H. Foster, Porphyrin bleaching and PDT-induced spectral changes are irradiance dependent in ALA-sensitized normal rat skin *in vivo*, *Photochem. Photobiol.*, 2001, **73**, 54–63.

102. H. Messmann, P. Mlkvy, G. Buonaccorsi, C.L. Davies, A.J. MacRobert and S.G. Bown, Enhancement of photodynamic therapy with 5-aminolaevulinic acid-induced porphyrin photosensitisation in normal rat colon by threshold and light fractionation studies, *Br. J. Cancer*, 1995, **72**, 589–594.

103. H. Messmann, R.M. Szeimies, W. Baumler, R. Knuchel, H. Zirngibl, J. Scholmerich and A. Holstege, Enhanced effectiveness of photodynamic therapy with laser light fractionation in patients with esophageal cancer, *Endoscopy*, 1997, **29**, 275–280.

104. P. Babilas, V. Schacht, G. Liebsch, O.S. Wolfbeis, M. Landthaler, R.M. Szeimies and C. Abels, Effects of light fractionation and different fluence rates on photodynamic therapy with 5-aminolaevulinic acid *in vivo*, *Br. J. Cancer*, 2003, **88**, 1462–1469.

105. E.A. Hryhorenko, A.R. Oseroff, J. Morgan and K. Rittenhouse-Diakun, Antigen specific and nonspecific modulation of the immune response by aminolevulinic acid based photodynamic therapy, *Immunopharmacology*, 1998, **40**, 231–240.

106. E.A. Hryhorenko, A.R. Oseroff, J. Morgan and K. Rittenhouse-Diakun, Deletion of alloantigen-activated cells by aminolevulinic acid-based photodynamic therapy, *Photochem. Photobiol.*, 1999, **69**, 560–565.

107. R.S. Boumedine and D.C. Roy, Elimination of alloreactive T cells using photodynamic therapy, *Cytotherapy*, 2005, **7**, 134–143.

108. S.O. Gollnick, L. Vaughan and B.W. Henderson, Generation of effective antitumor vaccines using photodynamic therapy, *Cancer Res.*, 2002, **62**, 1604–1608.

109. J.T. Dalton, M.C. Meyer and A.L. Golub, Pharmacokinetics of aminolevulinic acid after oral and intravenous administration in dogs, *Drug Metab. Dispos.*, 1999, **27**, 432–435.

110. P. Mustajoki, K. Timonen, A. Gorchein, A.M. Seppalainen, E. Matikainen and R. Tenhunen, Sustained high plasma 5-aminolaevulinic acid concentration in a volunteer: no porphyric symptoms, *Eur. J. Clin. Invest.*, 1992, **22**, 407–411.

111. J.T. Dalton, C.R. Yates, D. Yin, A. Straughn, S.L. Marcus, A.L. Golub and M.C. Meyer, Clinical pharmacokinetics of 5-aminolevulinic acid in healthy volunteers and patients at high risk for recurrent bladder cancer, *J. Pharmacol. Exp. Ther.*, 2002, **301**, 507–512.

112. R. Sroka, W. Beyer, L. Gossner, T. Sassy, S. Stocker and R. Baumgartner, Pharmacokinetics of 5-aminolevulinic-acid-induced porphyrins in tumour-bearing mice, *J. Photochem. Photobiol. B*, 1996, **34**, 13–19.

113. J. van den Boogert, R. Van Hillegersberg, F.W. de Rooij, R.W. de Bruin, A. Edixhoven-Bosdijk, A.B. Houtsmuller, P.D. Siersema, J.H. Wilson and H.W. Tilanus, 5-Aminolaevulinic acid-induced protoporphyrin IX accumulation in tissues: pharmacokinetics after oral or intravenous administration, *J. Photochem. Photobiol. B*, 1998, **44**, 29–38.

114. R.L. Prosst, L. Schroeter and J. Gahlen, Kinetics of intraoperative fluorescence diagnosis of parathyroid glands, *Eur. J. Endocrinol.*, 2004, **150**, 743–747.

115. R.L. Prosst, L. Schroeter and J. Gahlen, Enhanced ALA-induced fluorescence in hyperparathyroidism, *J. Photochem. Photobiol. B*, 2005, **79**, 79–82.

116. L. Lilge and B.C. Wilson, Photodynamic therapy of intracranial tissues: a preclinical comparative study of four different photosensitizers, *J. Clin. Laser Med. Surg.*, 1998, **16**, 81–91.

117. S.A. Friesen, G.O. Hjortland, S.J. Madsen, H. Hirschberg, O. Engebraten, J.M. Nesland and Q. Peng, 5-Aminolevulinic acid-based photodynamic detection and therapy of brain tumors (Review), *Int. J. Oncol.*, 2002, **21**, 577–582.

118. W. Stummer, H.J. Reulen, A. Novotny, H. Stepp and J.C. Tonn, Fluorescence-guided resections of malignant gliomas – an overview, *Acta Neurochir. Suppl.*, 2003, **88**, 9–12.

119. H. Stepp, T. Beck, W. Beyer, T. Pongratz, R. Sroka, R. Baumgartner, W. Stummer, B. Olzowy, J. Mehrkens, J.C. Tonn and H.J. Reulen, Fluorescence-guided resections and photodynamic therapy for malignant gliomas using 5-aminolevulinic acid, *Proceedings of the SPIE*, SPIE, Bellingham, Washington, 2005, **5686**, 547–557.

120. C. Perotti, A. Casas, H. Fukuda, P. Sacca and A. Batlle, ALA and ALA hexyl ester induction of porphyrins after their systemic administration to tumour bearing mice, *Br. J. Cancer*, 2002, **87**, 790–795.

121. S. Gerscher, J.P. Connelly, J. Griffiths, S.B. Brown, A.J. MacRobert, G. Wong and L.E. Rhodes, Comparison of the pharmacokinetics and phototoxicity of protoporphyrin IX metabolized from 5-aminolevulinic acid and two derivatives in human skin *in vivo*, *Photochem. Photobiol.*, 2000, **72**, 569–574.

122. J. Regula, A.J. MacRobert, A. Gorchein, G.A. Buonaccorsi, S.M. Thorpe, G.M. Spencer, A.R. Hatfield and S.G. Bown, Photosensitisation and photodynamic therapy of oesophageal, duodenal, and colorectal tumours using 5 aminolaevulinic acid induced protoporphyrin IX – a pilot study, *Gut*, 1995, **36**, 67–75.

123. K. Rick, R. Sroka, H. Stepp, M. Kriegmair, R.M. Huber, K. Jacob and R. Baumgartner, Pharmacokinetics of 5-aminolevulinic acid-induced protoporphyrin IX in skin and blood, *J. Photochem. Photobiol. B*, 1997, **40**, 313–319.

124. J. Webber, D. Kessel and D. Fromm, On-line fluorescence of human tissues after oral administration of 5-aminolevulinic acid, *J. Photochem. Photobiol. B*, 1997, **38**, 209–214.

125. J. Webber, D. Kessel and D. Fromm, Plasma levels of protoporphyrin IX in humans after oral administration of 5-aminolevulinic acid, *J. Photochem. Photobiol. B*, 1997, **37**, 151–153.

126. M. Doss, Normal ranges of porphyrins and precursors in human tissue, urine and feces, in *Chemical porphynia in man*, J.J.T.W.A. Strik and J.H. Koeman (eds), Elsevier, Biomedical press, Amsterdam, 1979, pp. 221.

127. J. Webber, D. Kessel and D. Fromm, Side effects and photosensitization of human tissues after aminolevulinic acid, *J. Surg. Res.*, 1997, **68**, 31–37.

128. M.A. Herman, J. Webber, D. Fromm and D. Kessel, Hemodynamic effects of 5-aminolevulinic acid in humans, *J. Photochem. Photobiol. B*, 1998, **43**, 61–65.

129. B.W. Henderson, L. Vaughan, D.A. Bellnier, H. Van Leengoed, P.G. Johnson and A.R. Oseroff, Photosensitization of murine tumor, vasculature and skin by 5-aminolevulinic acid-induced porphyrin, *Photochem. Photobiol.*, 1995, **62**, 780–789.

130. A.L. Golub, E.F. Gudgin-Dickson, J.C. Kennedy, S.L. Marcus, Y. Park and R.H. Pottier, The monitoring of ALA-induced protoprophyrin IX accumulation and clearance in patients with skin lesions by *in vivo* surface-detected fluorescence spectroscopy, *Lasers Med. Sci.*, 1999, **14**, 112–122.

131. J.C. Kennedy, S.L. Marcus and R.H. Pottier, Photodynamic therapy (PDT) and photodiagnosis (PD) using endogenous photosensitization induced by 5-aminolevulinic acid (ALA): mechanisms and clinical results, *J. Clin. Laser Med. Surg.*, 1996, **14**, 289–304.

132. J.C. Kennedy, P. Nadeau, Z.J. Petryka, R.H. Pottier and G. Weagle, Clearance times of porphyrin derivatives from mice as measured by *in vivo* fluorescence spectroscopy, *Photochem. Photobiol.*, 1992, **55**, 729–734.

133. R.J. Riopelle and J.C. Kennedy, Some aspects of porphyrin neurotoxicity *in vitro*, *Can. J. Physiol. Pharmacol.*, 1982, **60**, 707–714.

134. H.L. Bonkovsky, J.F. Healey, P.R. Sinclair and J.F. Sinclair, Conversion of 5-aminolaevulinate into haem by homogenates of human liver. Comparison with rat and chick-embryo liver homogenates, *Biochem. J.*, 1985, **227**, 893–901.

135. L. Brancaleon and H. Moseley, Laser and non-laser light sources for photodynamic therapy, *Lasers Med. Sci.*, 2002, **17**, 173–186.

136. W. Beyer, Systems for light application and dosimetry in photodynamic therapy, *J. Photochem. Photobiol. B*, 1996, **36**, 153–156.

137. T. Stepinac, P. Grosjean, A. Woodtli, P. Monnier, B.H. van den and G. Wagnieres, Optimization of the diameter of a radial irradiation device for photodynamic therapy in the esophagus, *Endoscopy*, 2002, **34**, 411–415.

138. H. van den and Bergh, On the evolution of some endoscopic light delivery systems for photodynamic therapy, *Endoscopy*, 1998, **30**, 392–407.

139. L.K. Lee, C. Whitehurst, M.L. Pantelides and J.V. Moore, An interstitial light assembly for photodynamic therapy in prostatic carcinoma, *BJU Int.*, 1999, **84**, 821–826.

140. T.C. Zhu, S.M. Hahn, A.S. Kapatkin, A. Dimofte, C.E. Rodriguez, T.G. Vulcan, E. Glatstein and R.A. Hsi, *In vivo* optical properties of normal canine prostate at 732 nm using motexafin lutetium-mediated photodynamic therapy, *Photochem. Photobiol.*, 2003, **77**, 81–88.

141. T. Johansson, M.S. Thompson, M. Stenberg, K.C. af, S. Andersson-Engels, S. Svanberg and K. Svanberg, Feasibility study of a system for combined light dosimetry and interstitial photodynamic treatment of massive tumors, *Appl. Opt.*, 2002, **41**, 1462–1468.

142. P. van Veen, J.H. Schouwink, W.M. Star, H.J. Sterenborg, D.S. van Jr, F.A. Stewart and P. Baas, Wedge-shaped applicator for additional light delivery and dosimetry in the diaphragmal sinus during photodynamic therapy for malignant pleural mesothelioma, *Phys. Med. Biol.*, 2001, **46**, 1873–1883.

143. D.J. Robinson, H.S. de Bruijn, N. Van der Veen, M.R. Stringer, S.B. Brown and W.M. Star, Fluorescence photobleaching of ALA-induced protoporphyrin IX during photodynamic therapy of normal hairless mouse skin: the effect of light dose and irradiance and the resulting biological effect, *Photochem. Photobiol.*, 1998, **67**, 140–149.

144. K. König, H. Schneckenburger, A. Ruck and R. Steiner, *In vivo* photoproduct formation during PDT with ALA-induced endogenous porphyrins, *J. Photochem. Photobiol. B*, 1993, **18**, 287–290.

145. M.A. Ortner, B. Ebert, E. Hein, K. Zumbusch, D. Nolte, U. Sukowski, J. Weber-Eibel, B. Fleige, M. Dietel, M. Stolte, G. Oberhuber, R. Porschen, B. Klump, H. Hortnagl, H. Lochs and H. Rinneberg, Time gated fluorescence spectroscopy in Barrett's oesophagus, *Gut*, 2003, **52**, 28–33.

146. S. Lam, T. Kennedy, M. Unger, Y.E. Miller, D. Gelmont, V. Rusch, B. Gipe, D. Howard, J.C. Leriche, A. Coldman and A.F. Gazdar, Localization of bronchial intraepithelial neoplastic lesions by fluorescence bronchoscopy, *Chest*, 1998, **113**, 696–702.

147. R. Drezek, C. Brookner, I. Pavlova, I. Boiko, A. Malpica, R. Lotan, M. Follen and R. Richards-Kortum, Autofluorescence microscopy of fresh cervical-tissue sections reveals alterations in tissue biochemistry with dysplasia, *Photochem. Photobiol.*, 2001, **73**, 636–641.

148. K. Badizadegan, V. Backman, C.W. Boone, C.P. Crum, R.R. Dasari, I. Georga-koudi, K. Keefe, K. Munger, S.M. Shapshay, E.E. Sheetse and M.S. Feld, Spectroscopic diagnosis and imaging of invisible pre-cancer, *Faraday Discuss.*, 2004, **126**, 265–279.

149. P.J. Tadrous, J. Siegel, P.M. French, S. Shousha, E. Lalani and G.W. Stamp, Fluorescence lifetime imaging of unstained tissues: early results in human breast cancer, *J. Pathol.*, 2003, **199**, 309–317.

150. H. Stepp, R. Baumgartner, W. Beyer, R. Knüchel, T.O. Körner, M. Kriegmair, K. Rick, P. Steinbach, H.G. Stepp and A. Hofstetter, *Fluorescence imaging and spectroscopy of ALA-induced protoporphyrin IX preferentially accumulated in tumour tissue*, 1995, **2627**, 13–24.

151. M. Kriegmair, H. Stepp, P. Steinbach, W. Lumper, A. Ehsan, H.G. Stepp, K. Rick, R. Knuchel, R. Baumgartner and A. Hofstetter, Fluorescence cystoscopy following intravesical instillation of 5-aminolevulinic acid: a new procedure with

high sensitivity for detection of hardly visible urothelial neoplasias, *Urol. Int.*, 1995, **55**(4), 190–196.

152. M.A. D'Hallewin, L. Bezdetnaya and F. Guillemin, Fluorescence detection of bladder cancer: a review, *Eur. Urol.*, 2002, **42**, 417–425.

153. D. Zaak, D. Frimberger, H. Stepp, S. Wagner, R. Baumgartner, P. Schneede, M. Siebels, R. Knuchel, M. Kriegmair and A. Hofstetter, Quantification of 5-aminolevulinic acid induced fluorescence improves the specificity of bladder cancer detection, *J. Urol.*, 2001, **166**, 1665–1668.

154. S. Andersson-Engels, G. Canti, R. Cubeddu, C. Eker, K.C. af, A. Pifferi, K. Svanberg, S. Svanberg, P. Taroni, G. Valentini and I. Wang, Preliminary evaluation of two fluorescence imaging methods for the detection and the delineation of basal cell carcinomas of the skin, *Lasers Surg. Med.*, 2000, **26**, 76–82.

155. H. Schneckenburger, K. Stock, M. Lyttek, W.S. Strauss and R. Sailer, Fluorescence lifetime imaging (FLIM) of rhodamine 123 in living cells, *Photochem. Photobiol. Sci.*, 2004, **3**, 127–131.

156. J.R. Lakowicz, H. Szmacinski, K. Nowaczyk, K.W. Berndt and M. Johnson, Fluorescence lifetime imaging, *Anal. Biochem.*, 1992, **202**, 316–330.

157. U. Mahmood and R. Weissleder, Near-infrared optical imaging of proteases in cancer, *Mol. Cancer Ther.*, 2003, **2**, 489–496.

158. F. de Braud, M. Maffezzini, V. Vitale, P. Bruzzi, G. Gatta, W.F. Hendry and C.N. Sternberg, Bladder cancer, *Crit. Rev. Oncol. Hematol.*, 2002, **41**, 89–106.

159. D. Zaak, E. Hungerhuber, P. Schneede, H. Stepp, D. Frimberger, S. Corvin, N. Schmeller, M. Kriegmair, A. Hofstetter, R. Knuechel and R. Knochel, Role of 5-aminolevulinic acid in the detection of urothelial premalignant lesions, *Cancer*, 2002, **95**, 1234–1238.

160. W.R. Potter, T.S. Mang and T.J. Dougherty, The theory of photodynamic therapy dosimetry: consequences of photo-destruction of sensitizer, *Photochem. Photobiol.*, 1987, **46**, 97–101.

161. S.L. Jacques, Laser-tissue interactions. Photochemical, photothermal, and photomechanical, *Surg. Clin. North Am.*, 1992, **72**, 531–558.

162. S.L. Jacques, Light distributions from point, line and plane sources for photochemical reactions and fluorescence in turbid biological tissues, *Photochem. Photobiol.*, 1998, **67**, 23–32.

163. W.M. Star, Light dosimetry *in vivo*, *Phys. Med. Biol.*, 1997, **42**, 763–787.

164. S.G. Bown and L.B. Lovat, The biology of photodynamic therapy in the gastrointestinal tract, *Gastrointest. Endosc. Clin. N. Am.*, 2000, **10**, 533–550.

165. K.K. Wang and P.K. Nijhawan, Complications of photodynamic therapy in gastrointestinal disease, *Gastrointest. Endosc. Clin. N. Am.*, 2000, **10**, 487–495.

# Chapter 3

# ALA/MAL-PDT in Dermatology

## Verena Schleyer and Rolf-Markus Szeimies

**Table of Contents**

## Abstract

Depending on the applied light dose and the concentration of the photosensitizer, photodynamic therapy (PDT) in Dermatology can be used either as high-dose PDT for dermatooncologic indications inducing rapid cell death by necrosis and apoptosis or as low-dose PDT for inflammatory dermatoses by stimulation of immunomodulatory effects. Topical PDT already plays a substantial role in the treatment of non-melanoma skin cancers or precancerous lesions such as superficial basal cell carcinoma, actinic keratoses, Bowen's disease, and squamous cell carcinoma *in situ*. However, a therapeutical benefit of PDT is also evident for inflammatory dermatoses like acne vulgaris, scleroderma, or viral warts.

## 3.1. Introduction

### 3.1.1. Photosensitizers

Since systemic photosensitizers such as hematoporphyrin derivative (HPD) induce prolonged and generalized cutaneous phototoxicity which lasts for several weeks,[1] their use for dermatologic indications is restricted. Furthermore, topical application of HPD or other large molecules with tetrapyrolic ring structures is impractical because those drugs do not penetrate the skin in sufficient amounts. Therefore, further work has been focused on the development of topical photosensitizers.

In 1990 the heme precursor 5-aminolevulinic acid (ALA) and later its methylester (MAL), which induce photoactive porphyrins, have been introduced by Kennedy and co-workers.[2] The low molecular weight, together with their pharmacochemical properties allow easy penetration into the skin[2] with preferential intracellular accumulation inside the altered cells of the epidermis, where they are metabolized in the heme biosynthesis to photosensitizing porphyrins like protoporphyrin IX (PpIX)[3–6] Normally within the next 24–48 h PpIX is transformed to the photodynamically inactive heme.[7] These approaches have been studied extensively with great success for epithelial cancers after topical application. In contrast, systemic application of ALA in the treatment of solid tumors is disappointing, probably due to a heterogeneous ALA distribution and metabolism in the diseased tissue, leading to an incomplete destruction of the tumor microcirculation and resulting in only insufficient tumor ischemia, which is crucial for the PDT effect in solid tumors.[7–9]

In the United States 5-ALA hydrochloride as a solution (Levulan Kerastick, DUSA, USA) has reached approval status for PDT of actinic keratoses (AK) in combination with blue light (Blu-U lamp, DUSA, USA).[8] In Europe and Australia methyl aminolevulinate MAL (trade name Metvix (Europe) or Metvixia (US); Photocure ASA, (Norway), and Galderma S.A., (France), 160 mg g$^{-1}$) was approved for PDT of basal cell carcinoma, AK and recently also Bowen's disease in combination with red light. Also, in some countries in

**Table 1.**  Protocol for topical PDT in epithelial cancers or precanceroses with different photosensitizers

| Photosensitizer | 20% ALA in custom made oil-in-water emulsions or gels | Levulan Kerastick solution | 16% Metvix/Metvixia cream |
|---|---|---|---|
| Application | Occlusive, light impermeable | Occlusive, light impermeable | Occlusive, light impermeable |
| Incubation time | 4–6 h | 14–18 h | 3–4 h |
| Light source | Blue, green or red light (indication-dependent) | Blue light (Blu-U 4170) | Red light from LED source (ActiLite) |
| Irradiation parameters | Light intensity 100–180 mW cm$^{-2}$<br>Light dose 120–180 J cm$^{-2}$ | Light intensity 10 mW cm$^{-2}$ | Light intensity approx. 60 mW cm$^{-2}$<br>Light dose 37 J cm$^{-2}$ |
| Indication | Epithelial precancerous lesions, superficial basal cell carcinoma | Actinic keratoses | Superficial and nodular basal cell carcinoma (BCC) and actinic keratoses (AK), Bowen's disease |
| Treatments | 1 (retreatment if needed) | 1 (retreatment if needed) | 2 sessions 7 days apart for BCC and Bowen's disease, session for AK (retreated if needed) |

Europe custom-made oil-in-water emulsions or gels in concentrations up to 20% ALA (Crawford Pharmaceuticals, UK; photodynamic GmbH, Germany), or in combination with penetration enhancers, such as 40% DMSO are used for the treatment of skin tumors.

Meanwhile, for the treatment of epithelial cancerous lesions well-defined guidelines regarding photosensitizer concentrations, incubation times, light sources, and irradiation parameters exist (Table 1). In contrast, inflammatory skin diseases are treated with more experimental protocols, very often just a "low-dose PDT" is performed. Nevertheless, since there is neither official approval for those indications nor sufficient data with large study populations, these therapeutic approaches are still considered as experimental.

### 3.1.2. Light Sources

For dermatological purposes in topical PDT, light sources with wide illumination fields, which accomplish the simultaneous irradiation of larger areas, are most appropriate. Therefore mainly incoherent light sources like lamps or LED arrays, which match the absorption maxima of ALA- or MAL-induced porphyrins, are used.[10–14] Lesions may also be treated with laser systems matching the specific absorption bands of the applied photosensitizers. However, lasers are generally costly and require more maintenance than incoherent light sources. As no single light source is ideal for every possible indication for

**Table 2.** List of possible lamps and lasers for PDT in Dermatology (for details see the appendix)

| Type | Specific type | Emission $\lambda$ (nm) | Fluence rate (mW cm$^{-2}$) | Maximum field diameter (cm) |
|---|---|---|---|---|
| Laser | Argon dye | 630 | 10–500 | 10 |
| | Copper vapor dye | 630 | 10–500 | 10 |
| | Nd:YAG-KTP dye | 630 | 10–500 | 10 |
| | Semiconductor diode | 630 ± 5 | 10–500 | 10 |
| LED array | PRP 100 | 630 ± 5 | <150 | 4 |
| Xenon arc | Paterson PTL | 630 ± 15 | 10–130 | 8 |
| Metal halide | Waldmann 1200L | 600–750 | 10–200 | 15 |
| Tungsten/ halogen | Projector (modified) | 570–1100 | <200 | ~15 |
| | Photocure Curelight | 570–670 | <150 | 5.5 |
| Fluorescent | DUSA Blu-U 4170 | 417 ± 5 | 10 | >20 |
| LED array | Aktilite R 16 | 633 ± 5 | 70–100 | 5 |
| | Aktilite R 128 | 633 ± 5 | 70–100 | ~21 |
| | Omnilux | 633 ± 5 | 50–100 | ~23 |

topical PDT, the choice should be made according to clinical indications, treatment times, flexibility, and costs (Table 2).

For inflammatory dermatoses the use of broad-spectrum red light (580–700 nm), usually with a light dose of 10–40 J cm$^{-2}$ and an intensity of 50–70 mW cm$^{-2}$ is reported in the literature. For oncologic indications light doses of 100–150 J cm$^{-2}$ (100–200 mW cm$^{-2}$) are chosen. For the more narrow emission spectra of the LED systems (bandwidth approx. 30 nm) the values are significantly lower (37–50 J cm$^{-2}$). This is also true for blue light. To avoid hyperthermic effects, the light intensity should not exceed 200 mW cm$^{-2}$.[11,12] A comparative trial demonstrated that light at shorter wavelengths was less effective in the treatment of Bowen's disease at a theoretically equivalent dose, and the use of red light is recommended mainly for oncologic PDT to achieve maximum tissue penetration.[5,15] However, due to the lower degree of lesion thickness associated with non-hypertrophic AK the combination of Levulan Kerastick with blue light has also reached approval status for photodynamic treatment.

With red light non-melanoma skin cancer up to a thickness of 2–3 mm can be treated; thicker lesions require multiple treatments or tissue preparation (de-bulking) prior to PDT.[16–18] The site of the lesion is usually illuminated for 5–20 min. During illumination, both the patient and clinic staff should wear protective goggles in order to avoid the risk of eye damage.[19]

### 3.1.3. Biological Mechanism

Irradiation with light of the appropriate wavelength leads to activation of a photosensitizer, followed by generation of reactive oxygen species (ROS).

Depending on the amount and localization in the target tissue these ROS either induce cell death by necrosis or apoptosis or modify cellular functions[5,8,20] (see Chapters 1 and 2), making PDT an effective treatment option in special oncologic and inflammatory skin conditions. Since ALA or MAL accumulate significantly in proliferating, relatively iron-deficient tumor cells of epithelial origin and in inflammatory cells, tissue damage or immunomodulation is largely restricted to the sensitized cells, almost omitting the surrounding tissue, especially those cells of mesenchymal origin such as fibroblasts. Although PDT seems to have the potential to promote genotoxic effects like chromosomal aberrations *in vitro*,[21,22] this has never been shown *in vivo*. Moreover, a recent study showed even a delay in photoinduced carcinogenesis in mice following repetitive treatments with ALA-PDT.[23] Therefore, apart from two case reports with possible coincidence no further report on the carcinogenic potential of ALA/MAL-PDT has yet been published. The first case involved an 82-year-old man who had received seven sessions of ALA-PDT over 4 years to treat AK and squamous cell carcinomas (SCCs). In the treated area, a melanoma developed 6 months after the last PDT.[24] The other report concerned a 38-year-old man with histologically controlled clearance of erythroplasia of Queyrat after ALA-PDT. Because of clinical doubt about the completeness of remission topical 5-fluorouracil (5-FU) was applied twice weekly for 4 months. Subsequently a nodule developed on the penis that was histologically confirmed as SCC.[25] The British Photodermatology Group consequently rated ALA-PDT as therapy with a low frequency of severe adverse effects, good cosmesis, and a low risk of carcinogenic potential (strength of recommendation B, quality of evidence IIiii).[6] (See Table 3 for explanation.)

**Table 3.** Strength of recommendations and quality of evidence[6]

*Strength of recommendations*
A: There is good evidence to support the use of the procedure
B: There is fair evidence to support the use of the procedure
C: There is poor evidence to support the use of the procedure
D: There is fair evidence to support the rejection of the use of the procedure
E: There is good evidence to support the rejection of the use of the procedure
*Quality of evidence*
I: Evidence obtained from at least one properly designed, randomized controlled study
II(i): Evidence obtained from well-designed controlled trials without randomization
II(ii): Evidence obtained from well-designed cohort or case-control analytical studies, preferably from more than one centre or research group
II(iii): Evidence obtained from multiple time series with or without the intervention. Dramatic results in uncontrolled experiments could also be regarded as this type of evidence
III: Opinions of respected authorities based on clinical experience, descriptive studies or reports of expert committees
IV: Evidence inadequate owing to problems of methodology
A-II(iii): A range of light sources is effective in promoting dermatological applications of 5-aminolaevulinic acid-photodynamic therapy (ALA-PDT)
B-II(iii): Topical ALA-PDT is a safe treatment with few side-effects and no evidence of carcinogenicity during a decade of clinical use

**Table 4.** Possible drug interactions during PDT

| Drugs | Interaction |
| --- | --- |
| Anticoagulants or aspirin | Possible bleeding in case of vigorous lesion preparation (curettage before PDT) |
| Nonsteroidal anti-inflammatory drugs | Risk of NSAIDs quenching a vital inflammatory reaction contributing to the phototoxic effect |
| Cytostatic drugs like hydroxyurea, azathioprine, *etc.* | Risk of prolonged healing after PDT |
| Antioxidants like vitamin C or E | Risk of antioxidants quenching the photodynamic reaction |
| Other photosensitizing drugs such as diuretics, antibiotics or antimalarials | Risk of amplification of the phototoxic effect |

### 3.1.4. Lesion Preparation

In case of prominent tumor parts or extensive hyperkeratosis, removal with a ring curettage or scalpel is required prior to incubation[†].[17–19,26] ALA preparations are usually applied to the lesions with little overlap of the surrounding tissue for 4–6 h prior to irradiation under occlusion and in addition with a light protective dressing or clothing[‡].[5] For the commercially available MAL ointment (Metvix or Metvixia), a shorter incubation time of 3 h is sufficient because of preferential uptake and higher selectivity,[27,28] but the entire area must also be covered with an occlusive dressing for better penetration.

### 3.1.5. Contraindications and Drug Interactions

There are only a few contraindications when treating a patient with topical ALA-PDT. Two single case reports showed allergic contact dermatitis after ALA[29] and MAL[30] following several PDT treatments. Patients developed acute eczema in the treated area and hyperreaction of the surrounding untreated skin. These findings were confirmed by a positive patch test. Other contraindications are pregnancy and porphyrias, to exclude the possible risk of inducing/enhancing porphyria by topical administration of 5-ALA. Also, possible drug interactions must be taken into account before performing ALA-PDT (Table 4).

### 3.1.6. Side Effects and Complications

The most important side effects of ALA/MAL-PDT are stinging pain and a burning sensation (Table 5). They are usually well tolerated when smaller areas are treated. However, in the case of larger areas (mostly when AK with coinciding severely sun damaged skin on the scalp or the face are treated) a significant discomfort (usually restricted to the time span of irradiation and a

---

[†] Overnight occlusive dressing with petrolatum in order to reduce crusts or to alleviate mechanical removal might also be effective.

[‡] Pure aluminium foil should never be used as light protective dressing on skin on top of cream, since the acidose cream can destroy the aluminium.

**Table 5.** Side effects of PDT

| Frequency | Side effects |
| --- | --- |
| Usually | Stinging or burning sensation during illumination, especially face and scalp |
| | Localized erythema and edema after illumination |
| | Erosion, crust formation and dry necrosis of the tumor |
| Occasionally | Stinging or burning some hours after illumination |
| | Follicular bound sterile pustules |
| Rarely | Minor scarring |
| | Ulceration |
| | Pigmentary changes (hypo- or hyperpigmentation) |
| | Alopecia |

couple of hours thereafter[6]), often limits the patients' compliance. Therefore, especially in an extended irradiation field, analgesia with metamizole or piritramide or even, rarely, general anesthesia can be necessary.[31] Pain perception can also be alleviated by concurrent cold air analgesia, using a fan or water spray or by pouring water on the treated area during illumination.[32] Application of topical anesthetics is not recommended, since eutectic mixtures of lidocaine/prilocaine (EMLA) prior to irradiation might interfere with the acidity of the ALA/MAL-preparation: its high pH leads to a chemical inactivation (condensation) of the photosensitizer. In addition, topically applied analgesics can also induce local vasoconstriction, which adversely affects the oxygen supply, thus reducing the generation of ROS. Besides, in a randomized, placebo-controlled, double-blind study application of a tetracaine gel 1 h before irradiation did not significantly reduce pain during and after ALA-PDT.[33]

After illumination localized erythema and edema usually develops in the treated area. This is followed by a dry necrosis sharply restricted to the tumor bearing areas over the next days, and occasionally small follicular bound sterile pustules (when areas with a high density of sebaceous glands are treated). Local application of an antiseptic or antibiotic, for example, fusidic acid or clioquinol may then be helpful. After 10–21 days, the formed crusts come off and usually complete re-epithelialization occurs. During this period, most patients report little or no discomfort. In general, the cosmetical outcome of the completely responding lesions is good to excellent, with only about 2% minor scarring, or pigmentary changes that often are only transient. Because of the significantly lower doses of both light and photosensitizer in the context of "low-dose ALA-PDT" for inflammatory dermatoses little or no side effects appear, although multiple treatments are necessary. Irreversible alopecia has not yet been observed in the vast majority of the treated patients: however, owing to the concomitant sensitization of the pilosebaceous units, this effect should be taken into account.[5–6,31] ALA-PDT can be repeated several times and even in areas with prior exposure to ionizing radiation, ALA-PDT is possible.[34]

## 3.2. Oncologic Indications

### 3.2.1. Epithelial Cancers or Precancerous Lesions

Regarding oncologic indications, AK and nodular or superficial basal cell carcinomas (BCCs) are already approved indications for MAL in combination with red light. ALA is approved in the United States in combination with blue light for AK. In addition, treatment of Bowen's disease is also possible for PDT with ALA/MAL-induced porphyrins as already recommended by evidence-based guidelines.[6] Meanwhile, MAL has also been registered for this indication. Several efficient alternative treatments are available for single lesions, for example, surgery or cryotherapy. In contrast, ALA-PDT is already a favored choice for the therapy of extensive, widespread or multiple low-risk superficial lesions, for example, nevoid basal cell carcinoma syndrome,[8] immunosuppressed patients after organ transplantation (who often suffer from multiple lesions) and elderly patients with multiple AK of the scalp, face, and hands (Table 6). Possible indications for topical ALA/MAL-PDT are discussed below, and guidelines from the British Photodermatology Group workshop to promote efficacy and safety of PDT are stated[6] (Table 3).

### 3.2.1.1. Basal Cell Carcinoma.
*3.2.1.1.1. Different clinical types of basal cell carcinomas.* Basal cell carcinomas (BCCs) are the commonest malignant tumors of the skin, arising from the basal cells of the epidermis. They are mainly located in sun-exposed areas like the face, the neck region, upper extremities, and the trunk and then rarely metastasize. Treatment of BCCs should be chosen according to clinical type, tumor size, and location. Due to the limited penetration of red light into tissue, tumor thickness is a determinant response parameter of BCCs to ALA/MAL-PDT and should not exceed 2–3 mm to achieve complete destruction[6] (Figures 1 and 2). Therefore, nodular BCCs with vertical growth and greater thickness should be preferentially treated by surgery since treatment of nodular BCCs by single ALA-PDT has resulted in low average cure rates. Better results by far are achieved using MAL-PDT (possibly due to higher lipophilicity, more

**Table 6.** Indication for PDT in epithelial tumors

- Multiple lesions
- Large, extensive or widespread lesions
- Lesions <2–3 mm thickness
- Nevoid basal cell carcinoma syndrome
- Immunosuppressed patients with multiple lesions
- Site of lesion with a risk of disfigurement or poor healing from conventional therapies
- Contraindications for surgery, *e.g.*, anticoagulants

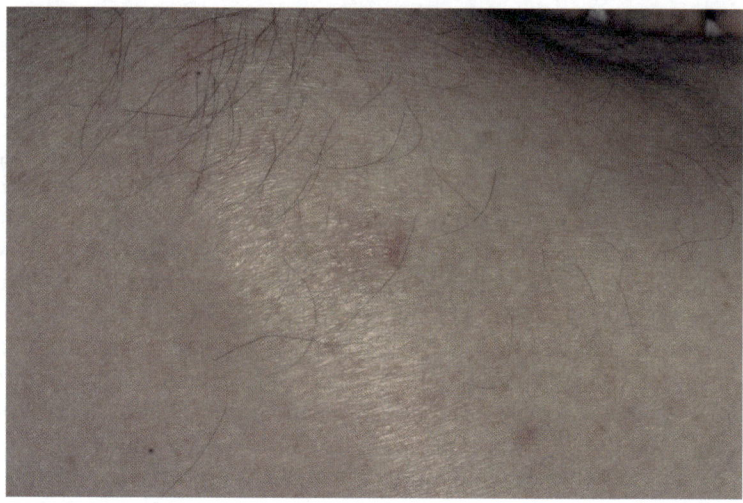

**Figure 1.**  Fifty-nine-year old man with superficial BCC on the back prior to ALA-PDT.

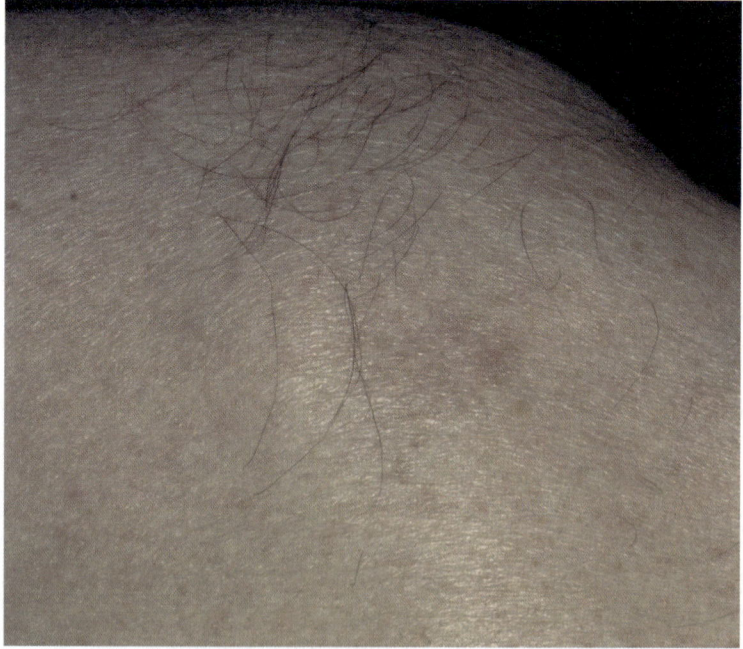

**Figure 2.**  Two months after a single course of ALA-PDT, excellent clinical result.

rapid skin penetration, and higher selectivity[8]) on the one hand, but also due to regular lesion preparation (debulking) prior to PDT. (See further studies below.)

In addition, pigmented BCCs do not allow an optimal penetration of the light and should therefore not be treated with PDT. This restriction is also suggested for morphoeic BCCs with diffuse and unpredictable borders. However, a recent report showed that ALA-PDT could be used for adjuvant therapy in combination with Moh's surgery.[35] In four patients who underwent Moh's micrographic surgery for extensive BCCs, the central infiltrating tumor part was first excised. After re-epithelialization, ALA-PDT was performed on the surrounding tumor rims (2–5 cm) bearing remaining superficial tumor parts. This led to a complete remission of the tumors with an excellent clinical and cosmetic result (with a follow-up period of up to 27 months).

Superficial BCCs usually occur on the trunk and are often multiple. A large number of treatment modalities exist, especially for single superficial BCCs, for example, excisional surgery, curettage and electrocautery, cryotherapy, cytotoxic agents, or radiotherapy. However, in the treatment of multiple lesions (*e.g.*, in the case of basal cell nevus syndrome[8] or in immunosuppressed patients after organ transplantation). ALA-PDT has the potential of becoming a first line therapy with an excellent cosmesis but without complications like scar formation, requirement for grafts, need of multiple treatments or pigmentary changes.

*3.2.1.1.2. Literature.* The above-mentioned conclusions are drawn from extensive studies in past years using ALA/MAL-PDT for nodular and superficial BCCs (Table 7). In a review published in 1997, the weighted average complete clearance rates in 12 studies treating 826 superficial and 208 nodular BCCs with ALA-PDT after a follow-up varying between 3 and 36 months were 87% and 53%, respectively.[36] Compiled data from other trials have shown comparable results with an average of 87–92% for superficial BCCs and 48–71% for nodular BCCs.[8] The exquisite results of ALA-PDT in superficial BCCs were also confirmed in two other trials in 61 and 87 patients.[12,13] At a 12 months follow-up a complete clearance rate of 82% and 97%, respectively, was determined.

A discrepancy between clinical and histological recurrence rates was seen in a prospective phase III trial comparing ALA-PDT with cryosurgery in 88 superficial and nodular BCCs.[37] In the PDT arm, a 20% ALA/water-in-oil cream was applied for 6 h under an occlusive dressing, followed by illumination with a laser at 635 nm (80 mW cm$^{-2}$, 60 J cm$^{-2}$). The other lesions were treated by cryosurgery using liquid nitrogen (open spray technique; two freeze–thaw cycles, each 25–30 s). After 3 months, punch biopsies were taken and revealed obviously higher histological recurrence rates of 25% in the PDT group and 15% in the cryosurgery group compared to clinical recurrence rates of only 5% and 13%, respectively. Besides better cosmetic outcome, healing time was also shorter in the PDT treated group.

**Table 7.**  Summary of results of selected clinical studies using topical ALA- or MAL-
PDT for the treatment of basal cell carcinoma

| Study | Indication | Sensitizer/ Procedure | Number of lesions/ (patients) | Complete remission | Follow-up |
|---|---|---|---|---|---|
| Morton[54] | Superficial BCCs <2 mm thickness | ALA | | 100% | |
| Thissen[18] | Nodular BCCs | ALA/debulking 3 weeks prior to PDT | 24 | 92% | 3 months (histological control) |
| Haller[16] | Superficial BCCs | ALA/double treatment within 7 days | 26 | 96% | 15–45 months |
| Wang[37] | Superficial and nodular BCCs | ALA vs. cryotherapy | 88 | BCCs: 75% (histologically) vs. 95% (clinically) Cryotherapy: 85 vs. 87% | 3 months (histological control) |
| Solèr[17] | Superficial and nodular BCCs | MAL/ debulking 3 weeks prior to PDT | 350 | 79% | 24–48 months |
| Varma[13] | Superficial BCCs | ALA | 61 | 82% | 12 months |
| Clark[12] | Superficial BCCs | ALA | 87 | 97% | 12 months |
| Horn[38] | Superficial and nodular BCCs | MAL | 94 patients | 87–92% (clinically) vs. 75–85% (histologically) | 3 months |
| Foley[27] | Nodular BCCs | MAL vs. placebo | 66 patients | 72% MAL 73% vs. placebo 21% | 24 months 6 months |
| Rhodes[39] | Nodular BCCs | MAL vs. surgery | 101 patients | 90% for MAL-PDT vs. 98% for surgery | 24 months |

In the treatment of nodular BCCs with ALA-PDT an improved outcome was achieved in 23 patients with 24 BCCs by performing prior debulking of the tumor 3 weeks before ALA-PDT (incoherent red light; 100 mW cm$^{-2}$, 120 J cm$^{-2}$) with a histological and clinical complete response in 92% of the treated lesions at a 3-month follow-up.[18]

Lesion preparation was also performed prior to incubation in a MAL-PDT study in 59 patients.[17] Nodular BCCs ($n$=350) were curetted before PDT, and MAL was applied to all tumors for 24 or 3 h prior to irradiation with a broadband halogen light source (50–200 J cm$^{-2}$). After a long-term follow-up for 24–48 months, the overall cure rate was 79% with an excellent or good cosmetic outcome in 98% of the completely responding lesions.

A better response rate of MAL-PDT compared to ALA-PDT in the treatment of nodular BCCs, even without debulking, was also seen in a recent open, prospective, uncontrolled, multicenter trial. Ninety-four patients with superficial or nodular BCCs at risk of complications, poor cosmetic outcome, disfigurement or recurrence using conventional therapy were treated with two sessions of MAL-PDT 1 week apart. At a follow-up visit at 3 months the clinical lesion remission rate was 92% for superficial BCC and 87% for nodular BCC, compared to a histological cure rate of 85% and 75%, respectively. Non-responders were retreated. At the 24-months follow-up, the overall lesion recurrence rate was 18%.[38]

A course of two sessions of MAL-PDT (75 J cm$^{-2}$, incoherent red light) 7 days apart was also performed in another European multicenter, open, randomized trial with 101 patients for nodular BCC in comparison with surgery. The clinical remission rate at 3 months was similar with MAL-PDT and surgery (91% vs. 98%) and the 24-months recurrence rate was 10% with MAL-PDT and 2% with surgery. A good or excellent cosmetic result was rated in 85% of the patients receiving PDT vs. 33% with surgery.[39]

The most recent study reported in the literature is a comparative, randomized, double-blind controlled trial in Australia for nodular BCCs. Lesions from 66 patients were treated with two sessions of either placebo or MAL-PDT. In cases where there was no complete response 3 months after initial treatment, lesions were excised. After 6 months, complete remission rate was 73% for MAL-PDT compared to 21% of placebo.[27]

The guidelines from the workshop of the British Photodermatology Group for topical photodynamic therapy (2002) postulated, before licensing MAL-PDT for BCC treatment, that topical ALA-PDT should be seen as an effective therapy for superficial (<2 mm thick) BCC, at least as effective as cryotherapy, but again with superior healing and cosmesis, and with particular advantages in large and multiple lesions (strength of recommendation A, quality of evidence I). Topical ALA-PDT was estimated as less effective for nodular BCC, and although additional therapy with prior curettage or with penetration enhancers or fractionated therapy may improve results, there was no published randomized evidence of their benefit (strength of recommendation C, quality of evidence IIiii)[6] (Table 3).

### 3.2.1.2. Squamous Cell Carcinoma

Squamous cell carcinoma (SCC) is a malignant, potentially metastasizing tumor arising from the keratinocytes of the epidermis, caused mainly by excessive lifetime exposure to sunlight. It often develops from precancerous lesions like AK or Bowen's disease, when atypical keratinocytes penetrate the basement membrane and invade the dermis. With a curative intention, histologically controlled surgery is still the treatment of choice, bearing in mind the metastatic potential of SCCs. Although initial response rates of 54–100% for superficial lesions could be seen in open studies with topical ALA-PDT, recurrence appeared in up to 69% after 3–47 months, so that caution is advised

by the British guidelines in the treatment with ALA-PDT (strength of recommendation D, quality of evidence IIiii)[6] (Table 3).

The high recurrence rate is probably due to the limited depth of tissue
penetration of ALA-PDT and the lack of histological control. Therefore, ALA-
PDT should be restricted to early invasive SCCs in patients in which surgery is
contraindicated or for immunosuppressed patients after organ transplant with
multiple initial lesions, when a regular follow-up can be guaranteed. However,
excellent results can be achieved in precancerous lesions like AK or Bowen's
disease, as reported below.


### 3.2.1.3. Actinic Keratoses

ALA-PDT has so far shown to be most effective for the treatment of AK
(Table 8). Clearance rates have ranged from 71% to 100% already after a single
treatment[6,40] (Figures 3 and 4). Irradiation has been performed either with red
(635 nm) or blue light (417 nm).[40,41] Although green light may theoretically also
be effective, one should always bear in mind that illumination with non-red
light should not be used for indications other than AK due to the lack of tissue
penetration.[5]

A recent study evaluated the Levulan Kerastick-preparation in a placebo-
controlled, uneven-parallel-group trial with 243 patients. The patients were
randomized to receive either vehicle or ALA, following irradiation with visible
blue light 14–18 h after incubation. After 8 weeks, a clearance rate of ≥ 75% of
the treated lesions was achieved in 77% of the ALA group *vs*. 18% of
the placebo group. At week 12 the outcome was 89% and 13%, respectively.
The 12-week clearing rates included 30% of patients who received a second
ALA-PDT course. Whereas moderate to severe discomfort during irradiation
was perceived by at least 90% of patients, only 3% of patients discontinued
therapy.[41] In order to lower the number of side effects of ALA-PDT, shorter
incubation periods (1, 2, and 3 h), in combination with (a) pretreatment with
40% urea to enhance ALA penetration and (b) the use of topical 3% lidocaine
hydrochloride to ease discomfort, were also evaluated. At a follow-up 5 months
after therapy up to 90% reduction of lesions in the target area was observed in
18 patients with at least four non-hypertrophic AKs. No difference was seen
between the three incubation periods, nor did pretreatment with urea or
lidocaine have an influence on the therapeutical outcome and pain.[31] As signs
of photoaging were also reduced with photodynamic therapy, this confirmed
the rationale for the use of a short-contact ALA-PDT regimen in photorejuvenation, to be discussed later.

In another trial, efficacy of ALA-PDT was compared to topical use of 5-FU
in a large area treatment using a short incubation time of 1 h followed by
irradiation with blue light or a pulsed dye laser.[42] Best results were achieved
with either blue light-ALA-PDT or 5-FU, but PDT was better tolerated. Two
other trials in 35 patients using 5-ALA (incubation time 3 or 14–18 h; illumination with a 585 nm vascular laser) showed clear differences in treatment
response depending on the site of the lesions. Overall, in 2561 lesions at

**Table 8.** Summary of results of selected clinical studies using topical ALA- or MAL-PDT for the treatment of actinic keratoses

| Study | Sensitizer | Number of lesions/ (patients) | Comments | Complete remission | Follow-up |
|---|---|---|---|---|---|
| Szeimies[28] | 16% MAL | 699 | MAL-PDT vs. cryotherapy | 69% MAL-PDT vs. 75% cryotherapy | 3 months |
| Alexiades-Armenakas[43,44] | ALA | 2561 | Comparison of different lesion sites | Face and scalp: 98.4% Torso: 74.4% Extremities: 49.1% | 2 months |
| Freeman[47] | 16% MAL | 204 patients | MAL-PDT vs. cryotherapy vs. placebo | 91% MAL-PDT vs. 68% cryotherapy vs. 30% placebo | 3 months |
| Pariser[48] | 16% MAL | 80 patients | MAL-PDT vs. placebo | 89% MAL-PDT vs. 38% placebo | 3 months |
| Piacquadio[41] | ALA-solution | 243 patients | Irradiation with incoherent blue light | Clearance rate ≥ 75% | 3 months |
| Touma[31] | ALA | 18 patients with at least 4 AK | ALA-PDT vs. placebo Pretreatment with urea and lidocaine, different incubation times | 89% ALA-PDT vs. 13% placebo 90% | Up to 5 months |
| Dragieva[45] | ALA | 40 patients | Immunocompetent (IC) vs. immunosuppressed (IS) patients | 94% (IC) vs. 86% (IS) 72% (IC) vs. 48% (IS) | 4 weeks 48 weeks |
| Dragieva[46] | MAL | 129 | MAL-PDT in immunosuppressed patients vs. placebo | 76% for MAL-PDT vs. 0% in placebo | 16 weeks |

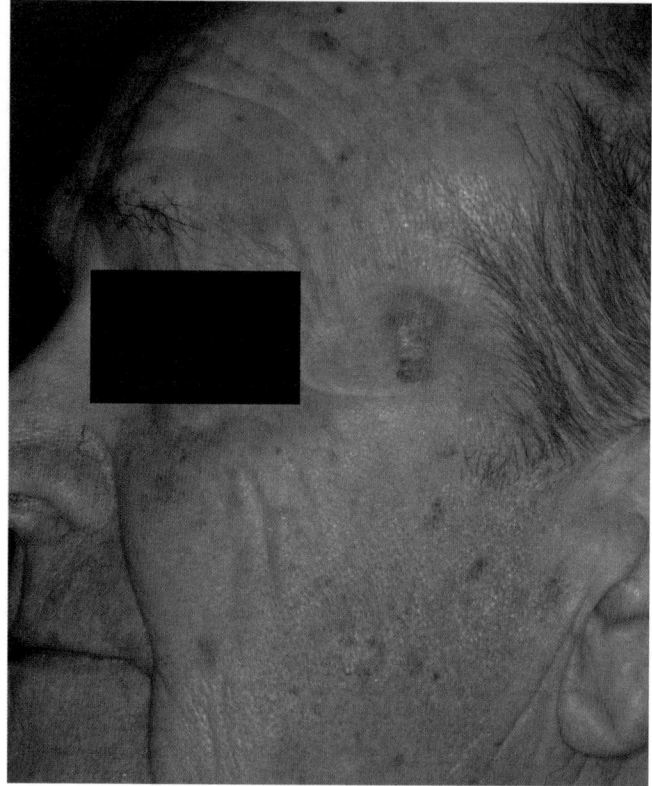

**Figure 3.** Seventy-eight-year old man with actinic keratoses on the left temple prior to MAL-PDT.

2-months follow-up AK showed a clearance rate of 98.4% on the face and scalp, of 74.4% on the trunk, and of 49.1% on the extremities.[43,44]

In immunosuppressed transplant recipients there is a dramatically increased risk for widespread epithelial tumors, which constrains standard treatments like cryosurgery or surgery. Therefore, ALA-PDT could be an ideal treatment modality, allowing for repeated treatments even in larger areas. In a recent study results at a short-term follow-up 4 weeks after ALA-PDT revealed comparable cure rates of 86% *vs.* 94% for immunosuppressed patients and immunocompetent control groups. However, after 48 weeks the therapeutic outcome in transplant recipients was significantly lower than in non-immuno-suppressed patients (48% *vs.* 72%).[45] This result perhaps arises from an increased disposition of immunosuppressed patients to develop new tumors and a critical host immune response regarding tumor clearing. In another placebo-controlled, double-blind trial the efficacy of MAL-PDT in transplant recipients was evaluated in the treatment of 129 lesions. At 16 weeks, a complete remission rate of 76% was found in the MAL-arm compared to 0% with placebo.[46]

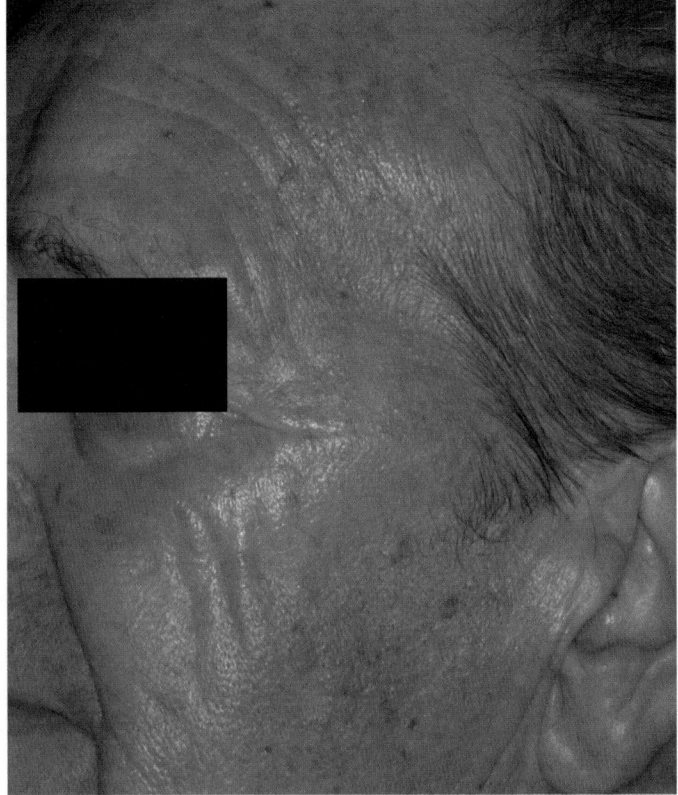

**Figure 4.** Six months after a single course of MAL-PDT; complete remission with only slight hypopigmentation.

The efficacy of MAL-PDT, compared to cryosurgery, for the treatment of AK in immunocompetent patients was assessed in a European, multicenter, randomized prospective study. Over 190 patients (95%) with 699 lesions completed the trial. Patients were randomized to either a single treatment with MAL-PDT (with repetition after 1 week in 8%) or a double freeze–thaw course of liquid nitrogen cryosurgery. MAL was applied for 3 h after slight lesion preparation, followed by irradiation with broad-spectrum red light (75 J cm$^{-2}$). At the follow-up visit 3 months post-treatment the efficacy of MAL-PDT (single application) was 69% *vs.* 75% for cryosurgery (no statistical significance). Highest response rates were seen in thin lesions on the scalp with 80% and 82% for PDT and cryosurgery, respectively. The cosmetic outcome, as judged by the investigator, was superior for MAL-PDT (96% *vs.* 81%).[28] A comparable study in Australia involved MAL-PDT with two treatment sessions 1 week apart, compared to a single course of cryosurgery or placebo in 204 patients. A follow-up at 3 months revealed a significantly higher complete remission rate with MAL-PDT (91%) than with cryosurgery (68%) and placebo (30%). An excellent cosmetic result was achieved in 81% of MAL-PDT patients vs. 51% in the cryosurgery arm.[47] Finally, a multicenter, randomized, double-blind, placebo-controlled study with two MAL-PDT cycles was performed in 80 patients with AK in the USA with similar PDT parameters to the above-mentioned trials. After 3 months, a complete lesion response rate of 89% for MAL-PDT vs. 38% for placebo was observed with an excellent or good cosmetic result in more than 90% of MAL-treated patients.[48] In the guidelines reported from a workshop of the British Photodermatology Group both ALA- and MAL-PDT were rated to be an effective treatment for non-hyperkeratotic AK on the face and scalp, with response rates comparable to topical 5-FU and cryotherapy but with a superior cosmetic outcome (strength of recommendation A, quality of evidence I)[6] (Table 3).

Additionally, encouraging results using ALA-PDT were also seen in actinic cheilitis in a case series of three subjects.[49]

### 3.2.1.4. Bowen's Disease

As extensively assessed so far in more than 14 open and three randomized comparison studies[6,15,50,51] for Bowen's disease, topical PDT using 20% ALA achieved the best cure rates (up to 100%) for all epithelial cancers or precursors (Table 9) (Figures 5 and 6). A recent bi-center, randomized, phase III trial in 40 patients compared ALA-PDT to topical 5-FU.[51] One to three lesions of previously untreated, histologically proven Bowen's disease received either PDT or 5-FU. ALA 20% in an oil/water-emulsion was applied 4 h prior to illumination with an incoherent light source (Paterson lamp, Photo therapeutics, UK, $\lambda_{em}$: 630 ± 15 nm; 50–90 mW cm$^{-2}$, 100 J cm$^{-2}$). Treatment with 5-FU was given once daily in week 1 and twice daily during weeks 2–4. When necessary, both ALA-PDT and 5-FU applications were repeated at the first follow-up at week 6, when 88% of the lesions (29 of 33) in the PDT arm showed complete response vs. 67% after 5-FU (22 of 33). At the 1-year follow-up,

**Table 9.** Summary of results of selected clinical studies using topical ALA- or MAL-PDT for the treatment of Bowen's disease

| Study | Sensitizer | Number of lesions/ (patients) | Comments | Complete remission | Follow-up |
|---|---|---|---|---|---|
| Reviews of 14 open and 3 randomized trials[6,15,50] | 20% ALA | | | 69–100% | 3–36 months |
| | | | | Average 90% | After 15 months |
| Salim[51] | 20% ALA | 40 patients | ALA-PDT vs. 5-FU | 88% PDT vs. 67% 5-FU | 6 weeks |
| | | | | 82% vs. 42% | 1 year |
| Morton[52] | MAL | 275 | MAL-PDT vs. cryotherapy vs. 5-FU vs. placebo | 86% PDT vs. 82% cryotherapy vs. 83% 5-FU vs. 17% placebo | 12 weeks |

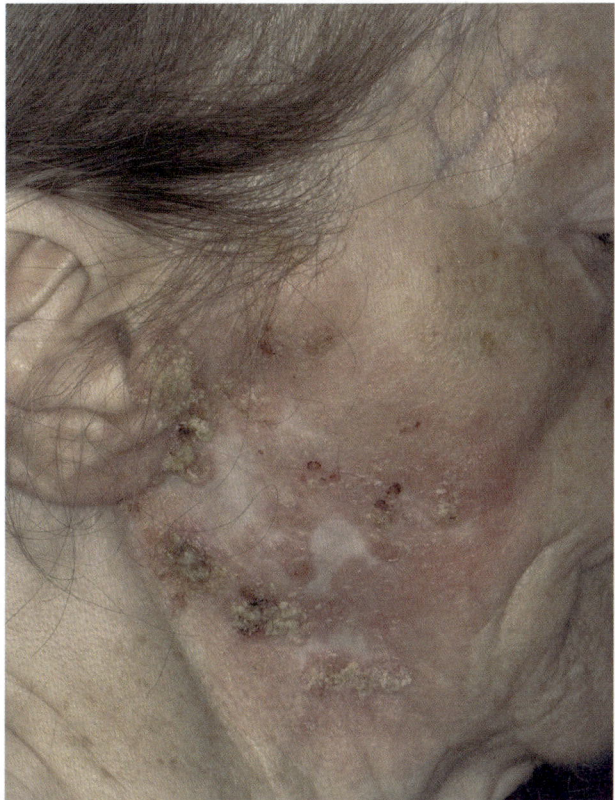

**Figure 5.** Eighty-four-year old woman with extensive Bowen's disease on the right mandibula prior to ALA-PDT (multiple surgical pretreatments).

further improvement of the complete clinical clearance rate after PDT was observed (82% *vs.* 42%).[51] Another phase III trial compared the results of MAL-PDT, placebo-PDT, cryotherapy with liquid nitrogen and 5-FU in 275 lesions. MAL-PDT (broadband red light, 75 J cm$^{-2}$) consisted of two treatment sessions 1 week apart: 5-FU treatment was performed for 4 weeks. Twelve weeks after the last treatment there was complete clearance in 107 of 124 lesions (86%) treated with MAL-PDT, 4 of 24 (17%) lesions with placebo-PDT, 75 of 91 (82%) with cryotherapy, and 30 of 36 (83%) with 5-FU, and the overall cosmetic outcome was best using MAL-PDT.[52]

In summary, in the guidelines of the British Photodermatology Group ALA-PDT was rated as an effective treatment in Bowen's disease, with at least as good results as in cryotherapy and topical application of 5-FU, but with a better cosmesis and fewer adverse events. It might especially have advantages with large or multiple lesions, lesions in sites with impaired healing such as lower legs, and for facial or digital lesions where conventional therapies show limitations (strength of recommendation A, quality of evidence I)[6] (Table 3).

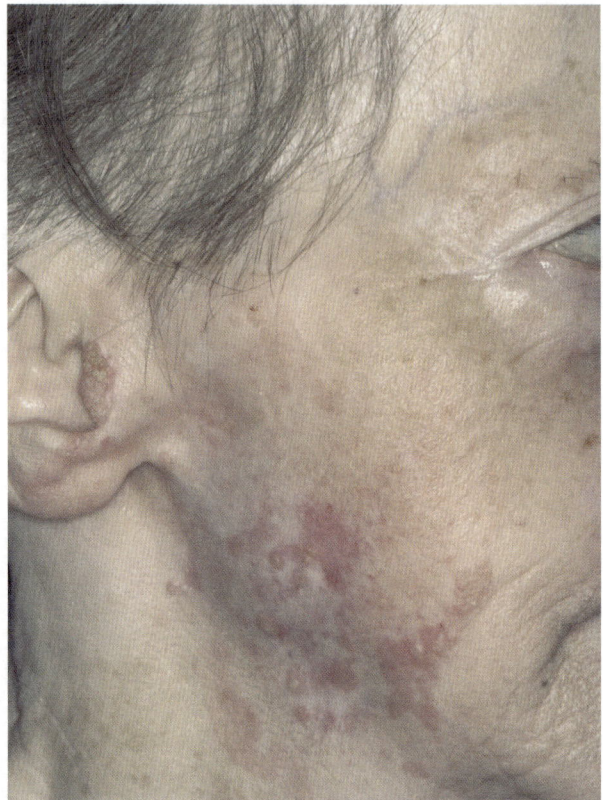

**Figure 6.**   Partial remission after the first course of PDT in a still ongoing treatment cycle. In extensive diseases like that several courses might be needed to achieve complete clearance.

In erythroplasia of Queyrat first case studies show an efficacy of ALA-PDT inferior to that in cutaneous Bowen's disease with less responsiveness, especially in more extensive disease.[53] Of concern is a recent report about development of a SCC after ALA-PDT and subsequent application of topical 5-FU in erythroplasia of Queyrat of the penis. Although it was likely related to the disease rather than to adverse events of PDT, this also raises concerns about the use of PDT in erythroplasia of Queyrat.[25]

*3.2.1.5.  Varia.*
*3.2.1.5.1.  Improvement of remission rates and retreatment.* Regarding the treatment of superficial oncologic epithelial lesions like Bowen's disease, AK or superficial BCCs (tumor thickness <2–3 mm) cure rates achieved by PDT are equal to the cure rates of the respective standard therapeutic procedures (*e.g.*, a clearance rate of 100% using a single PDT course with ALA application 6 h prior to irradiation for BCCs less than 2 mm in thickness[54]). To improve remission rates in thicker lesions, it may be necessary to prepare the tissue

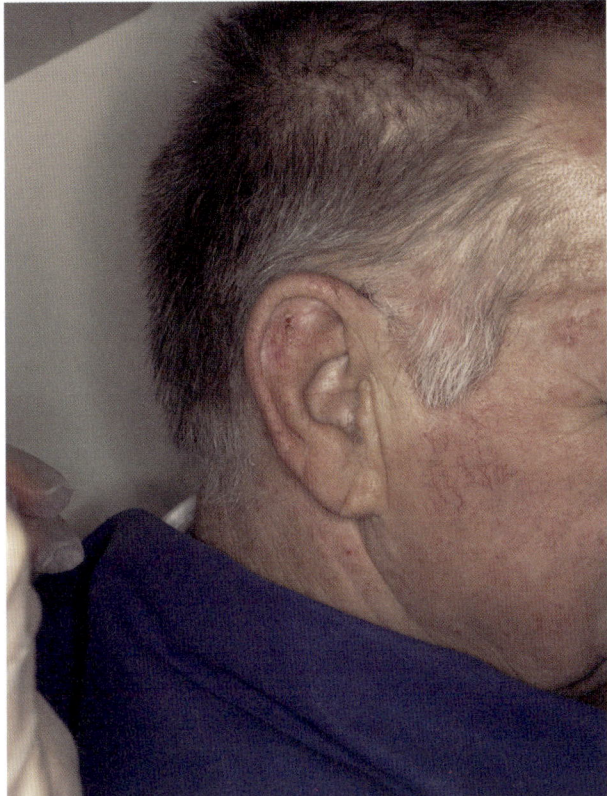

**Figure 7.** Selection of area to be treated (hypertrophic actinic keratoses on the right ear).

(debulking) prior to PDT with a slight curettage without major bleeding (Figures 7 and 8), or to add penetration enhancers like DMSO or EDTA. Several researchers also propose double or even multiple ALA-PDT treatments to improve therapeutic outcome. Repeated ALA-PDT every alternate day (to a maximum of three treatments) until no tumor was apparent, achieved 100% complete remission at 30 days, with a recurrence rate of 13% over a follow-up period of 24–36 months.[55] Recurrence-free cure rates of 60%, 80%, and 100% after a follow-up of 12–24 months were reported for lesions less than 1 cm in diameter, with respectively one, two, and three treatments at 1 month intervals until complete remission was achieved.[56] A routine double treatment at a 7-day interval, to provide time for maximum photodynamic damage and for some healing between treatments, achieved a complete remission of 96% after a median follow-up of 27 months.[16] However, the optimal timing of repeated treatments is still an open question, whether it is preferable to give a stand-ardized double treatment to all patients, resulting in some patients receiving a second PDT without medical need, or else to observe patients regularly and only retreat when necessary.

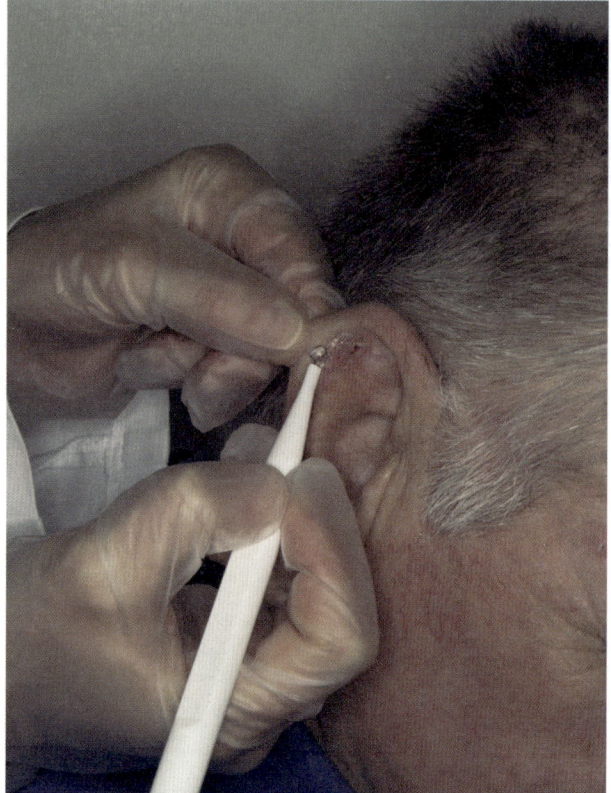

**Figure 8.** Curettage of hyperkeratotic parts of the tumor.

*3.2.1.5.2. Follow-up.* A general need for regular follow-ups can be seen in some studies with a decrease of the cure rate to 50–60% after long-term follow-up (with recurrences even after 3 years in single patients, personal observation), despite very high clinical as well as histological controlled cure rates at the short-term follow-ups. Therefore, long-term follow-up of at least 12 months for AK and up to 5 years after PDT for BCC is strongly recommended to ensure early recognition of recurrent tumors.

*3.2.1.5.3. Benefits of PDT.* The benefits of ALA-PDT compared to other treatments are the low level of invasiveness, the excellent cosmetic results after treatment, comparable clinical outcome to standard treatment in superficial lesions, the simultaneous treatment of multiple tumors or larger areas in one session, the relatively short healing time, tumor control in immunosuppressed patients (*i.e.*, transplant recipients) and an overall good patient tolerance (Tables 6 and 10). Besides clinical efficacy, cost-effectiveness is also an important aspect in the evaluation of PDT. Cost analysis indicates that with relatively low costs for permanent equipment, topical ALA-PDT is comparable in cost

**Table 10.** PDT in comparison to other treatment modalities in epithelial tumors

| Treatment | Advantages | Disadvantages |
|---|---|---|
| Topical PDT | Excellent cosmesis | Pain during PDT |
| | Non-invasive | Time consuming |
| | Safe | No histology |
| | Tumor-selectivity | |
| | Simultaneous treatment of multiple tumors | |
| | Short healing time | |
| | Cost effective | |
| Surgery | Safe | Invasive |
| | Histology | Local anesthesia |
| | | Scarring |
| Cryotherapy | Cheap and fast to perform | Pain during treatment |
| | | Blistering |
| | | Delayed wound healing |
| | | Scarring |
| | | Hypopigmentation |
| | | Skin atrophy |
| | | No histology |
| Curettage | Cheap and easy to perform | Local anesthesia |
| | Histology | Scarring |
| | | Recurrence rate |

with other therapies when morbidity costs in standard treatments are included, becoming more economical in patients where multiple tumors can be treated in a single PDT session.[6]

### 3.2.2. Nonepithelial Tumors of the Skin

### 3.2.2.1. Malignant Melanoma, Kaposi's Sarcoma and Cutaneous Metastases
Other indications for topical ALA-PDT, apart from epithelial tumors in dermato-oncology, are rare at present. Whereas systemic PDT with photosensitizers other than ALA achieved promising results in the treatment of cutaneous breast cancer metastases,[57] clinical response to topical application of ALA was poor, probably due to the depth of the metastatic nodules,[58] so that this treatment regimen has not been pursued. The same findings could be derived using ALA-PDT in the therapy of both amelanotic and melanotic metastasis from malignant melanoma.[59] In pigmented lesions this could be a result of poor light penetration caused by shading effects of melanin, so that trials *in vivo* have been performed to eliminate pigment prior to PDT by a Nd:YAG laser treatment.[60] Considering the aggressiveness of malignant melanoma and the disappointing results with PDT so far, further studies do not seem ethical now. For Kaposi's sarcoma there are several reports about successful systemic PDT with photosensitizers other than ALA.[61,62] In all of these indications, topical ALA does not seem to penetrate sufficiently through the skin into tumor cells and systemic application is not suitable because of heterogeneous ALA distribution and metabolism resulting in insufficient PDT-induced tumor ischemia and consecutive necrosis.

*3.2.2.2. Cutaneous T-Cell Lymphoma*

A selective photosensitizer accumulation in lymphocyte cell lines in cutaneous T-cell lymphoma (CTCL)[63–65] and B- and T-cell toxicity,[64,66] as well as lymphoblast mutations[67,68] when performing PDT *in vitro,* was the rationale for the use of PDT in CTCL. There are only isolated clinical reports in patients with different stages of CTCL. Two patients showed complete response in patch lesions using a single treatment of PDT with 20% ALA and irradiation with a pulsed frequency doubled Nd:YAG dye laser (630 nm, $<110$ mW cm$^{-2}$, 60 J cm$^{-2}$).[69,70] In plaque-stage mycosis fungoides clinical as well as histological response with remission for 8 and 14 months, respectively, was seen after five sessions of PDT with 20% ALA and a modified slide projector as light source (44 mW cm$^{-2}$, 40 J cm$^{-2}$).[60] Similarly good results were achieved in a study of 10 patients when using a Waldmann PDT 1200L lamp (600–730 nm, 88–180 J cm$^{-2}$, 20–265 mW cm$^{-2}$, 20% ALA),[71] whereas there were controversial results in tumor stage CTCL.[71,72] Thus, results so far indicate that topical PDT could act as an alternative therapy to common treatments in selected cases of patients with patch- and plaque-stage mycosis fungoides who are resistant to or who can not tolerate other treatment modalities. In addition PDT could be used in combination with radiotherapy to reduce tumor volume prior to conventional therapy, which was shown in the treatment of a localized lesion of large cell anaplastic cutaneous lymphoma (LACL).[73] This "pretreatment" could contribute to minimizing the possible adverse effects of ionizing radiation and chemotherapy. In summary, as the optimal regimen for treatment has yet to be established, the guidelines of the British Photodermatology Group currently rate strength of recommendation C, quality of evidence IIiii)[6] (Table 3).

## 3.3. Non-Oncologic Indications

Specific ALA uptake in diseased tissue and conversion to the active species, PpIX, is mainly facilitated by an increased permeability of abnormal stratum corneum and alterations in porphyrin enzyme profiles in diseased tissue.[36] As accumulation of photosensitizers in altered tissue is not specific for neoplastic or preneoplastic diseases, but has also been found in benign skin disorders,[74] topical ALA-PDT has been used in the treatment of various benign proliferative skin conditions. However, relating publications are mostly restricted to small case series or case reports and methodology has not been standardized, so that results must therefore be interpreted with caution as long as studies of adequate size and design are lacking.

A recent randomized controlled trial has demonstrated lack of efficacy of ALA-PDT in alopecia areata. More promising, single case reports document benefit in the treatment of epidermodysplasia verruciformis, penile lichen planus, nevus sebaceous, hirsutism,[75] Hailey–Hailey disease,[76] and hidradenitis suppurativa.[77] In light of limited evidence, no recommendations are proposed concerning the aforementioned indications.

*3.3.1. Small Case Series and Initial Trials*

Some interesting initial results have been seen in small case series in inflammatory and infectious diseases like scleroderma, lichen sclerosus or condylomata acuminata.

*3.3.1.1. Inflammatory Dermatoses.*
*3.3.1.1.1. Scleroderma.* In the treatment of localized scleroderma, psoralen+UVA-light (PUVA-therapy) is a very effective modality. Possible advantages in using PDT as opposed to PUVA are a deeper tissue penetration and the use of light with longer wavelengths without clear signs of carcinogenicity.

Promising results of ALA-PDT using a 3% gel (6 h incubation) and irradiation with incoherent light (Waldmann PDT 1200L, 40 mW cm$^{-2}$, 12 J cm$^{-2}$) were seen in a pilot study in five subjects previously resistant to common therapies.[78] Treatment was delivered one to two times weekly for 3–6 months, with a median number of $26 \pm 8$ treatments. Significant improvement in clinical skin scores and quantitative durometry scores were seen in all subjects. Minimal side effects consisted of slight burning or pruritus and transient hyperpigmentation. A follow-up of 2 years did not show further progression in significantly improved or nearly cleared lesions. However development of further lesions at remote sites was not prevented.

The pathogenesis of localized scleroderma reveals excessive collagen deposit in the dermis of lesional skin and a reduction of matrix metalloproteinases responsible for collagen turnover. Recent *in vitro* studies have shown that the antisclerotic effects of ALA-PDT are mediated by direct induction of the collagen-degrading matrix metalloproteinase (MMP)-1 and MMP-3 and a reduction of collagen type I in fibroblasts.[79] Further observations also show that PDT can trigger MMP production not only directly, but also by an indirect paracrine loop mediated by cytokines which are released by epidermal keratinocytes.[80]

*3.3.1.1.2. Darier's disease (keratosis follicularis).* Darier's disease can be treated with various topical ointments, retinoids, oral antibiotics or surgical modalities like excision, electrodessication, dermabrasion, and ablation lasers. In a small, uncontrolled study six patients with an unsatisfactory response to these conventional treatments were treated with ALA-PDT in addition to stable doses of retinoids (apart from one patient who was not on systemic treatment). One to two treatments at 4–12week intervals were performed with 5-ALA 20% w/w ointment applied 4 h prior to irradiation with an incoherent light source (PDT 1200L, Waldmann, 580–740 nm, mean fluence rates of 110–150 mW cm$^{-2}$, 150 J cm$^{-2}$). After PDT the treated areas were dressed with fusidic acid 2% ointment twice daily. This regimen resulted in sustained clearance or improvement in four of the five remaining patients during a follow-up period of 6 months to 3 years, while three patients remained on systemic retinoids and one had never been on systemic treatment. The fifth patient showed initial improvement followed by recurrence when retinoids were stopped. One patient did not tolerate treatment because of adverse events like erythema, exudation, and discomfort, which where related to the inflammatory reaction and lasted up to 2 and 3 weeks.[81]

*3.3.1.1.3. Lichen sclerosus.* In a pilot study in 12 subjects with vulval lichen sclerosus, PDT (application of 20% ALA solution 4–5 h prior to irradiation with an argon ion pulsed dye laser at 635 nm, 40–70 mW cm$^{-2}$, 80 J cm$^{-2}$) produced symptomatic relief with reduction of pruritus for several months after one to three treatment sessions in 10 of 12 subjects. However, only two premenopausal women showed objective signs of improvement, and no alterations of skin morphology was seen in 10 of the 12 women. (Eight were postmenopausal with advanced disease.)[82]

A newly designed bioadhesive patch containing 38 mg cm$^{-2}$ ALA was used in the treatment of lichen sclerosus and squamous hyperplasia. Ten patients were either treated with either one (three cases) or two (seven cases) sessions of PDT using a nonlaser light source (Paterson lamp, 630 nm) after incubation with ALA for 4–6 h. Pretreatment fluorescence was performed using a Wood's lamp. At 6 weeks post-treatment six of nine patients who completed follow-up reported a significant symptomatic relief (decrease from 8.1 to 4.6) on a visual analogue scale. Again no changes in skin morphology and architecture could be evaluated in post-treatment biopsies.[83] Whether more treatment cycles of PDT could improve efficacy, especially in premenopausal women, still needs to be clarified. However, a clear symptomatic relief of pruritus suggests PDT as a reasonable concomitant treatment modality in combination with other therapies.

*3.3.1.2. Infectious Cutaneous Diseases due to Microorganisms.*
*3.3.1.2.1. Viral infections: Condylomata acuminata.* Anogenital warts present with a high recurrence rate independent of the treatment modality chosen. Most destructive therapies like laser vaporization or cauterization only destroy the visible part of the warts, whereas subclinical lesions are not targeted. As selective photosensitizer accumulation in condylomata acuminata after ALA-application has been shown[84] (probably due to enhanced stratum corneum permeability), this could be a promising therapy to reduce recurrence rates especially for subclinical lesions.

In a pilot study in seven patients 20% ALA gel in addition with lidocaine was applied for 14 h prior to irradiation with an argon dye laser ($\lambda = 630$ nm, 100 J cm$^{-2}$, 75–150 mW cm$^{-2}$). In one patient very marked swelling of the penis developed and two patients withdrew because of severe pain. The remaining patients showed response rates of 75–100%, although one patient developed recurrence at 4 weeks.[85] This treatment regimen did not seem to be optimal, as a strong burning sensation was the limiting factor of the therapy. Also an incubation time of 14 h needed to be questioned, since *in vivo* fluorescence imaging and microscopy revealed the optimal irradiation time 30 min–3 h after ALA application, and longer application causes loss of tissue specificity also leading to sensitization of normal tissue.[84,86] Another trial in 16 patients with vulvar and vaginal lesions compared PDT with 10% ALA gel (incubation of 2–4 h, 80–125 J cm$^{-2}$ laser light at wavelength of 635 nm) to $CO_2$ laser vaporization. Postoperative discomfort lasted about 5 days. A complete clearance rate of 66% for PDT appeared to be as effective as conventional treatments, but with shorter healing time and an excellent cosmesis.[87] Despite difficult treatment sites, good clinical results were achieved in patients suffering from 164 urethral condylomata acuminata. Topical ALA-PDT with intraurethral irradiation using a cylindrical fiber showed complete response of 95% and a recurrence rate of 5% after a 6–24 months follow-up. Mechanisms active in PDT for condylomata might be triggering both apoptosis and necrosis in human papillomavirus-infected keratinocytes.[88] Still further trials are needed to consider pain as the main adverse event and treating condylomata in regions difficult to reach by light like the intraanal and vaginal cavity.

*3.3.1.2.2. Mycotic infections: Interdigital mycoses of the feet.* In vitro studies showing photosensitization of dermatophytes and yeasts after topical application of ALA[89] were the rationale for a trial in nine patients suffering from clinically and microbiologically confirmed interdigital mycoses of the feet. After occlusive application of 20% ALA cream for 4 h, lesions were irradiated with 75 J cm$^{-2}$ of broadband red light (range 575–700 nm, with a cut off filter blocking emission of wavelengths with $\lambda < 600$ nm, 100mW cm$^{-2}$). Treatments with only light or ALA were performed as controls on interdigital lesions of the other foot. One week after the first treatment, clinical and microscopical results were controlled. If inspection was still suggestive of mycoses, therapy was continued with three additional weekly treatments. Although a clinical and microbiological recovery in six of nine patients after one (four cases) or four (two cases) treatments was achieved and the overall tolerability was good, later results turned out to be disappointing with recurrences 1 month after the last treatment in four patients. Possible reasons for a low cellular uptake of the photosensitizer may be related to altered environmental conditions like humidity, temperature, and pH of interdigital skin, as well as a deficient biosynthesis of PpIX. In addition, occlusive application of ALA in this peculiar area is hard to perform and better results might be achieved treating superficial skin mycosis.[90]

*3.3.1.2.3. Protozoic infections: Cutaneous leishmaniasis.* A recent case report documented efficacy of PDT in a single patient suffering from leishmaniasis.[91] As therapy with PDT started several months after pretreatment with systemic sodium stibogluconate and paromomycin sulfate ointment, there was discussion about possible self-termination of the disease and delayed effect of pretreatment on the outcome.[92] In confirmation of the first report, the same group successfully treated a series of 11 patients with 32 lesions of cutaneous leishmaniasis. All but one lesion became amastigote negative (verification of clearing by direct smears) after only two treatments, with a mean reduction in lesion size of 67% at that time, an excellent cosmetic result and no relapse within a 6-months follow-up.[93] Although PDT may not be an important option for the majority of patients in endemic areas, since it is a comparatively time consuming and expensive treatment modality that requires a special equipment, it might be a rapid therapeutic alternative for some travelers or for immigrants.

### 3.3.2. Larger Clinical Trials

Most studies on PDT for inflammatory indications concern psoriasis vulgaris, acne vulgaris, and recalcitrant warts.

### 3.3.2.1. Psoriasis Vulgaris

The rationale for the use of ALA-PDT in psoriasis vulgaris is the selectivity of ALA accumulation in psoriatic plaques and photobleaching during PDT.[74] Whereas very early reports showed partial remission, mostly after application of ALA in higher concentrations (20–30%)[94] and after repeated treatments,[95] and also rated therapeutic outcome of topical PDT with 10% ALA cream to be comparable with dithranol,[96] more recent studies displayed disappointing results, as shown below. Limiting factors were inconsistencies in photosensitizer accumulation, unpredictability in treatment response, patient discomfort and the need for multiple treatments.[97–99] In one study with 22 subjects, 10 of 36 sites (2 cm$^2$) treated within psoriasis plaques cleared, but all relapsed within 2 weeks.[98] PDT was performed using 4 h incubation time (20% ALA) and subsequent illumination with a modified slide projector (400–650 nm, 25 mW cm$^{-2}$, up to 16 J cm$^{-2}$).

More recently, a study of 10 subjects with plaque psoriasis showed that multiple treatments (using the same incubation parameters, but a reduced illumination intensity of 15 mW cm$^{-2}$ from a slide projector) up to three times a week, with a maximum of 12 treatments and an illumination dose of 8 J cm$^{-2}$ at each treatment, improved clinical response.[99] Although 8 of the 10 patients showed a positive response, only 4–19 treated sites cleared. Variations of PpIX fluorescence intensity were detected before treatment between sites on the same patient as well as between different patients and between fluorescence and clinical outcome. Fluorescence was found mainly in the epidermis and stratum corneum and interestingly also at psoriasis plaques distant to those treated with ALA cream. However, no evidence of systemic absorption was found.[97,99] Although the clinical appearance improved after multiple treatments, significant pain, and unpredictable response argues against this treatment modality. A lack of efficacy and tolerability was also confirmed in another randomized intra-patient comparison study on topical ALA-PDT[100] as well as when compared with narrowband UVB phototherapy,[101] so that ALA-PDT presented as an inadequate treatment option for psoriasis. Also Koebnerization (*i.e.*, flare up of psoriatic lesions owing to isomorphic stimulus) of psoriasis plaques was reported after PDT of AK and SCCss.[102]

Therefore the role of topical ALA-PDT in the treatment of psoriasis remains unclear. Optimization of the treatment regimen is required in order to reduce the unpredictable nature of response and patient discomfort. Psoriasis-PDT was evaluated in the British guidelines as follows: strength of recommendation C, quality of evidence IIiii[6] (Table 3). In future, the development of a systemic photosensitizer without associated prolonged photosensitivity would potentially facilitate the use of PDT as an alternative therapy to PUVA, and preliminary results are encouraging.[103]

### 3.3.2.2. Acne Vulgaris

PDT in the treatment of acne vulgaris is based on two observations targeting both *Propionibacterium acnes* and the pilosebaceous units. First, there is a selectivity of ALA-induced porphyrin fluorescence for pilosebaceous units shown in animal models[104] and secondly, *P. acnes* contain endogenous porphyrins, in particular coproporphyrin III.[105] There are several reports on the efficacy of visible and blue light phototherapy in the treatment of acne[106–109] with a non-significant trend to higher efficacy for a combined red and blue light treatment.[110] However, the mode of action of visible light phototherapy is likely to be multifactorial and, apart from a possible endogenous PDT, may derive from immunomodulation and anti-inflammatory effects.[75] "Exogenous" PDT with ALA using a quite intense regimen was conducted in an open, randomized, controlled study in 22 subjects with moderate acne on the trunk.[111] After incubation with 20% ALA cream for 3 h a broadband illumination (550–700 nm) at 150 J cm$^{-2}$ was performed. Intra-individual comparisons were made and test areas in each subject received either one or four ALA-PDT treatments, with the appropriate control sites. A single PDT treatment achieved a statistically significant reduction of inflammatory acne for 10 weeks along with significant reduction in sebum excretion, *P. acnes* fluorescence and sebaceous gland size. However, there was also clear damage to sebaceous glands. From these findings three possible modes of action were concluded: direct photodynamic injury of sebaceous glands leading to inhibition of sebum production, reduction of follicular obstruction by an effect on keratinocyte shedding and hyperkeratosis, and photodynamic killing of *P. acnes* with sterilization of the sebaceous follicle. With four treatments, changes in these parameters lasted for at least 20 weeks. However, significant adverse effects like folliculitis, discomfort, desquamation, blistering, crusting and long lasting pigmentation changes could be seen, which suggested this treatment regimen to be unsuitable for routine use.

A low-dose PDT with markedly reduced illumination parameters to diminish adverse events was performed in a recent clinical study in 10 patients with mild to moderate acne of their backs.[112] Each patient's back was marked with four areas of equal acne severity. Each site was then either treated with ALA-PDT, ALA alone, light alone or kept as untreated control site, using 20% ALA cream incubated under occlusion for 3 h and red light from a diode laser (635 nm, 25 mW cm$^{-2}$, 15 J cm$^{-2}$) for 10 min. To assess clinical efficacy, baseline numbers of inflammatory and non-inflammatory acne lesions were counted, the sebum excretion measured and surface *P. acnes* swabs performed. After three courses of ALA-PDT within 3 weeks, a follow-up visit 3 weeks later showed a statistically significant reduction in inflammatory acne lesion counts from 11.6 at baseline to 3.6 for the PDT-treated site, but not at any of the other sites. However, no statistically significant reduction of *P. acnes* fluorescence or sebum excretion was detected. The authors therefore conclude other modes of action than damage to sebaceous glands or reduction of *P. acnes* counts. Good clinical results were also achieved with reduced illumination parameters in 13 patients with facial acne.[113] After a single treatment with 20% ALA cream and irradiation with a polychromatic, visible light source from a halogen lamp (600–700 nm, 17 mW cm$^{-2}$, 13 J cm$^{-2}$), improvement persisted for up to 6 months.

Recent studies also show successful results using a topical ALA solution (Levulan Kerastick). Twenty patients with moderate to severe acne vulgaris were incubated with topical ALA 1 h prior to illumination with a novel intense pulsed light (IPL) and heat source that emits 430–1100 nm radiation fluences of 3–9 J cm$^{-2}$.[114] Twelve of 15 patients completing the trial showed an average reduction of 50.1% in active inflammatory acne lesions at the end of a 4-week treatment period, with even improved results of 68.5% 4 weeks after the final treatment and of 71.8% 12 weeks after final treatment. The regimen was well tolerated by all patients and no treated lesion recurred at the end of follow-up period. Additional benefit in combination with peeling was seen in a study of 18 patients with moderate to severe inflammatory acne.[115] Two to four treatments over 4–8 weeks or two cycles of ALA-PDT (weeks 2, 4) preceded by a salicylic acid peel (weeks 1, 3) over 4 weeks were performed using a topical ALA solution applied 15–30 min before exposure to blue light or a combination of optical and radiofrequency energy. On a scale of 0.0–4.0 the average acne grade improvement among 12 patients who experienced a benefit was 1.75 at an average follow-up time of 4 months. Eleven showed at least 50% improvement and 5 more than 75% improvement. Adverse events included mainly erythema and peeling for up to 5 days: one episode of impetiginization of the treated area was reported. Although these findings provide encouraging evidence that ALA-PDT may be a useful adjunct in acne, pain during treatment, crust formation, erythema, and pigmentation for up to 4 weeks after treatment may limit patient acceptance for this therapy (strength of recommendation according to the British guidelines B, quality of evidence I)[6] (Table 3).


### 3.3.2.3. Recalcitrant Warts

In recalcitrant viral warts, the rationale for the use of topical ALA-PDT is based on their hyperproliferative and inflammatory nature thus accumulating porphyrins inside the lesions. PpIX was found to enrich in warts after ALA application.[116]

A report of six patients with viral warts showed that one single treatment with ALA-PDT (20% ALA applied for 5–6 h prior to irradiation with a slide projector for 30 min) without surface preparation before therapy failed to be effective in four of them.[117] Paring of the lesions down to the level of visible blood vessels prior to therapy with the same formulation (12 h incubation) and again illumination with a slide projector at a dose of 50 J $cm^{-2}$ led to a complete response after two or three cycles without recurrence.[118] The only adverse effect reported was moderate burning sensation. The results of ALA-PDT treatment in 62 patients with recalcitrant warts were retrospectively analyzed by the same authors.[119] Again, paring was performed prior to PDT (20% ALA cream, similar illumination parameters), but incubation times were reduced to 4–5 h. Treatment was repeated three times at weekly intervals and in the follow-up period there was additional treatment of warts by paring and application of Verucid (salicylic acid and lactic acid) once or twice a week. Marked pain caused 10 patients to refuse further treatment, and 50% of the patients completing the trial reported substantial discomfort. At the end of the study, 58% of lesions had cleared, with no recurrence in up to 17 months of follow-up. Disappointingly, only one of five patients with immunosuppression responded to treatment.

The same group also performed a comparative pilot study in 30 subjects with 250 recalcitrant warts.[120] Again, aforementioned treatment parameters were used besides illumination, which was performed with white, blue or red light (total dose 40 J $cm^{-2}$ in each group). PDT was repeated three times in 10 days. In addition, paring twice per week and application of Verucid were carried out. These treatment modalities were compared to conventional cryotherapy (up to four times in a 2 months period). Overall, 73% of the lesions in the white light PDT group cleared, compared with 42% in the red light PDT group, 28% in the blue light PDT group and 20% in the cryotherapy group. Unfortunately, treatment was accompanied by mild to strong burning within a few minutes of light exposure, with some patients reporting persistent discomfort up to 48 h after treatment. Three patients stopped treatment because of intolerable pain.

Higher light doses of 70 J $cm^{-2}$ using a broadband source ($\lambda = 590$–700 nm, Waldmann PDT 1200L, 50 mW $cm^{-2}$) were applied in a double-blind, placebo-controlled study in 45 patients with 232 recalcitrant hand and foot warts.[116] Again, paring before and within the three PDT cycles (with either 20% ALA cream or placebo) at weekly intervals was performed as well as treatment with Verucid. In case of persistent warts, they were retreated for a second 3-weeks-course and then reassessed at weeks 14 and 18. Patients with immunosuppression were excluded from the study. A complete clearance of warts by week 18 was seen in 56% of patients in the active treatment group, compared to a surprisingly high number of 42% in the placebo-treated group. The latter could be due to additional treatment with paring and application of salicylic acid and lactic acid, and possibly due to PDT effects mediated by endogenous porphyrins, as small peaks of fluorescence at 630 nm were seen in some placebo-treated warts not incubated with ALA. In accordance with the findings in prior studies, a significantly higher proportion of ALA-PDT patients compared to placebo experienced moderate to severe pain immediately and up to 24 h after light exposure.

More impressive results were achieved using topical ALA-PDT (20% ALA, 5 h incubation, illumination with $\lambda = 400$–$700$ nm, 50 mW cm$^{-2}$, 50 J cm$^{-2}$) weekly for 3 weeks compared with vehicle-PDT in the treatment of 67 patients with 121 warts,[121] in combination with a keratolytic and gentle curettage. Seventy-five per cent of warts resolved with ALA-PDT compared with 22.8% of the vehicle treated group. Fewer adverse events than in previous studies were reported with only a mild burning sensation.

Although ALA-PDT, whether as monotherapy or with additional treatment, was superior to the corresponding treatment regimens with cryotherapy or placebo, the overall potential in the therapy of viral warts seems to be limited. Children, because of remarkable pain during treatment, and immunosuppressed patients with recalcitrant warts on hands and feet do not profit from this therapeutical approach. Further optimization of the treatment regimen in order to minimize adverse effects, (particularly pain), is needed. The guidelines of the British Photodermatology Group state strength of recommendation B, quality of evidence I[6] (Table 3).

### 3.3.3. Conclusion

In conclusion, the development in ALA/MAL-PDT is continuously advancing. Well-documented indications for ALA-PDT are superficial basal cell carcinoma, AK and Bowen's disease. Also, ample data already exists for the possible usefulness of ALA-PDT in inflammatory dermatoses. However, larger, properly designed studies are required to clarify whether ALA-PDT for inflammatory dermatoses can demonstrate superiority over existing approved therapeutic modalities.

## 3.4. Perspectives

### 3.4.1. Tumor Surveillance

In transplant patients under chronic immunosuppression the risk of widespread epithelial tumors is dramatically increased. Standard treatments like cryosurgery, curettage, topical application of cytotoxic agents, radiotherapy or surgery work well, but are limited because of their invasiveness and the extent of disease. PDT for the treatment of epithelial superficial skin tumors in immunocompetent patients revealed many of advantages like a low level of invasiveness, the possibility of simultaneous treatment of multiple tumors or larger areas in one session, a comparable clinical outcome to standard therapy, a relatively short healing time, an excellent cosmetic result after treatment and an overall good patient tolerance. PDT also targets lesions still *in situ*, which cannot be diagnosed with the unaided eye. Therefore, PDT was estimated to be an ideal treatment modality especially in transplant patients with multiple lesions. Although on a long-term follow-up clearing rates in immunosuppressed patients after ALA-PDT in AK are significantly lower compared to immunocompetent patients (48% *vs.* 72% after 48 weeks[45]), this is a general problem in nearly all treatment modalities, probably because of an increased disposition of immunosuppressed patients to develop new tumors and a critical host immune response regarding tumor clearing.

In a placebo-controlled trial, AK in transplant recipients after MAL-PDT showed a complete remission rate of 76% *vs.* 0% for placebo at 16 weeks.[46] Therefore, despite lower clearance rates compared to immunocompetent patients, in the treatment of immunosuppressed patients PDT may have the potential of being a first-line treatment modality for tumor control especially in multiple lesions. This is true because the therapy targets large areas, can be repeated several times and can even be performed in areas with prior exposure to ionizing irradiation.

### 3.4.2. Skin Rejuvenation

As cosmetic aspects reach significant importance in dermatology today, an increasing demand exists for alternative methods of photorejuvenation. The rationale for the use of PDT could be a minimal, controlled phototoxic response to improve surface quality using low concentrations of a photosensitizer.

Following a dose-ranging trial on the forearm skin assessing different ALA concentrations (between 5% and 20%) and contact times (from 30 to 120 min) for optimum results, a pilot single treatment study of the periorbital crow's feet area was performed in six patients. Prior to the application of the photosensitizer, the stratum corneum was removed by tape stripping to enhance its absorption. Then ALA-PDT was performed according to the dose-ranging trial with a single application of 5% ALA for 30 min and subsequent irradiation with the OmniLux LED (Photo therapeutics, narrow band visible light, 633 nm, 105 mW cm$^{-2}$, 96 J cm$^{-2}$) for 20 min. A follow-up period of 12 weeks revealed a significant improvement in 67% (4 of 6 subjects) with a reduction in fine lines and wrinkles and smoothing of the periorbital skin.[122]

Short-contact PDT (30 min–1 h incubation) with higher photosensitizer concentrations of 20% ALA solution (Levulan Kerastick) and IPL was performed in 10 patients with sun damaged skin and AK's as full-face therapy. After three single treatments in 1 month intervals with follow-ups at 1 and 3 months more than 85% of the AK responded and 90% of patients achieved a greater than 75% improvement in tactile roughness and crow's feet, a 90% improvement in mottled hyperpigmentation and 50% improvement in facial erythema.[123]

Another trial in 32 patients with moderate photodamage and multiple AK also using short-time incubation (1 h) in combination with blue light irradiation for one single treatment showed a 90% clearance of AK, an improvement in skin texture in 72% and skin pigmentation changes of 59% at the end of 6 months. In nearly two-thirds of the patients treatment was judged to be less painful than cryotherapy,[124] with a postinflammatory erythema at least resolving within 7 days as most common adverse event.[122] Poorer results after one single treatment were seen using an IPL source, with 55% improvement in teleangiectasias, 48% improvement in pigmentary irregularities and 25% improvement in coarseness of skin texture.[125]

Longer incubation periods of 4 h prior to irradiation with an IPL in two treatment sessions showed more promising results in 17 patients with AK and additional signs of photodamage. An excellent cosmesis was found with an 87% improvement in wrinkling, skin texture, pigmentary changes, and teleangiectasias in addition to improvement of AK.[126] Another sign of photoaging, sebaceous gland hyperplasia, was also reduced after PDT with application of an ALA solution 30–60 min prior to illumination with blue light or an IPL over 15 min. PDT was performed once per month for 4 months. Twelve weeks after the final treatment a larger than 50% reduction in the number of lesions in both treatment arms without lesional recurrence was achieved Apart from a mild, transient erythema in two patients and bullous eruption in one patient, treatment was well tolerated.[127]

Another approach has been published recently. In a dose-response study a combination of an IPL, a radiofrequency device and topical ALA solution was used. The minimal erythema dose in Fitzpatrick skin types I/II and III, respectively, was achieved either by ALA applied for 1–2 h and an IPL fluence of 24–26 or 26–28 J cm$^{-2}$, or ALA application for 2–3 h and an IPL fluence of 26–28 or 28–30 J cm$^{-2}$. Based on these results two photorejuvenation protocols were developed.[128]

As possible mode of action in PDT-mediated photorejuvenation, a controlled trauma with increased wound healing response was discussed. Utilizing ALA-PDT, the quick peak of the inflammatory (phototoxic) reaction could lead to better conditions for the subsequent proliferative phase, optimizing the remodeling of the cellular and collagen matrix.[122] However, although ALA-PDT might be effective in photorejuvenation of the skin, the optimal treatment protocol still has to be established to promote PDT as an alternative method of photorejuvenation that is safe, easy to perform and with a minimum of side effects.

### 3.4.3. Treatment Algorithm

The following steps should be carefully taken into account when an ALA-PDT procedure is performed for the treatment of dermatologic conditions.

### 3.4.3.1. Patient and Lesion Selection

- Check indication (number, size, site of lesions, need of pretreatment, alternative therapies), concomitant diseases and medication, exclusion criteria (pregnancy, allergy to components of the ALA/MAL-preparation, porphyria), and if necessary perform biopsy to confirm diagnosis
- Obtain written informed consent, according to the guidelines of the country
- Fully document lesions (*e.g.*, photodocumentation)

### 3.4.3.2. Pretreatment

- Select the area to be treated (Figure 7)
- Perform lesion preparation if needed (curettage of crusts or prominent tumor parts, application of gauze soaked in saline or ointment, removal with forceps) (Figure 8)
- Apply the photosensitizer formulation to the lesion with little overlap (about 0.5–1 cm) to the surrounding tissue (Figure 9)

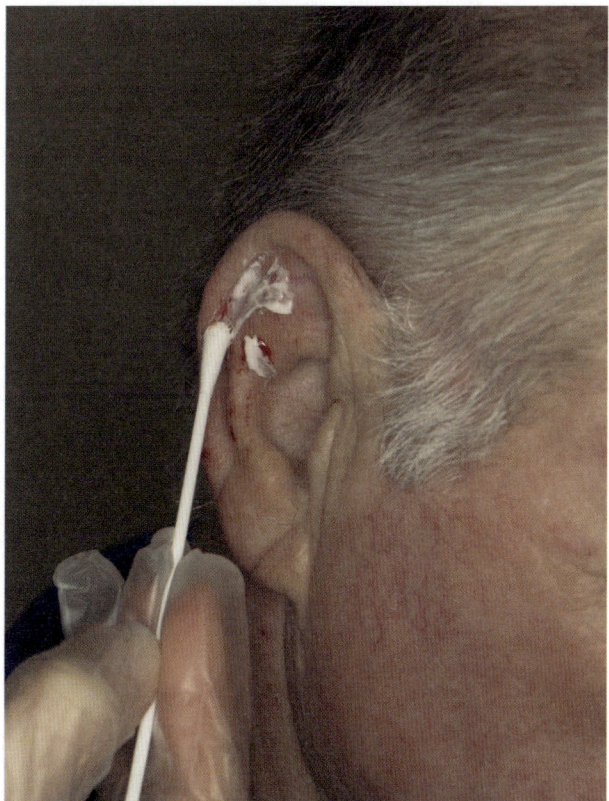

**Figure 9.** Application of ALA cream (20% ALA-hydrochloride in unguentum emulsificans) with little overlap using a cotton tip.

- Cover the entire area with an occlusive foil to allow a better penetration and with a light-protecting dressing or clothing to avoid photobleaching (Figures 10–12)
- Advise the patient to return after the incubation period (3–16 h, depending on the photosensitizer used (Metvix, Levulan Kerastick)
- If necessary give systemic analgesia (sometimes required in case of extensive disease/illumination fields)

### 3.4.3.3. Fluorescence Diagnosis and Illumination

- Remove the dressing and the emulsion. Optionally check surface fluorescence with an ultraviolet Wood's lamp or computerized diagnostic device (Dyaderm, Biocam)
- If necessary, apply local anesthetic (be sure not to induce drug incompatibilities (pH!))
- Protect the patient's and your eyes with goggles
- Place the light source within the right distance from the lesion (Figure 13)
- Double check for correct distance, especially when curved surfaces are treated!

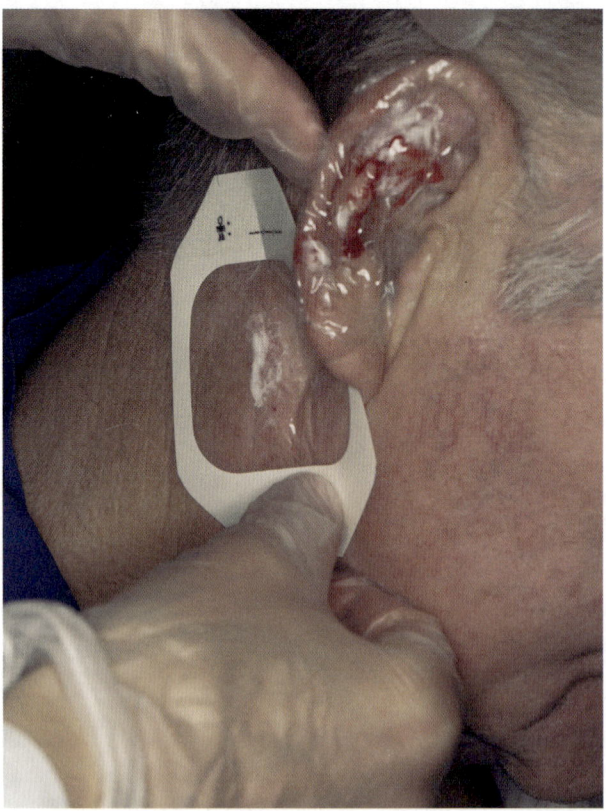

**Figure 10.** Application of an occlusive foil (Tegaderm, 3M Medica, St Louis, USA).

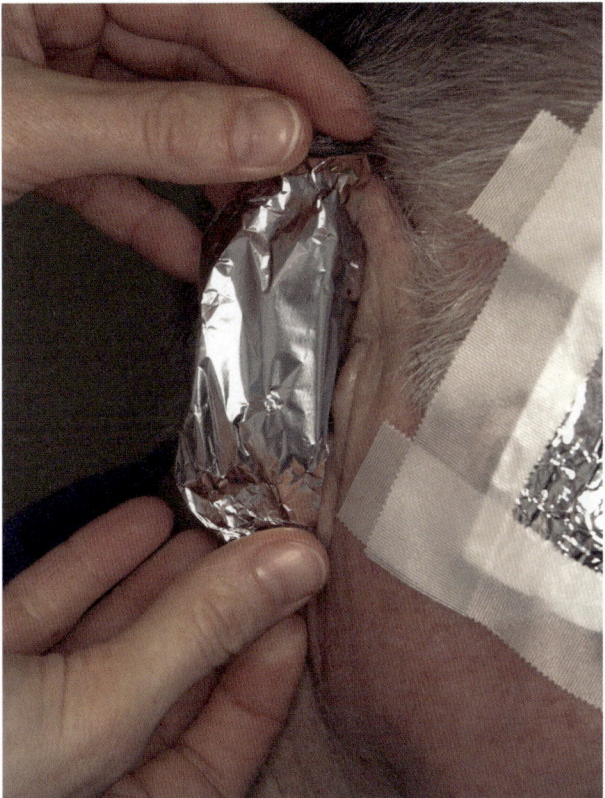

**Figure 11.**  Application of light-protecting dressings (aluminum foil).

- Irradiate with the advised parameters for total light dose and intensity
- If necessary, perform cold air analgesia, cooling by fan, *etc.* during illumination
- During illumination avoid misalignment of the light source owing to patient's movements

*3.4.3.4. After Care and Follow-Up*
- Suggest further cooling, for example, with cool pack or wet towel
- After illumination protect treatment site to direct sunlight for 24–48 h
- Avoid exposure to sunlight for the next 4–6 weeks (risk of postinflammatory hyperpigmentation)
- Inform patient again about frequent crust formation in the target area after 2–3 days (resolving within the next 2 weeks) and tell the patient to return if infection of the treated area or any other problem occurs
- Make mandatory follow-ups to ensure early recognition of recurrence, for example, at 4 weeks, 3, 6, 12, 18, 24, and 36 months after PDT (depending on diagnosis of lesions treated)

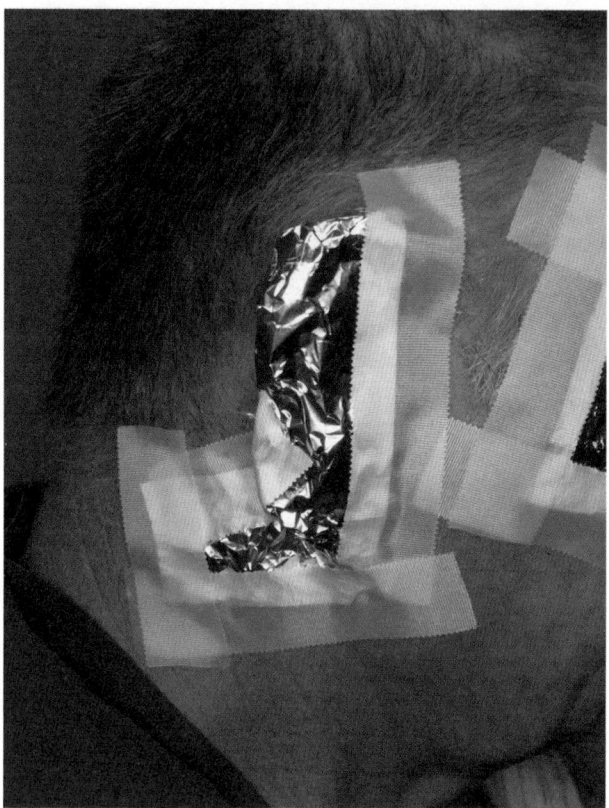

**Figure 12.**  Application of light-protecting dressings (aluminum foil).

• Perform a second treatment session (1–3 weeks apart) for thicker
  tumors (BCC, Bowen's disease) were treated, to improve the therapeutic
  outcome

*3.4.4.  Installation of a PDT Business Center (Derived from the Guidelines of the
British Photodermatology Group[6])*

  (i) Define the clinical need: estimate number of suitable patients and
current management approach (growing prevalence of non-melanoma
skin cancers, special situations such as transplant patients, complications and patient perceptions of existing therapies)

 (ii) Describe topical PDT and its uses, and its potential advantages over
existing therapies, including estimated savings, and treatment protocol

(iii) Costs:Location: isolated area for treatment, perhaps specific adapta-
tions for laser-source
Staff: medical (diagnostic and follow-up visits, supervision of service)nursing (ability to perform entire procedure, management of ap-
pointments)

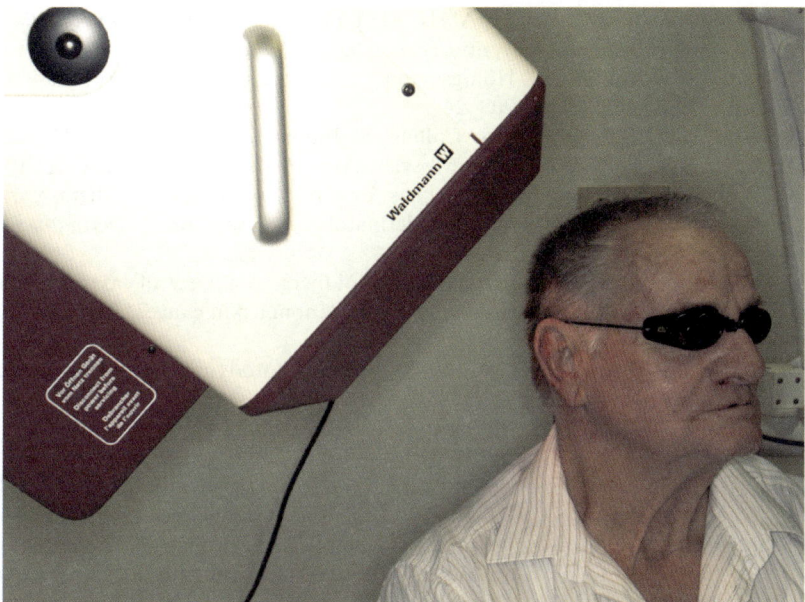

**Figure 13.**   Illumination with incoherent light source after incubation for 6 h.

Training: special accredited and certified courses like Euro-PDT (euro-pdt.com) comprehensive courses (how to perform treatment, to be aware of possible adverse events and how to treat them)
Equipment: light source (purchase *vs.* lease, maintenance costs); photosensitizer (licensed drug *vs.* self-made formulation); refrigerator for storage of photosensitizer, cold air analgesia, goggles for irradiation, disposables, for example, dressings, photographic documentation.

# References

1. V.G. Schweitzer, Photofrin-mediated photodynamic therapy for treatment of aggressive head and neck nonmelanomatous skin tumors in elderly patients, *Laryngoscope*, 2001, **111**, 1091–1098.
2. J.C. Kennedy, R.H. Pottier and D.C. Pross, Photodynamic therapy with endogenous protoporphyrin IX:basic principles and present clinical experience, *J. Photochem. Photobiol.*, 1990, **6**, 143–148.
3. C. Fritsch, B. Homey, W. Stahl, P. Lehmann, T. Ruzicka and H. Sies, Preferential relative porphyrin enrichment in solar keratoses upon topical application of D-aminolevulinic acid methylester, *Photochem. Photobiol.*, 1998, **68**, 218–221.
4. G. Ackermann, C. Abels, W. Bäumler, S. Langer, M. Landthaler, E.W. Lang and R.M. Szeimies, Simulations on the selectivity of 5-aminolevulinic acid-induced fluorescence *in vivo*, *J. Photochem. Photobiol. B Biol.*, 1998, **47**, 121–128.

5. R.M. Szeimies, S. Karrer, C. Abels, M. Landthaler and C.A. Elmets, *Photodynamic therapy in dermatology*, in: *Dermatological Phototherapy and Photodiagnostic Methods*, J. Krutmann, H. Hönigsmann, C.A. Elmets and P.R. Bergstresser (eds), Springer, Berlin, 2001, pp. 209–247.

6. C.A. Morton, S.B. Brown, S. Collins, S. Ibbotson, H. Jenkinson, H. Kurwa, K. Langmack, K. McKenna, H. Moseley, A.D. Pearse, M. Stringer, D.K. Taylor, G. Wong and L.R. Rhodes, Guidelines for topical photodynamic therapy: report of a workshop of the British Photodermatology Group, *Br. J. Dermatol.*, 2002, **146**, 552–567.

7. E.S. Marmur, C.D. Schmults and D.J. Goldberg, A review of laser and photodynamic therapy for the treatment of nonmelanoma skin cancer, *Dermatol. Surg.*, 2004, **30**, 264–271.

8. N.C. Zeitouni, A.R. Oseroff and S. Shieh, Photodynamic therapy for nonmelanoma skin cancers, *Mol. Immunol.*, 2003, **39**, 1133–1136.

9. V. Schacht, P. Weiderer, R.M. Szeimies and C. Abels, The microvascular effects are more pronounced following PDT with topical as compared with systemical application of 5-aminolevulinic acid (ALA) *in vivo*, *Arch. Dermatol. Res.*, 2002, **294**, 50.

10. G. Dragieva, J. Hafner, R. Dummer, P. Schmid-Grendelmeier, M. Roos, B.M. Prinz, G. Burg, U. Binswanger and W. Kempf, Topical photodynamic therapy in the treatment of actinic keratoses and Bowen's disease in transplant recipients, *Transplantation*, 2004, **77**, 115–121.

11. S.B. Brown, The role of light in the treatment of non-melanoma skin cancer using methyl aminolevulinate, *J. Dermatol. Treat.*, 2003, **14**, 11–14.

12. C. Clark, A. Bryden, R. Dawe, H. Moseley, J. Ferguson and S.H. Ibbotson, Topical 5-aminolaevulinic acid photodynamic therapy for cutaneous lesions: outcome and comparison of light sources, *Photodermatol. Photoimmunol. Photomed.*, 2003, **19**, 134–141.

13. S. Varma, H. Wilson, H.A. Kurwa, B. Gambles, C. Charman, A.D. Pearse, D. Taylor and A.V. Anstey, Bowen's disease, solar keratoses and superficial basal cell carcinomas treated by photodynamic therapy using a large-field incoherent light source, *Br. J. Dermatol.*, 2001, **144**, 567–574.

14. C.H. Yang, J.C. Lee, C.H. Chen, C.Y. Hui, H.S. Hong and H.W. Kuo, Photodynamic therapy for bowenoid papulosis using a novel incoherent light-emitting diode device, *Br. J. Dermatol.*, 2003, **149**, 1297–1299.

15. C.A. Morton, C. Whitehurst, J.V. Moore and R.M. MacKie, Comparison of red and green light in the treatment of Bowen's disease by photodynamic therapy, *Br. J. Dermatol.*, 2000, **143**, 767–772.

16. J.C. Haller, F. Cairnduff, G. Slack, J. Schofield, C. Whitehurst, R. Tunstall, S.B. Brown and D.J.H. Roberts, Routine double treatments of superficial basal cell carcinomas using aminolaevulinic acid-based photodynamic therapy, *Br. J. Dermatol.*, 2000, **143**, 1270–1274.

17. A.M. Solèr, T. Warloe, A. Berner and K.E. Giercksky, A follow-up study of recurrence and cosmesis in completely responding superficial and nodular basal cell carcinomas treated with methyl 5-aminolaevulinate-based photodynamic therapy alone and with prior curettage, *Br. J. Dermatol.*, 2001, **145**, 467–471.

18. M.R.T.M. Thissen, C.A. Schroeter and H.A.M. Neumann, Photodynamic therapy with delta-aminolaevulinic acid for nodular basal cell carcinomas using a prior debulking technique, *Br. J. Dermatol.*, 2000, **142**, 338–339.

19. C.A. Morton, Methyl aminolevulinate (Metvix(r)) photodynamic therapy-practical pearls, *J. Dermatol. Treat.*, 2003, **14**, 23–26.

20. C.A. Morton, Photodynamic therapy for nonmelanoma skin cancer-and more?, *Arch. Dermatol.*, 2004, **140**, 116–120.

21. J. Moan, H. Waksvik and T. Christensen, DNA single stranded breaks and sister chromatid exchanges induced by treatment with hematoporphyrin and light or by X-rays in human NHIK 3025 cells, *Cancer Res.*, 1980, **40**, 2915–2918.

22. D.M. Fiedler, P.M. Eckl and B. Krammer, Does D-aminolaevulinic acid induce genotoxic effects?, *J. Photochem. Photobiol. B Biol.*, 1996, **33**, 39–44.

23. I.M. Stender, N. Bech-Thomsen, T. Poulsen and H.C. Wulf, Photodynamic therapy with topical delta-aminolevulinic acid delays UV photocarcinogenesis in hairless mice, *Photochem. Photobiol.*, 1997, **66**, 493–496.

24. P. Wolf, R. Fink-Puches, A. Reimann-Weber and H. Kerl, Development of malignant melanoma after repeated topical photodynamic therapy with 5-amino-levulinic acid at the exposed site, *Dermatology*, 1997, **194**, 53–54.

25. S. Varma, P.J.A. Holt and A.V. Anstey, Erythroplasia of Queyrat treated by topical aminolaevulinic acid photodynamic therapy: a cautionary tale, *Br. J. Dermatol.*, 2000, **142**, 825–826.

26. J.V. Moore and E. Allan, Pulsed ultrasound measurements of depth and regression of basal cell carcinomas after photodynamic therapy: relationship to probability of 1-year local control, *Br. J. Dermatol.*, 2003, **149**, 1035–1040.

27. P. Foley, Clinical efficacy of methyl aminolevulinate (Metvix(r)) photodynamic therapy, *J. Dermatol. Treat.*, 2003, **14**, 15–22.

28. R.M. Szeimies, S. Karrer, S. Radakovic-Fijan, A. Tanew, P.G. Calzavara-Pinton, C. Zane, A. Sidoroff, M. Hempel, J. Ulrich, T. Proebstle, H. Meffert, M. Mulder, D. Salomon, H.C. Dittmar, J.W. Bauer, K. Kernland and L.R. Braathen, Photodynamic therapy using topical methyl 5-aminolevulinate compared with cryotherapy for actinic keratosis: a prospective randomized study, *J. Am. Acad. Dermatol.*, 2002, **47**, 258–262.

29. B. Gniazdowska, F. Rueff, P. Hillemanns and B. Pryzbilla, Allergic contact dermatitis from delta-aminolevulinic acid used in photodynamic therapy, *Contact Dermatitis*, 1998, **38**, 348–349.

30. H.C. Wulf and P. Philipsen, Allergic contact dermatitis to 5-aminolaevulinic acid methylester but not 5-aminolaevulinic acid after photodynamic therapy, *Br. J. Dermatol.*, 2004, **150**, 143–145.

31. D. Touma, M. Yaar, S. Whitehead, Konnikov and B.A. Gilchrest, A trial of short incubation, broad-area photodynamic therapy for facial actinic keratoses and diffuse photodamage, *Arch. Dermatol.*, 2004, **140**, 33–40.

32. J. Pagliaro, T. Elliott, M. Bulsara, C. King and C. Vinciullo, Cold air analgesia in photodynamic therapy of basal cell carcinomas and Bowen's disease: an effective addition to treatment: a pilot study, *Dermatol. Surg.*, 2004, **30**, 63–66.

33. M.V. Holmes, R.S. Dawe, J. Ferguson and S.H. Ibbotson, A randomised, double-blind, placebo-controlled study of the efficacy of tetracaine gel (Ametop) for pain relief during topical photodynamic therapy, *Br. J. Dermatol.*, 2004, **150**, 337–340.

34. C. Guillen, O. Sanmartin, A. Escudero, R. Botella-Estrada, A. Svila and P. Castejon, Photodynamic therapy for *in situ* squamous cell carcinoma on chronic radiation dermatitis after photosensitization with 5-aminolevulinic acid, *J. Eur. Acad. Dermatol.*, 2000, **14**, 298–300.

35. D.I.M. Kuijpers, N.W.J. Smeets, G.A.M. Krekels and M.R.T.M. Thissen, Photodynamic therapy as adjuvant treatment of extensive basal cell carcinoma treated with Mohs micrographic surgery, *Dermatol. Surg.*, 2004, **30**, 794–798.

36. Q. Peng, T. Warloe, K. Berg, J. Moan, M. Kongshaug, K.E. Giercksky and J.M. Nesland, 5-ALA based photodynamic therapy. clinical research and future challenges, *Cancer*, 1997, **79**, 2282–2308.

37. I. Wang, N. Bendsoe, C.A.F. Klinteberg, A.M.K. Enejder, S. Andersson-Engels, S. Svanberg and K. Svanberg, Photodynamic therapy *vs.* cryosurgery of basal cell carcinomas: results of a phase III clinical trial, *Br. J. Dermatol.*, 2001, **144**, 832–840.

38. M. Horn, P. Wolf, H.C. Wulf, T. Warloe, C. Fritsch, L.E. Rhodes, R. Kaufmann, M. de Rie, F.J. Legat, I.M. Stender, A.M. Solér, A.M. Wennberg, G.A.E. Wong and O. Larkö, Topical methyl aminolevulinate photodynamic therapy in patients with basal cell carcinoma prone to complications and poor cosmetic outcome with conventional treatment, *Br. J. Dermatol.*, 2003, **149**, 1242–1249.

39. L.E. Rhodes, M. de Rie, Y. Enström, R. Groves, T. Morken, V. Goulden, G.A.E. Wong, J.J. Grob, S. Varma and P. Wolf, Photodynamic therapy using topical methyl aminolevulinate *vs.* surgery for nodular basal cell carcinoma, *Arch. Dermatol.*, 2004, **140**, 17–23.

40. A. Sidoroff, *Actinic keratosis*, in: *Photodynamic Therapy and Fluorescence Diagnosis in Dermatology*, P.G. Calzavara. –Pinton, R.M. Szeimies and B. Ortel (eds), Elsevier, Amsterdam, 2001, pp. 199–216.

41. D.J. Piacquadio, D.M. Chen, H.F. Farber, J.F. Fowler, S.D. Glazer, J.J. Goodman, L.L. Hruza, E.W.B. Jeffes, M.R. Ling, T.J. Phillips, T.M. Rallis, R.K. Scher, C.R. Taylor and G.D. Weinstein, Photodynamic therapy with aminolevulinic acid topical solution and visible blue light in the treatment of multiple actinic keratoses of the face and scalp, *Arch. Dermatol.*, 2004, **140**, 41–46.

42. S. Smith, D. Piacquadio, V. Morhenn, D. Atkin and R. Fitzpatrick, Short incubation PDT versus 5-FU in treating actinic keratoses, *J. Drugs Dermatol.*, 2003, **2**, 629–635.

43. M. Alexiades-Armenakas, A.N.B. Kauvar, C.H. Bernstein, E.A. Maforng and R.G. Geronemus, Laser assisted photodynamic therapy of actinic keratoses, *J. Lasers Med. Surg.*, 2002, **14**, 24.

44. M. Alexiades-Armenakas and R.G. Geronemus, Laser-mediated photodynamic therapy of actinic keratoses, *Arch. Dermatol.*, 2003, **139**, 1313–1320.

45. G. Dragieva, J. Hafner, R. Dummer, P. Schmid-Grendelmeier, M. Roos, B.M. Prinz, G. Burg, U. Binswanger and W. Kempf, Topical photodynamic therapy in the treatment of actinic keratoses and Bowen's disease in transplant recipients, *Transplantation*, 2004, **77**, 115–121.

46. G. Dragieva, L. Schärer, R. Dummer and W. Kempf, Photodynamic therapy-a new treatment option for epithelial malignancies of the skin, *Onkologie.*, 2004, **27**, 407–411.

47. M. Freeman, C. Vinciullo, D. Francis, L. Spelman, R. Nguyen, P. Fergin, K.E. Thai, D. Murrell, W. Weightman, C. Anderson, C. Reid, A. Watson and P. Foley, A comparison of photodynamic therapy using topical methyl aminolevulinate (Metvix(r)) with single cycle cryotherapy in patients with actinic keratosis: a prospective randomized study, *J. Dermatol. Treat.*, 2003, **14**, 99–106.

48. D.M. Pariser, N.J. Lowe, D.M. Stewart, M.T. Jarratt, A.W. Lucky, R.J. Pariser and P.S. Yamauchi, Photodynamic therapy with topical methyl aminolevulinate for actinic keratosis: Results of a prospective randomized multicenter trial, *J. Am. Acad. Dermatol.*, 2003, **48**, 227–232.

49. I.M. Stender and H.C. Wulf, Photodynamic therapy with 5-aminolevulinic acid in the treatment of actinic cheilitis, *Br. J. Dermatol.*, 1996, **135**, 454–456.

50. C.A. Morton, C. Whitehurst, J.H. McColl, J.V. Moore and R.M. MacKie, Photodynamic therapy for large or multiple patches of Bowen's disease and basal cell carcinoma, *Arch. Dermatol.*, 2001, **137**, 319–324.

51. A. Salim, J.A. Leman, J.H. McColl, R. Chapman and C.A. Morton, Randomized comparison of photodynamic therapy with topical 5-fluorouracil in Bowen's disease, *Br. J. Dermatol.*, 2003, **148**, 539–543.

52. C.A. Morton, Photodynamic therapy for nonmelanoma skin cancer-and more?, *Arch. Dermatol.*, 2004, **140**, 116–120.

53. G.I. Stables, M.R. Stringer, D.J. Robinson and D.V. Ash, Erythroplasia of Queyrat treated by topical aminolaevulinic acid photodynamic therapy, *Br. J. Dermatol.*, 1999, **140**, 514–517.

54. C.A. Morton, R.M. MacKie, C. Whitehurst, J.V. Moore and J.H. McColl, Photodynamic therapy for basal cell carcinoma: effect of tumor thickness and duration of photosensitizer application on response, *Arch. Dermatol.*, 1998, **134**, 248–249.

55. P.G. Calzavara-Pinton, Repetitive photodynamic therapy with topical delta-aminolevulinic acid as an appropriate approach to the route treatment of superficial non-melanoma skin tumors, *J. Photochem. Photobiol. B*, 1995, **29**, 53–57.

56. C. Fritsch, G. Goerz and T. Ruzicka, Photodynamic therapy in dermatology, *Arch. Dermatol.*, 1998, **134**, 207–214.

57. S.A. Khan, T.J. Dougherty and T.S. Mang, An evaluation of photodynamic therapy in the management of cutaneous metastases of breast cancer, *Eur. J. Cancer*, 1993, **29A**, 1686–1690.

58. J.C. Kennedy, R.H. Pottier and D.C. Pross, Photodynamic therapy with endogenous protoporphyrin IX: basic principles and present clinical experience, *J. Photochem. Photobiol. B*, 1990, **6**, 143–148.

59. P. Wolf, E. Rieger and H. Kerl, Topical photodynamic therapy with endogenous porphyrins after application of 5-aminolevulinic acid, *J. Am. Acad. Dermatol.*, 1993, **28**, 17–21.

60. N.J. Petrelli, J.A. Cebollero, M. Rodriguez-Bigas and T. Mang, Photodynamic therapy in the management of neoplasms of the perianal skin, *Arch. Surg.*, 1992, **127**, 1436–1438.

61. R.R. Allison, T.S. Mang and B.D. Wilson, Photodynamic therapy for the treatment of nonmelanomatous cutaneous malignancies, *Semin. Cut. Med. Surg.*, 1998, **17**, 153–163.

62. R.M. Szeimies, T. Lorenzen, S. Karrer, C. Abels and A. Plettenberg, Photochemotherapy of cutaneous AIDS-associated Kaposi sarcoma with indocyanine green and laser light, *Hautarzt*, 2001, **52**, 322–326.

63. W.H. Boehncke, K. Konig, A. Ruck, R. Kaufmann and W. Sterry, *In vitro* and *in vivo* effects of photodynamic therapy in cutaneous T cell lymphoma, *Acta Dermato. –Venereol.*, 1994, **74**, 201–205.

64. D. Grebenova, H. Cajthamlova, J. Bartosova, J. Marinov, H. Klamova, O. Fuchs and Z. Hrkal, Selective destruction of leukaemic cells by photo-activation of 5-aminolaevulinic acid-induced protoporphyrin-IX, *J. Photochem. Photobiol. B*, 1998, **47**, 74–81.

65. K. Rittenhouse-Diakun, H. van Leengoed, J. Morgan, E. Hryhorenko, G. Paszkiewicz, J.E. Whitaker and A.R. Oseroff, The role of transferrin receptor (CD71) in photodynamic therapy of activated and malignant lymphocytes using the heme precursor delta-aminolevulinic acid (ALA), *Photochem. Photobiol.*, 1995, **61**, 523–528.

66. W.H. Boehncke, W. Sterry and R. Kaufmann, Treatment of psoriasis by topical photodynamic therapy with polychromatic light, *Lancet*, 1994, **343**, 801.
67. J.T. Deahl, N.L. Oleinick and H.H. Evans, Large mutagenic lesions are induced by photodynamic therapy in murine L5178Y lymphoblasts, *Photochem. Photobiol.*, 1993, **58**, 259–264.
68. W.H. Boehncke, A. Ruck, J. Naumann, W. Sterry and R. Kaufmann, Comparison of sensitivity towards photodynamic therapy of cutaneous resident and infiltrating cell types in vitro, *Lasers Surg. Med.*, 1996, **19**, 451–457.
69. A. Orenstein, J. Halik, J. Tamir, E. Winkler, H. Trau, Z. Malik and G. Kostenich, Photodynamic therapy of cutaneous lymphoma using 5-aminolevulinic acid topical application, *Dermatol. Surg.*, 2000, **26**, 765–769.
70. P. Wolf, R. Fink-Puches, L. Cerroni and H. Kerl, Photodynamic therapy for mycosis fungoides after topical photosensitization with 5-aminolevulinic acid, *J. Am. Acad. Dermatol.*, 1994, **31**, 678–680.
71. D. Edstrom, A. Porwit and A. Ros, Photodynamic therapy with topical 5-amino-levulinic acid for mycosis fungoides: clinical and histological response, *Acta Derm. Venereol.*, 2001, **81**, 184–188.
72. T. Markham, K. Sheahan and P. Collins, Topical 5-aminolaevulinic acid photo-dynamic therapy for tumour-stage mycosis fungoides, *Br. J. Dermatol.*, 2001, **144**, 1262–1263.
73. N. Umegaki, R. Moritsugu, S. Katoh, K. Harada, K. Nakano, K. Tamai, K. Hanada and M. Tanaka, Photodynamic therapy may be useful in debulking cutaneous lymphoma prior to radiotherapy, *Clin. Exp. Dermatol.*, 2004, **29**, 42–45.
74. W.H. Boehncke, K. Konig, R. Kaufmann, W. Scheffold, O. Prummer and W. Sterry, Photodynamic therapy in psoriasis: suppression of cytokine production in vitro and recording of fluorescence modification during treatment *in vivo*, *Arch. Dermatol. Res.*, 1994, **286**, 300–303.
75. S.H. Ibbotson, Topical 5-aminolaevulinic acid photodynamic therapy for the treatment of skin conditions other than non-melanoma skin cancer, *Br. J. Dermatol.*, 2002, **146**, 178–188.
76. R. Ruiz-Rodriguez, J.G. Alvarez, P. Jaen, A. Acevedo and S. Cordoba, Phtot-dynamic therapy with 5-aminolevulinic acid for recalcitrant familial benign pe-mphigus (Hailey–Hailey disease), *J. Am. Acad. Dermatol.*, 2002, **47**, 740–742.
77. M. Gold, T.M. Bridges, V.L. Bradshaw and M. Boring, ALA-PDT and blue light therapy for hidradenitis suppurativa, *J. Drugs Dermatol.*, 2004, **3**, S32–S35.
78. S. Karrer, C. Abels, M. Landthaler and R.M. Szeimies, Topical photodynamic therapy for localized scleroderma, *Acta Derm. Venereol.*, 2000, **80**, 26–27.
79. S. Karrer, A.K. Bosserhoff, P. Weiderer, M. Landthaler and R.M. Szeimies, Influence of 5-aminolevulinic acid and red light on collagen metabolism of human dermal fibroblasts, *J. Invest Dermatol.*, 2003, **120**, 325–331.
80. S. Karrer, A.K. Bosserhoff, P. Weiderer, M. Landthaler and R.M. Szeimies, Keratinocyte-derived cytokines after photodynamic therapy and their paracrine induction of matrix metalloproteinases in fibroblasts, *Br. J. Dermatol.*, 2004, **151**, 776–783.
81. D. Exadaktylou, H.A. Kurwa, E. Calonje and R. Barlow, Treatment of Darier's disease with photodynamic therapy, *Br. J. Dermatol.*, 2003, **149**, 606–610.
82. P. Hillemanns, M. Untch, F. Prove, R. Baumgartner, M. Hillemanns and M. Korell, Photodynamic therapy of vulvar lichen sclerosus with 5-aminolevulinic acid, *Obstet. Gynecol.*, 1999, **93**, 71–74.

83. P.A. McCarron, R.F. Donnelly, A.D. Woolfson, A. Zawislak, J.H. Price and P. Maxwell, Photodynamic treatment of lichen sclerosus and squamous hyperplasia using sustained topical delivery of aminolaevulinic acid from a novel bioadhesive, *Br. J. Dermatol.*, 2004, **151**(Suppl. 68), 105–106.

84. M.K. Fehr, C.F. Chapman and T. Krasieva, Selective photosensitizer distribution in vulvar condyloma acuminatum after topical application of 5-aminolaevulinic acid, *Am. J. Obstet. Gynaecol.*, 1996, **174**, 951–957.

85. R.G. Frank, J.D. Bos, F.W. Meulen and H.J.C.M. Sterenborg, Photodynamic therapy for condylomata acuminata with local application of 5-aminolaevulinic acid, *Genitourin. Med.*, 1996, **72**, 70–71.

86. E.V. Ross, R. Romero, N. Kollias, C. Crum and R.R. Anderson, Selectivity of protoporphyrin IX fluorescence for condylomata after topical application of 5-aminolaevulinic acid: implications for photodynamic treatment, *Br. J. Dermatol.*, 1997, **137**, 736–742.

87. M.K. Fehr, R. Hornung, A. Degen, V.A. Schwarz, D. Fink, U. Haller and P. Wyss, Photodynamic therapy of vulvar and vaginal condyloma and intraepithelial neoplasia using topically applied 5-aminolevulinic acid, *Lasers Surg. Med.*, 2002, **30**, 273–279.

88. X.L. Wang, H.W. Wang, H.S. Wang, S.Z. Xu, K.H. Liao and P. Hillemanns, Topical 5-aminolaevulinic acid-photodynamic therapy for the treatment of urethral condylomata acuminata, *Br. J. Dermatol.*, 2004, **151**, 880–885.

89. M.G. Strakhovskaya, A.O. Shumarina, G.Y. Fraikin and A.B. Rubin, Synthesis of PPIX induced by 5-aminolevulinic acid in yeast cells in the presence of 2,2;-dipyridyl, *Biochemistry (Mosc.)*, 1998, **63**, 725–728.

90. P.G. Calzavara-Pinton, M. Venturini, R. Capezzera, R. Sala and C. Zane, Photodynamic therapy of interdigital mycoses of the feet with topical application of 5-aminolevulinic acid, *Photodermatol. Photoimmunol. Photomed.*, 2004, **20**, 144–147.

91. K. Gardlo, Z. Horska, C.D. Enk, L. Rauch, M. Megahed, T. Ruzicka and C. Fritsch, Treatment of cutaneous leishmaniasis by photodynamic therapy, *J. Am. Acad. Dermatol.*, 2002, **48**, 893–896.

92. J. El On, M. Katz and L. Weinrauch, Letter to: treatment of cutaneous leishmaniasis by photodynamic therapy, *J. Am. Acad. Dermatol.*, 2004, **50**, e–12.

93. C.D. Enk, K. Gardlo and T. Ruzicka, Reply to letter, *J. Am. Acad. Dermatol.*, 2004, **50**, e–13.

94. G.D. Weinstein, J.L. McCullough, E.W. Jeffes, J.S. Nelson, N.L. Fong and A.J. McCormick, Photodynamic therapy of psoriasis with topical delta aminolevulinic acid a pilot dose ranging study, *Photodermatol. Photoimmunol. Photomed.*, 1994, 10, abstract P92.

95. A.F. Hürlimann, R.A. Panizzon and G. Burg, Topical photodynamic treatment of skin tumors and dermatoses, *Dermatology*, 1994, 3, Abstract P327.

96. W.H. Boehncke, W. Sterry and R. Kaufmann, Treatment of psoriasis by topical photodynamic therapy with polychromatic light, *Lancet*, 1994, **343**, 801.

97. M.R. Stringer, P. Collins, D.J. Robinson, G.I. Stables and R.A. Sheehan-Dare, The accumulation of protoporphyrin IX in plaque psoriasis after topical application of 5-aminolevulinic acid indicates a potential for superficial photodynamic therapy, *J. Invest Dermatol.*, 1996, **107**, 76–81.

98. P. Collins, D.J. Robinson, M.R. Stringer, G.I. Stables and R.A. Sheehan-Dare, The variable response of plaque psoriasis after a single treatment with topical 5-aminolaevulinic acid photodynamic therapy, *Br. J. Dermatol.*, 1997, **137**, 743–749.

99. D.J. Robinson, P. Collins, M.R. Stringer, D.I. Vernon, G.I. Stables, S.B. Brown and R.A. Sheehan-Dare, Improved response of plaque psoriasis after multiple treatments with topical 5-aminolaevulinic acid photodynamic therapy, *Acta Derm. Venereol.*, 1999, **79**, 451–455.

100. S. Radakovic-Fijan, U. Blecha-Thalhammer, V. Schleyer, R.M. Szeimies, T. Zwingers, H. Hönigsmann and A. Tanew, Topical aminolaevulinic acid-based photodynamic therapy as a treatment option for psoriasis? Results of a randomized observed-blinded study, *Br. J. Dermatol.*, 2005, **152**, 279–283.

101. P.E. Beattie, R.S. Dawe, J. Ferguson and S.H. Ibbotson, Lack of efficacy and tolerability of topical PDT for psoriasis in comparison with narrowband UVB phototherapy, *Clin. Exp. Dermatol.*, 2004, **29**, 560–562.

102. I.M. Stender and H.C. Wulf, Köbner reaction induced by photodynamic therapy using delta-aminolevulinic acid, *Acta Derm. Venereol.*, 1996, **76**, 392–393.

103. W.H. Boehncke, T. Elshorst-Schmidt and R. Kaufmann, Systemic photodynamic therapy is a safe and effective treatment of psoriasis, *Arch. Dermatol.*, 2000, **136**, 271–272.

104. D.X. Divaris, J.C. Kennedy and R.H. Pottier, Phototoxic damage to sebaceous glands and hair follicles of mice after systemic administration of 5-aminolevulinic acid correlates with localized protoporphyrin IX fluorescence, *Am. J. Pathol.*, 1990, **136**, 891–897.

105. W.L. Lee, A.R. Shalita and M.B. Poh-Fitzpatrick, Comparative studies of porphyrin production in Propionibacterium acnes and Propionibacterium granulosum, *J. Bacter.*, 1978, **133**, 811–815.

106. H. Meffert, H.P. Scherf and N. Sönnichsen, Behandlung von Akne vulgaris mit sichtbarem Licht, *Dermatol. Monatsschr.*, 1987, **173**, 678–679.

107. H. Meffert, K. Gaunitz, T. Gutewort and U.J. Amlong, Acnetherapie mit sichbarem Licht, *Dermatol. Montasschr.*, 1990, **176**, 597–603.

108. V. Sigurdsson, A.C. Knulst and H. van Weelden, Phototherapy of acne vulgaris with visible light, *Dermatology*, 1997, **194**, 256–260.

109. W.J. Cunliffe and V. Goulden, Phototherapy and acne vulgaris, *Br. J. Dermatol.*, 2000, **142**, 855–856.

110. P. Papageorgiou, A. Katsambas and A. Chu, Phototherapy with blue (415 nm) and red (660 nm) light in the treatment of acne vulgaris, *Br. J. Dermatol.*, 2000, **142**, 973–978.

111. W. Hongcharu, C.R. Taylor, Y. Chang, D. Aghassi, K. Suthamjariya and R.R. Anderson, Topical ALA-photodynamic therapy for the treatment of acne vulgaris, *J. Invest Dermatol.*, 2000, **115**, 183–192.

112. B. Pollock, D. Turner, M.R. Stringer, R.A. Bojar, V. Goulden, G.I. Stables and W.J. Cunliffe, Topical aminolaevulinic acid-photodynamic therapy for the treatment of acne vulgaris: a study of clinical efficacy and mechanism of action, *Br. J. Dermatol.*, 2004, **151**, 616–622.

113. Y. Itoh, Y. Ninomiya, S. Tajima and A. Ishibashi, Photodynamic therapy of acne vulgaris with topical delta-aminolaevulinic acid and incoherent light in Japanese patients, *Br. J. Dermatol.*, 2001, **144**, 575–579.

114. M.H. Gold, V.L. Bradshaw, M.M. Boring, T.M. Bridges, J.A. Biron and L.N. Carter, The use of a novel intense pulsed light and heat source and ALA-PDT in the treatment of moderate to severe inflammatory acne vulgaris, *J. Drugs Dermatol.*, 2004, **3**, S15–S19.

115. A.F. Taub, Photodynamic therapy for the treatment of acne: a pilot study, *J. Drugs Dermatol.*, 2004, **3**, S10–S14.

116. I.M. Stender, R. Na, H. Fogh, C. Gluud and H.C. Wulf, Photodynamic therapy with 5-aminolaevulinic acid or placebo for recalcitrant foot and hand: warts randomised double-blind trial, *Lancet*, 2000, **355**, 963–966.
117. R. Ammann, T. Hunziker and L.R. Braathen, Topical photodynamic therapy in verrucae. A pilot study, *Dermatology*, 1995, **191**, 346–347.
118. I.M. Stender and C.H. Wulf, Treatment of recalcitrant verrucae by photodynamic therapy with topical application of delta-aminolaevulinic acid, *Clin. Exp. Dermatol.*, 1996, **21**, 390.
119. I.M. Stender and H.C. Wulf, Photodynamic therapy of recalcitrant warts with 5-aminolaevulinic acid: a retrospective analysis, *Acta Dermato. -Venereol.*, 1999, **79**, 400–401.
120. I.M. Stender, J. Lock-Anderson and H.C. Wulf, Recalcitrant hand and foot warts successfully treated with photodynamic therapy with topical 5-aminolaevulinic acid: a pilot study, *Clin. Exp. Dermatol.*, 1999, **24**, 154–159.
121. G. Fabbrocini, M. Costanzo, A. Riccardo, M. Quarto, A. Colasanti, G. Roberti and G. Monfrecola, Photodynamic therapy with topical delta-aminolaevulinic acid for the treatment of plantar warts, *J. Photochem. Photobiol. B Biol.*, 2001, **61**, 30–34.
122. C. Whitehurst, LED clinical study focuses on PDT photorejuvenation, *Aesthetic Buyers. Guide January./February.*, 2004.
123. M.H. Gold, Intense pulsed light therapy for photorejuvenation enhanced with 20% aminolevulinic acid photodynamic therapy, *J. Lasers Med. Surg.*, 2003, **15**, 47.
124. M.D. Goldman, D. Atkin and S. Kincad, PDT/ALA in the treatment of actinic damage: real world experience, *J. Lasers Med. Surg.*, 2002, **14**, 24.
125. D.K. Avram and M.P. Goldman, Effectiveness and safety of ALA-IPL in treating actinic keratoses and photodamage, *J. Drugs Dermatol.*, 2004, **3**, S–39.
126. R. Ruiz-Rodriguez, T. Sanz-Sanchez and S. Cordoba, Photodynamic photorejuvenation, *Dermatol. Surg.*, 2002, **28**, 742–744.
127. M.H. Gold, V.L. Bradshaw, M.M. Boring, T.M. Bridges, J.A. Biron and T.L. Lewis, Treatment of sebaceous gland hyperplasia by photodynamic therapy with aminolevulinic acid and a blue light source or intense pulsed light source, *J. Drugs Dermatol.*, 2004, **3**, S6–S9.
128. J.A. Hall, P.J. Keller and G.S. Keller, Dose response of combination photorejuvenation using intense pulsed light-activated photodynamic therapy and radiofrequency energy, *Arch. Facial. Plast. Surg.*, 2004, **6**, 374–378.

*Chapter 4*

# ALA-PDT and ALA-FD in Urology

## Raphaela Waidelich and Herbert Stepp

**Table of Contents**

## Abstract

Apart from skin lesions, the first clinical application of ALA was for the detection of bladder cancer in 1992. The high recurrence rate of superficial bladder cancer after standard endoscopic resection had always been indicative of a diagnostic deficit. Today, ALA has proven its superior sensitivity in the detection and precise localization of superficial bladder cancer, especially for the highly malignant carcinoma *in situ*. In this chapter, the current knowledge about implications for recurrence and progression rates and for the treatment strategy will be reviewed. Fluorescence diagnosis for bladder cancer is approved in Europe and it is on the verge of defining the new standard for endoscopic diagnosis of multifocal bladder cancer. The photosensitising properties of ALA-induced PpIX are exploited in clinical studies for PDT of the entire bladder wall. ALA-PDT promises to be an efficient treatment modality for carcinoma *in situ* and very superficial lesions, but the appropriate selection of patients and irradiation parameters still have to be clearly defined. ALA fluorescence diagnosis and ALA-PDT are also studied in connection with condylomas in the urethra, penile cancer, renal cell cancer and prostate cancer. Endoscopic fluorescence detection is as simple as standard white light endoscopy and not significantly more expensive. A widespread use may thus be possible even in the near future. Equipment for ALA-PDT of the bladder is still less elaborate and more expensive. Cheaper light sources and a simple light application are definitely required before ALA-PDT may experience a breakthrough in urology.

## 4.1. Introduction

In urology, ALA has gained widespread use, mainly in Germany, in endoscopic fluorescence diagnosis (FD) of urothelial carcinoma of the bladder. ALA-PDT of bladder cancer, cancer in the upper urinary tract and prostate cancer as well as FD of tumours of the external genitalia and renal cell cancer have been investigated in clinical pilot studies. The diagnostic and therapeutic potentials of ALA and its hexyl-ester (HAL) are being extensively explored for applications in urology.

The standard treatment of non-invasive bladder cancer is endoscopic resection. The main concern with these neoplasias is the high recurrence rate, which is currently between 50% and 70%.[1] The complete transurethral resection (TUR) or destruction of all bladder tumours is supposed to be a crucial factor in preventing recurrent disease.[2] Using white light endoscopy (WLE), the detection of neoplasms of the urinary bladder is limited to morphologically conspicious patterns. Flat urothelial lesions with no epithelial thickening, such as carcinoma *in situ* (CIS) and dysplasia, and small papillary tumours are barely visible during conventional WLE and can easily be missed during TUR.[3] CIS, however, is an aggressive high grade lesion with a 5-year progression rate of more than 50% if left untreated.[4] After TUR of superficial bladder cancer, residual tumour has been identified in up to 55% of cases at a second resection performed 1–2 weeks later.[5–7] Since the sixties, therefore, in order to decrease the risk of overlooking tumours urologists have sought methods of *in vivo* labelling for barely visible neoplastic lesions with the help of an additional colour contrast. Diagnostic methods based on detection of the fluorescence of systemically administered tetracycline, systemic porphyrin mixtures or fluorescein have only been tested in a few patients and have been abandoned.[8-10] Intravesical instillation of methylene blue has also proven unsuitable, since 70% of CIS and 84% of dysplastic lesions were not stained.[11]

Urethral condylomata are genital warts caused by several human papilloma virus (HPV). This virus-induced sexually-transmitted disease is highly contagious and is strongly associated with cervical dysplasia and cervical carcinoma. Urethral condylomata can sometimes only be reached by endoscopy and are generally very susceptible to recurrence. They must therefore be considered as a therapeutic problem still unsolved. ALA-induced fluorescence diagnostics enhances the effectiveness of Nd:YAG laser therapy of these human papilloma-virus lesions.

Early diagnosis and treatment are crucial in penile carcinoma, since the 5-year survival rate of 66% for patients with organ-confined disease is reduced by half for patients with metastatic spread to the inguinal lymph nodes. Additionally, patients with systemic disease have a poor prognosis, since radiation therapy and chemotherapy show partial or complete remissions in only a few cases.[12] As the naked eye is often not able to reliably distinguish the tumours from surrounding healthy tissue, treatment of penile cancer may be incomplete or too radical. There is still a diagnostic dilemma between over treatment of lymph node-negative patients and the missing of occult metastases by watchful waiting. Palpable inguinal lymph nodes are positive for cancer in only 50% of the cases, while cancer can be found in 15% of clinically unsuspicious nodes after standard inguinal dissection.[13] Neither preoperative imaging with computed tomography (CT) nor magnetic resonance imaging (MRI) scan are sensitive enough to use as a basis for deciding whether to remove the lymphatic tissue.[14] Therefore, the only reliable staging information is the surgical exposure of the inguinal area. Additionally, radical removal of all affected tissue provides the best chances for long-term survival. Even for patients with tumour-positive lymph nodes, cure can be achieved in 40-75% with radical surgery if only one or two nodes are involved.[15] However, the side effects and the complications of a radical lymphadenectomy can be disastrous for patients.[16] Recently, several groups examined the value of lymphoscintigraphic sentinel node detection in penile cancer for clinically insignificant disease.[17] Fluorescence detection of ALA-induced PpIX has been shown to be helpful in the diagnosis of the primary tumour and also of the metastatic lesions.[18,19]

In cases of renal cell cancer, the affected kidney cannot always be removed completely, but the tumour can be excised from the normal parenchyma. During this kidney-preserving partial nephrectomy, it is crucial to judge the resection border for residual tumour, which is usually done by submitting a slice of the tissue at the resection border to the pathology unit to perform instantaneous section diagnosis. At this stage, fluorescence imaging certainly cannot replace histopathology, but might be suitable as an intraoperative tool to pre-select the tissue submitted for histopathology. It may thus enable a more tissue-sparing resection without impairing safety.

Observations that bladder cancer may be a multifocal diathesis have led to a search for an effective treatment modality to reliably eradicate bladder cancer cells when they are not yet endoscopically visible via a method based on the systemic and diffuse effects on abnormal areas of the bladder mucosa. As a result, integral treatment of the whole bladder mucosa with adjuvant intravesical chemotherapy or intravesical immunotherapy is indicated in patients at high risk for tumour recurrence and progression. The failure of initial intravesical treatment clearly identifies a group of patients at high risk for subsequent tumour progression. The standard treatment of non-invasive urothelial carcinoma of the bladder refractory to intravesical chemotherapy and immunotherapy is radical cystectomy. In patients with upper tract urothelial carcinoma, nephroureterectomy must be performed. Alternatively, patients who are not candidates for an open operation may benefit from PDT. Although it was first applied clinically to bladder cancer in 1975,[20] PDT has not become an established treatment. Local and systemic toxicity caused by the commonly-used synthetic porphyrins that occasionally caused bladder shrinkage and regularly led to long term cutaneous photosensitivity, have prevented wide clinical application.[21-24] In the last two decades an intensive search for new photosensitizers overcoming these side effects has been underway. Among these so-called "second generation photosensitizers" ALA is one of the most promising.

PDT of prostate cancer is an emerging experimental field. Standard treatment of prostate cancer (surgical prostatectomy or brachytherapy) shows a considerable morbidity. PDT lends itself as a potentially less invasive alternative for patients who are not amenable to standard treatment.

## 4.2. Fluorescence Cystoscopy: Fluorescence Diagnosis of Bladder Cancer

Fluorescence cystoscopy following intravesical instillation of ALA has gained broad application in Germany because of its proven high sensitivity for the detection of barely visible urothelial neoplasias. Hexyl aminolevulenate (HAL) was approved for FD of bladder cancer by the Swedish regulatory authorities in September of 2004. Approval from another 25 European countries was obtained in March of 2005. The "EAU guidelines on the Diagnosis and Treatment of Urothelial Carcinoma in Situ" issued by the European Association of Urology recommend: "Fluorescence cystoscopy should be considered because it has a greater sensitivity than white light cystoscopy".[25]

The procedure for fluorescence cystoscopy involves application of ALA (usually 1.5 g in 50 mL isotonic saline, buffered to pH 5) or HAL (8 mM) by intravesical instillation about 2 h prior to examination. Following application of ALA or HAL there is a selective accumulation of PpIX in urothelial cancer, providing an intense colour contrast between red fluorescing malignant lesions and the nonfluorescing normal mucosa, against blue backscattered light[26] (see Chapter 2). *In vivo* spectral measurements showed a more than 10-fold higher intensity of fluorescence of urothelial cancer in comparison to normal urothelium. Because of the topical application of ALA and its fast metabolism, only minor side effects such as urgency and alginuresis were observed.[27] Initially, a krypton ion laser was used for fluorescence excitation. When large-scale laser technology was replaced by a specially designed incoherent light system, the equipment met all the requirements for a daily routine procedure and can now be used in combination with only slightly modified urological endoscopes (see Chapter 2). Based on a biopsy-related evaluation, with the additional evaluation of the porphyrin fluorescence, a significant increase in sensitivity was observed for the diagnosis of neoplastic urothelial lesions such as dysplasia and CIS as well as for papillary tumours. FD with ALA or HAL is characterized by a high sensitivity of 87–97% and a specificity of around 65%.[27–37] The observed sensitivity was significantly higher than that observed in standard cystoscopy in all studies, whereas the observed specificity was not increased.

In order to clarify the clinical significance of FD, the following questions have to be addressed:

(i) Is FD superior to standard cystoscopy in the detection of high risk lesions?
(ii) Does FD-guided resection lead to a reduced rate of residual tumour?
(iii) Does FD-guided resection reduce recurrence and progression rates?
(iv) Does FD influence therapeutic strategy?

These questions are discussed in detail below.

### 4.2.1. High-Risk Lesions

In the largest study published to date, Zaak *et al.*,[38] on the basis of 1414 procedures on 713 patients, reported that FD more than doubled the detection rate of CIS: standard cystoscopy had missed 52.8% of CIS and 30.3% of moderate dysplasia. The incidence of both high-risk findings combined was as high as 24.3%. Interestingly 77 (47.2%) CIS lesions occurred synchronously with other tumours (18 TaG1-2, 7 TaG3, 27 dys II, 4 T1G1-2, 15 T1G3, 4 T2). These data have recently been updated and are summarized in Table 1.

It is a common finding in all studies of FD for bladder cancer that FD is especially advantageous for the detection of high-grade lesions.[39] The number of patients with CIS detected only by FD versus the number of patients with CIS detected only by random biopsy or standard white light cystoscopy was 6:0 (Landry *et al.*[31]), 9:1 (Jichlinski *et al.*[33]), 18:3 (Schmidbauer *et al.*[40]) or, based on biopsies, 2:1 (König *et al.*[32]) and 29:4 (Grimbergen *et al.*[41]).

**Table 1.** List of all tumour entities and their incidence found in 1713 cystoscopies on 875 patients

| Tumour | No. of biopsies | % of all tumours | FD positive | % FD positive | FD negative | % FD overlooked | WLE positive | % WLE positive | WLE negative | % WLE overlooked |
|---|---|---|---|---|---|---|---|---|---|---|
| pTxGI | 47 | 2.9 | 42 | 89.4 | 5 | 10.6 | 40 | 85.1 | 7 | 14.9 |
| pTxGII | 28 | 1.7 | 26 | 92.9 | 2 | 7.1 | 23 | 82.1 | 5 | 17.9 |
| pTxGIII | 18 | 1.1 | 15 | 83.3 | 3 | 16.7 | 15 | 83.3 | 3 | 16.7 |
| pTaGI | 495 | 30.7 | 466 | 94.1 | 29 | 5.9 | 408 | 82.4 | 87 | 17.6 |
| pTaGII | 205 | 12.7 | 196 | 95.6 | 9 | 4.4 | 175 | 85.4 | 30 | 14.6 |
| pTaGIII | 45 | 2.8 | 42 | 93.3 | 3 | 6.7 | 34 | 75.6 | 11 | 24.4 |
| pT1GI | 12 | 0.7 | 11 | 91.7 | 1 | 8.3 | 8 | 66.7 | 4 | 33.3 |
| pT1GII | 53 | 3.3 | 52 | 98.1 | 1 | 1.9 | 38 | 71.7 | 15 | 28.3 |
| pT1GIII | 149 | 9.2 | 144 | 96.6 | 5 | 3.4 | 119 | 79.9 | 30 | 20.1 |
| >=T2GIII | 101 | 6.3 | 90 | 89.1 | 11 | 10.9 | 87 | 86.1 | 14 | 13.9 |
| Cis | 274 | 17.0 | 254 | 92.7 | 20 | 7.3 | 155 | 56.6 | 119 | 43.4 |
| Dysplasia II | 186 | 11.5 | 146 | 78.5 | 40 | 21.5 | 129 | 69.4 | 57 | 30.6 |
| Total | 1613 | 100.0 | 1484 | 92.0 | 129 | 8.0 | 1231 | 76.3 | 382 | 23.7 |

The detection rates found by fluorescence cystoscopy (FD) and white light endoscopy (WLE) are shown. Courtesy of Dirk Zaak.

### 4.2.2. Residual Tumour

The question of residual tumour was specifically addressed in three studies, which randomized patients for tumour resection into a white light and a FD arm which involved a second look resection 10 days–6 weeks later.

In a multicentre, parallel group, phase III study[42] statistical analysis revealed that the risk of residual tumour after TUR of urothelial carcinoma of the bladder is significantly decreased by ALA-FD. In the WLE group 40.6% of cases were resected tumour-free at primary resection, whereas with 5-amino-levulinic acid fluorescence endoscopy (ALA-FD)-guided TUR, 61.5% were resected tumour-free ($p < 0.014$). Secondary resection was with white light.

Riedl et al.[43] reported on 102 patients enrolled in a study to compare residual tumour rates 6 weeks after initial resection with standard cystoscopy (51 pts.) or FD (51 pts.). Residual tumours were found in 39% of the white light arm versus 16% in the FD arm ($p = 0.005$). The secondary resection in this study was with FD in both arms.

In another prospective, single institution, randomized trial on 301 patients,[44] the residual tumour rate was 25.2% in the white light arm versus 4.5% in the ALA-FD arm ($p < 0.0001$). Recurrence-free survival in the ALA-FD group was 89.6% after 12 and 24 months compared with 73.8% and 65.9%, respectively, in the white light group ($p = 0.004$). This superiority proved to be independent of risk group.

Although the residual tumour rates were very different in these studies, FD proved to be superior to the current gold standard in each of them.

### 4.2.3. Recurrence and Progression Rates

While it is valuable for the surgeon to identify malignant lesions more clearly, especially high grade ones, and while it is comfortable for the patient to have a reduced risk of suffering from extended tissue resections in control cystoscopies, a real clinical benefit of FD would have to be assessed on more long-term effects. Currently, there are only two published follow-up studies comparing TUR guided by either FD or WLE.

Filbeck et al.[44] conducted a randomized trial with follow-up visits every 3 months using WLE and cytology in both groups. The median follow-up of the 191 available patients was 21.2 months for the FD-arm and 20.5 months for the WLE-arm, respectively. There were no statistical differences with regard to either the risk profiles or the intravesical therapy. Recurrence-free survival in the FD-arm after 12 and 24 months was 89.6% compared to 73.8% and 65.9% in the WLE arm, respectively ($p = 0.004$). The superiority proved to be an independent prognostic parameter and the adjusted hazard ratio for FD-TUR compared to WLE-TUR was 0.33 (95% confidence interval, 0.16–0.67). Recently these authors have updated the follow-up with significantly different recurrence-free survivals after 48 months of 85% for FD and 60.7% for WLE[45] (Figure 1).

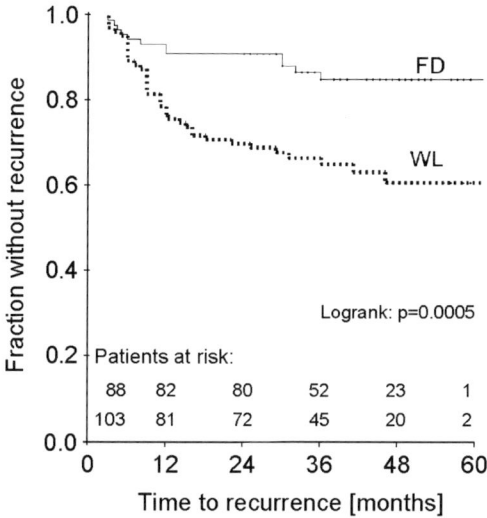

**Figure 1.** Kaplan–Meier plot of time to recurrence after transurethral resections of bladder cancer performed under either standard white light endoscopy (lower curve) or fluorescence endoscopy (upper curve). (Courtesy of Filbeck *et al.*, adapted from ref. 45.)

Babjuk *et al.*[46] have reported their data comparing recurrence rates after TUR using FD in one arm and WLE in the other. A first follow-up cystoscopy with white light 10–15 weeks after TUR revealed recurrent tumours in only five patients who had received TUR guided by FD (8%), whereas 23 patients (37%) of the WLE group had developed a recurrence. The median time to first recurrence was 17 months for the FD-group and 8 months for the WLE-group ($p = 0.008$) with the most significant benefit for patients with multiple or recurrent tumours.

### 4.2.4. Therapeutic Strategy

It is evident that an increased rate of CIS detected will influence the kind of therapeutic procedure suggested for the patients concerned. A randomized phase III multicentre study explicitly evaluated HAL-FD in comparison to WLE on the basis of the rate of changes to the therapeutic concept.[47] For 17% of 146 patients, a blinded independent urologist designed an improved treatment plan when he assessed the findings obtained with FD compared to the findings obtained with WLE for the same patient. Thus, a direct impact on the patient's intraoperative or postoperative treatment by HAL-FD was stated for every sixth patient.

Since 1994, FD has proven its superior sensitivity and clinical relevance for the management of early stage bladder cancer in a series of independent studies. There is one drawback, however; the low specificity caused by frequent false-positive fluorescence findings. About one third of the lesions that show PpIX fluorescence are histologically benign as assessed by traditional histology. Most frequently, cystitis is responsible for false-positive findings, but there is also squamous metaplasia, hyperplasia or normal urothelium among the histological findings. An increased rate of false-positive findings has been found in the areas of previous resections when the examination takes place within 6 weeks after resection,[48] and after intravesical instillation therapies within 6 months prior to FD.[41] However, some of the false-positive findings may, in fact, be serious precursors of malignancies. Using a fluorescence *in situ* hybridization (FISH) technique on microdissected tissue, Hartmann *et al.*[49] documented genetic alteration patterns in fluorescent hyperplasias that were also found in concomitant papillary tumours of the same patients in seven of eight cases. Even in 6 out of 12 samples of fluorescent tissue judged "normal", premalignant genetic alterations (deletions of chromosome 9) were found.

From a practical point of view, low specificity is seldom a significant restriction, since small areas of inflammatory tissue may undergo a complete "diagnostic resection" without additional risk, and larger areas can be probed with a directed biopsy.

### 4.2.5. Detailed Procedure for Fluorescence Cystoscopy

In order to integrate fluorescence cystoscopy into the daily routine of a urological clinic or practice, only a few logistical steps are necessary. (Currently, only HAL has been approved in 26 European countries.)

- ALA or HAL must have been ordered from the pharmacy and sterile preparation of the instillation liquid must be established according to product information provided by the drug suppliers (see Section 8).
- Instillation with a sterile, single-use catheter after complete voiding of the bladder. Air must be prevented from entering the bladder by pre-rinsing the catheter.
- Patient is asked not to void the instillation liquid and is advised to move freely in order to guarantee a thorough contact of the whole bladder wall with the instillation liquid.
- The instillation liquid must remain inside the bladder prior to investigation for a minimum of a half-hour (optimally, 2–3 h). If the patient is unable to prevent micturition, less than 1 h should have been passed prior to cystoscopy.
- Auxillary equipment required in the operating room includes a special light source, a special light guide and special cystoscopes. Also, a special camera is useful. Approximate total cost is 20,000 € (plus camera cost), but this extra instrumentation can also be used for WLE. (Ensure that all parts are suitable for fluorescence cystoscopy!) Most experience has been obtained with rigid cystoscopes. Flexible cystoscopes provide significantly less brightness in fluorescence mode, but the sensitivity obtained may still be equivalent to standard white light rigid cystoscopy.[50,51]

- When inserting the cystoscope, it is important to completely void any urine prior to using the fluorescence mode, since urine gives a diffuse green fluorescence which obscures a clear view of the bladder wall. Perform barbotage cytology as desired.
- Usually bladder wall inspection starts with a standard orientation in white light, and the location of visible tumours is identified. Light exposure of the bladder wall, however, should not be extensive, in order to avoid photochemical bleaching of the PpIX.
- An initial check on an efficient instillation of ALA/HAL as well as on the functionality of the equipment can be performed by examination for fluorescence at the bladder outlet, which always shows – usually unspecific – positive fluorescence.
- During fluorescence inspection of the bladder wall, excitation light exposure should be kept as low as possible. PpIX will undergo photochemical degradation, leading to gradual fading of the fluorescence. This, of course, only affects the area exposed to light and any light-induced fluorescence fading occurs faster when exposed to UV light than when exposed to white light. Small or flat lesions, not easily identified in white light, should in any case first be resected or biopsied and documented. Depending on the camera system used, image repetition rate may be reduced, resulting in blurring of the images during movement of the endoscope. Performing biopsies or resection will usually be under white light, but can also be done in fluorescence mode, provided the repetition rate is sufficiently high or the movements slow enough. After resection, the resected margin is checked for residual fluorescence. It must be realized that the deeper layers of the resected area cannot be checked for residual tumour by fluorescence, as PpIX-staining is restricted to the very superficial layer of any lesion.
- Fluorescence judgement is simple, since the decision is based on the colour observed, rather than on the relative intensity as described in Chapter 2. However, some caveats have to be considered.

  (i) Bleaching must always be avoided.
 (ii) The bladder wall must be kept unfolded so as to avoid false positive-appearing wrinkles.
(iii) One should avoid a tangential viewing of the mucosa, since there is a little PpIX in normal mucosa that causes red fluorescence under small observation angles.
(iv) A clear and urine-free rinsing liquid should be maintained at all times.
 (v) It should be kept in mind that especially single-chip cameras tend to produce a reddish glare in overexposed situations, usually in the centre of the image.

In the event of unusual results, remove the camera, check visually for a positive bladder outlet/prostatic urethra. If there is no visible red fluorescence, either the drug has not been applied sufficiently or the endoscope has no observation filter.

*4.2.6. Images*

The images below are typical examples.

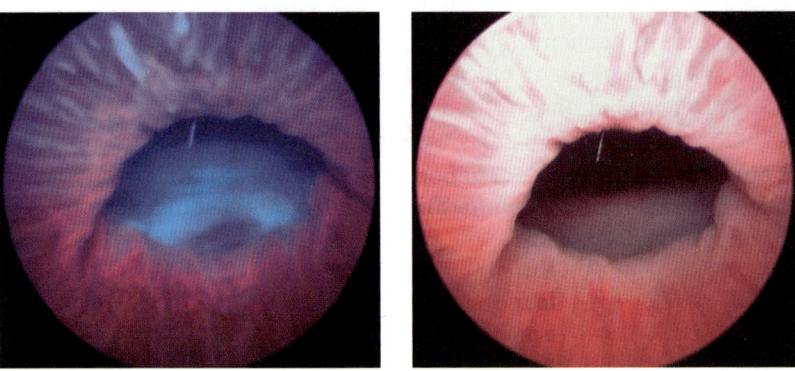

Left: Unspecific red PpIX-fluorescence at the bladder outlet is indicative of functional equipment and efficient ALA-application. Right: corresponding white light image.

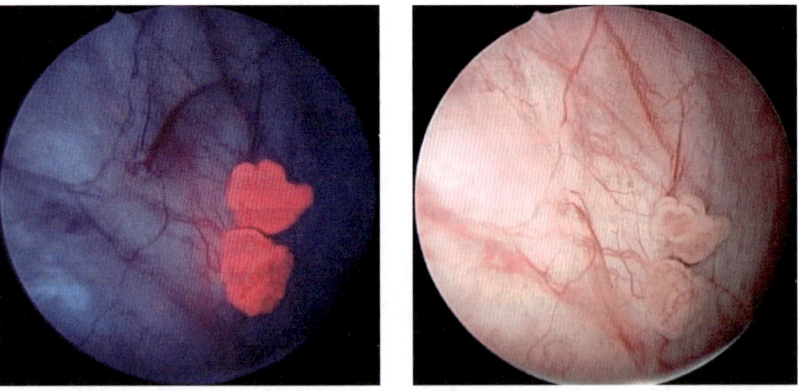

Two small papillary tumours, visible both in fluorescence and white light. Fluorescence confirms that there is no flat peritumoural dysplastic tissue. Note the tiny jet of blood flowing out of a small injured vessel, displayed black in fluorescence mode.

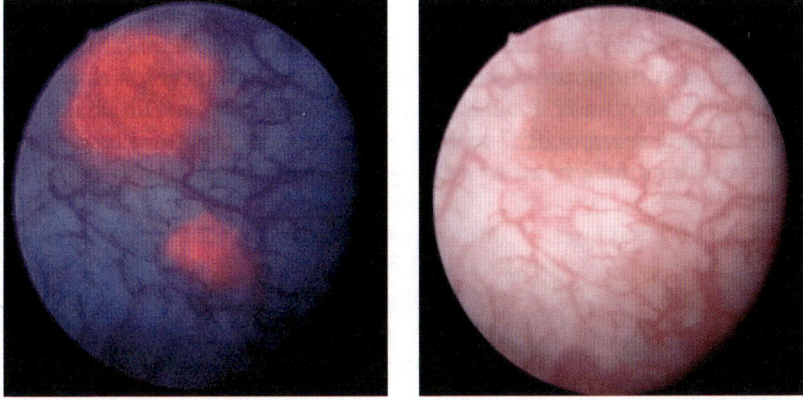

Left: A flat lesion (pTaG1) close to a papillary tumour is readily identified by FD, but is invisible in white light (right).

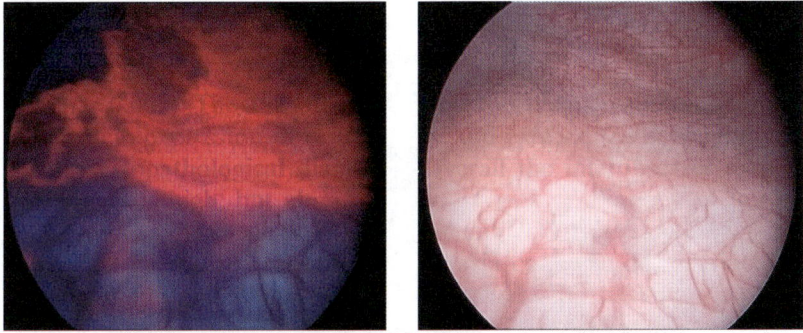

A CIS-lesion in fluorescence (left) and white light (right) modes.

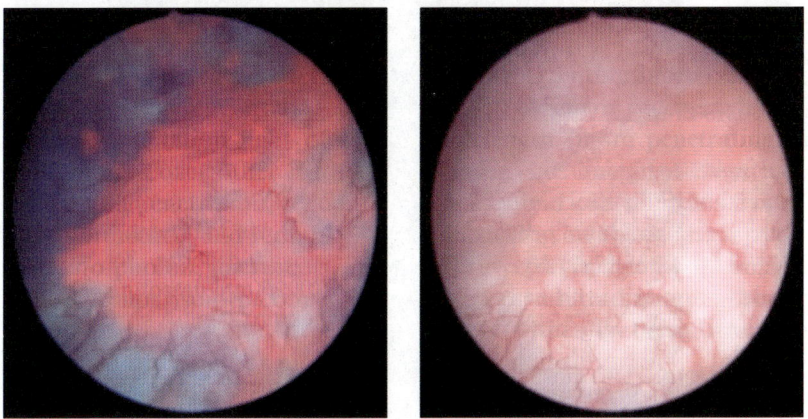

Left: Example of a false-positive lesion, histologically judged as normal, although a suspicious alteration of tissue texture is also visible in white light (right).
Images courtesy of Dirk Zaak.

### 4.2.7. Fluorescence Quantification

Since the false-positive fluorescence originating from inflammatory or scarred tissue can be attributed mainly to infiltrating immune cells, as shown in fluorescence microscopy,[52] it has been suggested that inflammations might be discernible from malignant lesions by a lower fluorescence intensity. This hypothesis has been investigated in a small study of 25 patients and 53 biopsies with fluorescence image quantification.[53] Without affecting sensitivity, the false-positive rate could be decreased by 30% in this study. However, it was also evident that the fluorescence intensities within the group of malignant findings varied as strongly as in the group of non-malignant fluorescence-positive findings. A rather limited benefit can therefore be expected only from quantitative fluorescence measurements.

### 4.2.8. Fluorescence Cytology

Some tumour cells can end up in the urine, and thus the appearance of red fluorescent tumour cells in a bladder lavage specimen of patients instilled with ALA is to be expected. Such fluorescent cells have indeed been observed (Figure 2).[54,55] There is an apparent contrast of cells positive for PpIX-fluorescence and cells exhibiting only green autofluorescence. The method is currently tested for its sensitivity and specificity for the diagnosis of tumours with different gradings. It has a potential of becoming a valuable complement for standard urinary cytology, especially where highly differentiated urinary tumours and flat premalignant dysplasias are concerned.

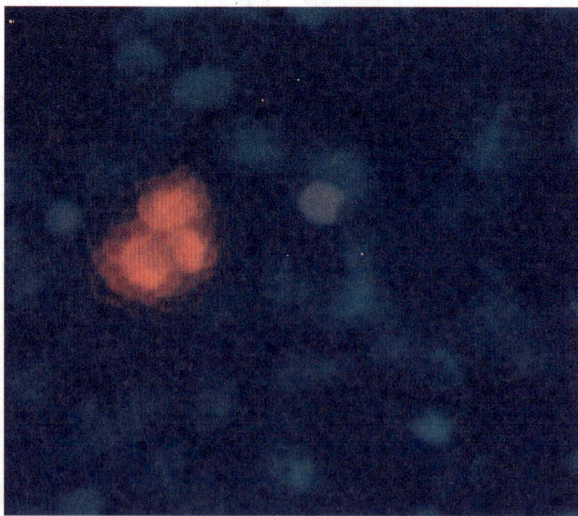

**Figure 2.**    Fluorescence microscopy (10x objective) of bladder lavage specimen showing cells positive for PpIX accumulation (red) and negative cells (green autofluorescence) as well as many erythrocytes (small dark spots). (Image courtesy of Stefan Tauber.)

## 4.3. Other Indications for FD in Urology

### 4.3.1. *Fluorescence Urethroscopy: FD of Condylomata Acuminata and Urothelial Carcinoma of the Urethra*

Early clinical experience with ALA-FD for the diagnosis of urothelial carcinoma of the urethra has been reported.[56] The results show that marking these tumours by fluorescence may improve TUR and thus preserve the urethra. Fluorescence urethroscopy is also a promising diagnostic procedure for detecting subtle clinical and subclinical HPV lesions of the urethra which are not normally visualized by conventional endoscopy (Figure 3). In a single-institution trial, fluorescence urethroscopy detected additional subclinical lesions in 13 of 43 men.[57] A study of Nd:YAG-laser coagulation of urethral condylomata following conventional white light urethroscopy alone and WLE in addition to fluorescence endoscopy after topical application of ALA (total: 168 patients), showed fewer recurrences (21.3% *vs.* 47.3%) in the fluorescence-controlled group.[58]

### 4.3.2. *ALA-FD of Penile Cancer*

Clinical studies have now shown that FD following topically applied ALA can assist urologists in the detection of neoplastic and preneoplastic lesions of the penis, thus ensuring a more reliable destruction of all tumour material in penile sparing surgery.[59]

Additionally, FD following orally applied ALA may be helpful to detect occult metastasis in squamous cell carcinoma of the penis. Preliminary results show that following oral administration of ALA, the fluorescent metabolite PpIX is accumulated in tumour-positive lymph nodes, making FD in penile cancer possible.[60]

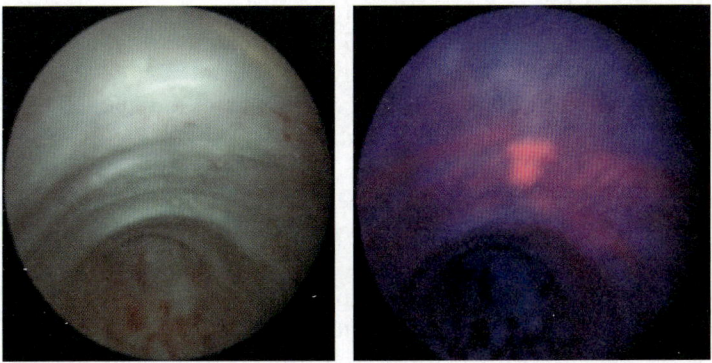

**Figure 3.**   Small HPV-positive lesion in the urethra.

*4.3.3. ALA-FD of Renal Cell Cancer*

There are preliminary clinical data published in 1999 on the fluorescence contrast obtained from kidney cancer versus normal surrounding parenchyma.[61] ALA had been applied systemically in a dose of 20 mg kg$^{-1}$ 4–6 h prior to organ preserving tumour resection. Although the authors reported no side effects and could clearly demarcate resection borders by PpIX-fluorescence imaging in eight of nine patients investigated, there has been no further clinical experience reported to date.

## 4.4. ALA-PDT of Urothelial Carcinoma

Initial clinical results by Waidelich *et al.*[62] and Berger *et al.*[63] have shown that ALA-PDT is effective as an organ preserving procedure for treating non-invasive bladder cancer, even in patients with multifocal, refractory urothelial carcinoma. ALA-PDT showed a good initial response rate (CR at 3 months: 19 of 24 patients in ref. 62). Long-term control is worse of course (CR: 7 of 24 in ref. 62 (median follow-up: 36 months) and 16 of 31 in ref. 63 (mean follow-up: 20 months)). Still, ALA-PDT of bladder cancer represents a reasonable additional treatment option especially for inoperable patients. Flat lesions responded significantly better than papillary tumours, even if the latter are small. Thus, at present, resection of visible papillary tumours is recommended rather than relying on PDT for their removal. As far as this preliminary experience is concerned, oral delivery of ALA (ref. 62, 40 mg kg$^{-1}$ bw) is not superior to intravesical instillation (ref. 63, 3% 50 mL). There may be considerable potential in optimising ALA-PDT parameters and devices (described below). A promising treatment option may be the combination of Mitomycin C instillation therapy and ALA-PDT.[64]

ALA-PDT has also been shown to be an effective organ preserving treatment modality of multifocal urothelial carcinoma of the upper urinary tract.[65] Two of four patients not eligible for nephroureterectomy responded to ALA-PDT alone (CR at 7 and 17 months follow-up), and the other two were disease-free at 24 months after additional Nd:YAG laser treatment of the small residual tumours. ALA had been applied orally with 40 mg kg$^{-1}$, then laser light (514 and/or 635 nm) was applied using 2 mm diameter radial diffusers, 200 mW cm$^{-1}$ for 471 s. The promising results obtained so far warrant further evaluation.

One should be aware of a risk of hemodynamic instability after the oral intake of ALA, especially in patients with cardiovascular comorbidity.[62,65] Therefore, after the oral administration of ALA patients should be monitored by measuring the blood pressure and pulse rate, and hospitalized for 24 h. Invasive cardiovascular monitoring should be included in the protocol of oral ALA-PDT in patients with elevated risk for complications associated with anaesthesia (ASA score 3 or greater). The side effects associated with oral

ALA, including hypotension, tachycardia, nausea and vomiting, may mimic porphyria because patients with acute intermittent porphyria and those who receive ALA orally have elevated plasma ALA. These side effects, however, can be avoided if ALA is delivered by intravesical instillation. ALA is the only agent for PDT currently available that causes reliable photosensitization when administered topically.

Using a white light source with a transurethral irradiation catheter following intravesical instillation of ALA or HAL makes whole-bladder wall ALA-PDT an easy procedure and avoids systemic side effects.[66,67] From the clinical results that can be reported so far,[66] there is no noticeable difference in outcome with white light irradiation versus the use of red or green laser light. Five of eleven patients were disease-free after a median follow-up of 18 months. With these patients biopsies had been taken from different sites soon after PDT and checked for immediate effects (stained sections and electron microscopy). Effects could be shown on papillary tumours, where the most superficial cells showed signs of moderate to severe damage, whereas deeper lying cells and healthy mucosa appeared to be unaffected or minimally affected by PDT. Areas with CIS prior to PDT contained no cells after PDT, with extremely damaged urothelial cells at the border of these denuded areas.

## 4.5. ALA-PDT of Prostate Cancer

Various photosensitizers are being tested as an alternative treatment modality for prostate cancer. Early clinical applications have been reported for Pd-Bacteriopheophorbide (TOOKAD, Negma-Lerads and Steba Biotech, France) and mTHPC (Foscan, BioLitec, Germany).[68,69] ALA has also been applied for PDT of human prostates by Zaak et al.[70] In this study, ALA had been applied orally with 20 mg kg$^{-1}$ bw, which led to a cancer cell-specific PpIX formation as shown with fluorescence microscopy (Figure 4). PDT was performed with radial diffusers with 5–10 insertions in three patients, and perineal introduction of the fibres in two more patients. 500 mW cm$^{-1}$ had been applied using a 633 diode laser as an irradiation source. A decreased individual PSA-value (prostate specific antigen) 6 weeks after ALA-PDT showed a measurable response. Further development of a reproducible light application and dosimetry, as well as more clinical studies are warranted to better define the optimized parameters and indication profile for ALA-PDT of prostate cancer.

## 4.6. Conclusions

Urology, a domain of endoscopic evaluation and treatment, provides many possibilities for the use for FD and PDT. Since the main targets for endoscopic approach are non-invasive tumours, ALA-induced PpIX is a suitable photosensitizer. PpIX is excited most effectively by violet light, but the absorption in

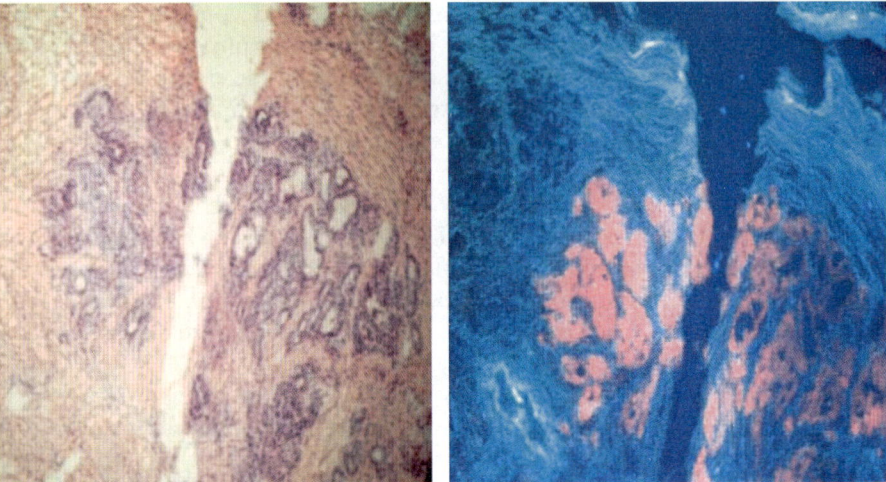

**Figure 4.** Localization of red PpIX-fluorescence in prostate cancer tissue (Gleason 6). Left: H&E-stain. Right: Frozen section fluorescence. Adjacent stroma shows autofluoresence (courtesy of Zaak and Sroka, URO and LFL Munich).

the red region is strong enough to release the full phototoxic capacity of accumulated PpIX with the light doses commonly used for PDT. To detect and treat superficial lesions, there is no need for the deeper penetrating red light, and blue light can be used for irradiation.

In recent years, ALA-FD has been shown to be an invaluable diagnostic tool for endoscopic detection of bladder carcinoma. The European approval of Hexyl aminolevulenate (HAL or Hexvix, Photocure, Norway) for that indication is a great milestone.

The early clinical results with ALA-FD in the diagnosis of intraurethral tumours and penile cancer are encouraging, but these approaches still need further evaluation by phase III trials. Avoiding systemic side effects and the risk of bladder shrinkage makes PDT following intravesically-applied ALA an attractive therapeutic alternative to radical surgery. Initial clinical results indicate that ALA-PDT is an effective organ-preserving procedure for treating multifocal superficial urothelial carcinoma of the upper urinary tract and bladder, even in patients with refractory urothelial carcinoma and it is effective in destroying flat neoplastic lesions like CIS. To avoid systemic side effects, the intravesical application of ALA is preferable. Whole bladder wall PDT, using a white light source and an irrigation catheter, is an easy procedure.

## 4.7. Future Perspectives

There are many potential applications for ALA-FD and ALA-PDT in urology that are in the early stages of clinical trials. Amongst these, whole bladder wall

PDT of bladder cancer, interstitial PDT of prostate cancer and FD or PDT of urethral condylomata may have the clearest and most promising indications. There is significant competition with other photosensitizers like hypericin for bladder cancer and FD, TOOKAD or Foscan for PDT of prostate cancer. As well, there is always competition with continuously improved standard procedures like radioactive seed implantation (brachytherapy) for prostate cancer.

Whether ALA-PDT will be acceptable, and where ALA-PDT will have the greatest impact, will in large part depend on the costs of the procedures. The development of cheaper light sources and the availability of cheap and reliable light application devices will thus be of great importance. Another determining factor is the practicality of the procedures. Long drug-light intervals or a requirement for general anaesthesia will certainly not contribute to a quick and broad dissemination of ALA-PDT, since competitive procedures do not require them. The decisive precondition, of course, is that ALA-PDT is at least as effective as the current standard. ALA meets at least one of the most important preconditions: it has very few side effects because of high tissue selectivity and fast metabolism.

## 4.8. Materials and Methods for Various Indications

It should be noted that, with the exception of bladder cancer detection using Hexvix, all of the applications described below are investigational procedures and should be performed only in clinical studies. Protocols and equipment described are as applied in the department of urology of the University Clinic of Munich, Germany. Other suppliers may recommend different procedures.

### 4.8.1. ALA-FD of Bladder Cancer

Two to three hours before endoscopy, 1.5 g ALA dissolved in 50 mL of 5.7% sodium monohydrogen phosphate, or 8 mM Hexvix as provided by the supplier, is instilled in patients intravesically. Before ALA-FD, all patients undergo WLE, and a bladder washing cytologic specimen is obtained. For fluorescence excitation of PpIX a special light source (D-Light, Karl Storz GmbH, Tuttlingen, Germany) provides blue light at 375–440 nm for fluorescence excitation. A yellow long pass filter fit into the eyepiece of the endoscope reduces the blue excitation light and enhances fluorescent contrast. Suspicious sites are identified by their red fluorescence contrasting against backscattered blue light when observed through the long pass filter.

### 4.8.2. ALA-FD of Tumours Within the Urethra

A 0.1% solution of ALA in lidocain jelly ((Instillagel), Farco Pharma, Cologne, Germany) is transurethrally instilled 1 h before urethroscopy using FD-cystoscopic equipment.

*4.8.3. ALA-FD of Penile Cancer: Local Treatment of Primary Tumour*

For local treatment of neoplastic lesions of the male external genitalia, 2–3 h before operation, a 1% solution of ALA in lidocain jelly (Instillagel, Farco Pharma, Cologne, Germany) is applied to the glans and the shaft of the penis. A condom is used in order to expose the entire epithelium of the penis to the ALA allowing for an even distribution of the agent, as previously described. In the operation theatre, the condom is removed and the penis is inspected under regular white light. Afterwards, the room lights are dimmed and the penis is illuminated with the blue excitation light for the FD.

Suspicious penile lesions are detected by their bright red fluorescence. Any detected lesions can be treated with Nd:YAG laser coagulation (Dornier, Germering, Germany) in continuous wave mode at 30–50 W console power. The treatment energy is dependent on the size of the lesion with an irradiation time of 100 s (range: 60–150 s). After treating the lesion, a safety margin of 3 mm around the lesion is coagulated. The laser energy is delivered by a 600 μm bare fibre (Dornier) with no tissue contact and a fibre–tissue distance between 0.5 and 1 cm (NA 0.34). The focal spot depends on the fibre–tissue distance and is characterised by a gaussian intensity profile. The laser-treated areas are ablated with a scalpel and biopsies are taken from the tumour base. After laser coagulation, the penis is again examined with FD to check for remaining fluorescence. Laser coagulated areas with residual positive fluorescence are biopsied and then laser-coagulated again. Before and after laser treatments, biopsies must be sent for histopathological examination.

*4.8.4. ALA-FD of Penile Cancer: Detection of Lymph Node Metastases*

Five hours before the scheduled surgery, patients receive 15–20 mg ALA (medac GmbH, Wedel, Germany) per kilogram bodyweight. The ALA is dissolved in water, and the agent is given orally. After the application of ALA, the patients are protected from direct artificial and natural light for 24 h to prevent skin irritations. In the operation theatre patients are prepared as usual for an inguinal dissection. For FD, the operation field is illuminated with a D-Light and observed with a cystoscope for FD to identify the PpIX marked lymph nodes. Lymph nodes suspicious for tumours are resected as usual.

*4.8.5. ALA-PDT of Urothelial Carcinoma*

*4.8.5.1. Photosensitization: ALA Applied Orally*
ALA is dissolved in 200 mL water 3–6 h before irradiation and is administered orally. The dosage is 40 mg $kg^{-1}$ body weight. In order to avoid phototoxic skin reaction following oral administration of ALA, the patients must be protected from sunlight and intensive indoor illumination for 48 h.

*4.8.5.2. Photosensitization: ALA Applied Intravesically*
The urothelial neoplasms are photosensitized by the intravesical instillation of 5 g ALA-hydrochloride dissolved in 52 mL sodium bicarbonate (pH 4.8) freshly prepared and sterile-filtered before instillation. The duration of intravesical retention should be 2–4 h.

*4.8.5.3. Whole Bladder Wall Irradiation Using Coherent Light*
For whole bladder wall irradiation using coherent light, an argon ion laser pumped dye laser was used for most of the treatments performed in the Urology Division of the University Clinic of Munich, Germany. Of course a diode laser operating at around 635 nm and an output power exceeding 1 W can be used as well. With the argon-dye system, we first applied 40 J cm$^{-2}$ using the 514 nm line of the argon laser. An additional 20 J cm$^{-2}$ were applied from the dye laser tuned to the wavelength of 635 nm. The spherical light diffuser consisted of a 400 µm quartz fibre with a 1.7 mm diameter spherical light emitter. Output power of this emitter was measured with a calibrated integrating sphere powermeter at 0.8–1.5 W (514 nm) and 1.7–2.6 W (635 nm). The bulb-shaped fibre tip is placed in a central region within the bladder through a constant flow cystoscope and positioned using a 5 French ureteral catheter marked at cm intervals. Continuous irrigation with isotonic saline solution had to be maintained during the whole irradiation procedure at a pressure that distended the bladder without folds.

*4.8.5.4. Whole Bladder Wall Irradiation Using a White Light Source*
Irradiation of 100 J cm$^{-2}$ is applied using a 500 W Xenon short arc lamp (T-Light System, K. Storz, Tuttlingen, Germany).[66,67] Two to six Watts are coupled into a glass fibre with a core diameter of 1.5 mm. The tip of the bare fibre is positioned approximately in the bladder centre by means of a special balloon catheter, which is constructed on the basis of a transurethral irrigation catheter. A rod of light scattering material in front of the fibre tip distributes the light over the inner bladder surface[66,67] (Figure 5). Both light source and catheter are currently being tested in a two-centre phase I study using Hexvix (Photocure, Oslo, Norway).

The aim of the irradiation is to exploit the photosensitising capacity of the porphyrin synthesized in the transformed bladder mucosa as completely as possible. This is achieved when all porphyrin has been bleached during irradiation (see Chapter 2). In the case of bladder wall PDT, this can be confirmed easily by fluorescence cystoscopy after PDT. Continuous irrigation with isotonic saline solution has to be maintained during the entire irradiation procedure with a pressure that distends the bladder without folds. Irrigation also provides cooling of the fibre tip and the scattering rod. During irradiation, transabdominal ultrasonography is used to help localize the fibre tip in the bladder and to control the filling volume. The bladder wall area is calculated from the filling volume, assuming a spherical shape.

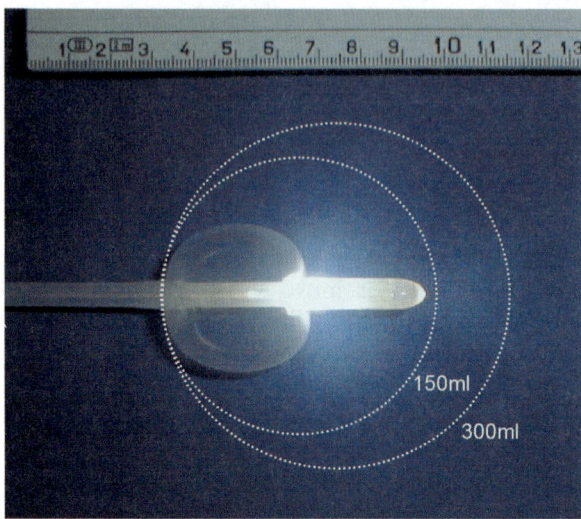

**Figure 5.**    Tip of white light catheter with indicated bladder filling volumes of 150 and 300 mL.

*4.8.5.5. Irradiation of the Upper Urinary Tract*

Three to six hours after oral administration of ALA (40 mg kg$^{-1}$ bw), a flexible quartz glass fibre with a cylindrical diffuser tip (length 1 or 2 cm, diameter 2 mm) is placed through a ureteroscope next to all visible tumours in the pyelocaliceal system and the ureters. Irradiation is performed using 514 nm of the argon ion laser or with 635 nm of the dye laser. Irradiation time is 471 s for each lesion, resulting in an irradiation of 50 J cm$^{-2}$ on an assumed tissue surface of 1.88 cm$^2$ cm$^{-1}$ length of diffuser (3 mm radius tube). Lesions protruding into the ureteral lumen by more than 50% of its radius are treated 2 or 3 times with appropriate lateral repositioning of the cylindrical diffuser tip after each irradiation. To avoid thermal damage, a percutaneous nephrostomy tube is placed before irradiation and irradiation is done under continuous lavage with isotonic saline solution. After irradiation the nephrostomy tube is replaced with a 7 French Double-J ureteral stent.

# References

1.  R. Lee and M.J. Droller, The natural history of bladder cancer. Implications for therapy, *Urol. Clin. North Am.*, 2000, **27**, 1–13.
2.  J.M. Holzbeierlein and J.A. Smith Jr., Surgical management of noninvasive bladder cancer (stages Ta/T1/CIS), *Urol. Clin. North Am.*, 2000, **27**, 15–24.
3.  M.S. Soloway, W. Murphy, M.K. Rao and C. Cox, Serial multiple-site biopsies in patients with bladder cancer, *J. Urol.*, 1978, **120**, 57–59.

4. J.A. Witjes, Bladder carcinoma *in situ* in 2003: state of the art, *Eur. Urol.*, 2004, **45**, 142–146.

5. R. Klan, V. Loy and H. Huland, Residual tumor discovered in routine second transurethral resection in patients with stage T1 transitional cell carcinoma of the bladder, *J. Urol.*, 1992, **146**, 316–318.

6. T.A. Vögeli, M.O. Grimm and R. Ackermann, Prospective study for quality control of TUR of bladder tumors by routine 2nd TUR (ReTUR), *J Urol.*, 1998, 159(Suppl.), 143-abstract 543.

7. K.U. Köhrmann, M. Woeste, J. Kappes, J. Rassweiler and P. Alken, Der Wert der transurethralen Nachresektion beim oberflächlichen Harnblasenkarzinom, *Akt Urol*, 1994, **25**, 208–213.

8. M. Devonec, P. Lenz, R. Bouvier, N. Blanc-Brunat and J.M. Dubernard, Clinically occult bladder cancer diagnosis. Trial using ultraviolet cystoscopy, *Cancer*, 1985, **55**, 468–471.

9. D. Jocham, R. Baumgartner, N. Fuchs, H. Lenz, H. Stepp and E. Unsold, Fluorescence diagnosis of porphyrin-marked urothelial tumors. Status of experimental development, *Urologe A*, 1989, **28**, 59–64.

10. W.F. Whitmore and I.M. Bush, Ultraviolet cystoscopy in patients with bladder cancer, *J Urol.*, 1966, **95**, 201–207.

11. J. Vicente, G. Chechile and F. Algaba, Value of *in vivo* mucosa-staining test with methylene blue in the diagnosis of pretumoral and tumoral lesions of the bladder, *Eur. Urol.*, 1987, **13**, 15–16.

12. A.M. Hussein, P. Benedetto and K.S. Sridhar, Chemotherapy with cisplatin and 5-fluorouracil for penile and urethral squamous cell carcinomas, *Cancer*, 1990, **65**, 433–438.

13. L. Persky and J. deKernion, Carcinoma of the penis, *CA Cancer J. Clin.*, 1986, **36**, 258–273.

14. S. Horenblas, H. Van Tinteren, J.F. Delemarre, L.M. Moonen, V. Lustig and R. Kroger, Squamous cell carcinoma of the penis: accuracy of tumor, nodes and metastasis classification system, and role of lymphangiography, computerized tomography scan and fine needle aspiration cytology, *J. Urol.*, 1991, **146**, 1279–1283.

15. S. Horenblas, H. Van Tinteren, J.F. Delemarre, L.M. Moonen, V. Lustig and E.W. van Waardenburg, Squamous cell carcinoma of the penis. III. Treatment of regional lymph nodes, *J. Urol.*, 1993, **149**, 492–497.

16. A.A. Ornellas, A.L. Seixas, A. Marota, A. Wisnescky, F. Campos and J.R. de Moraes, Surgical treatment of invasive squamous cell carcinoma of the penis: retrospective analysis of 350 cases, *J. Urol.*, 1994, **151**, 1244–1249.

17. S. Horenblas, L. Jansen, W. Meinhardt, C.A. Hoefnagel, D. de Jong and O.E. Nieweg, Detection of occult metastasis in squamous cell carcinoma of the penis using a dynamic sentinel node procedure, *J. Urol.*, 2000, **163**, 100–104.

18. D. Frimberger, R. Linke, H. Meissner, H. Stepp, D. Zaak, E. Hungerhuber, R. Waidelich, N. Schmeller, A. Hofstetter and P. Schneede, Fluorescence diagnosis: a novel method to guide radical inguinal lymph node dissection in penile cancer, *World J. Urol.*, 2004, **22**, 150–154.

19. D. Frimberger, P. Schneede, E. Hungerhuber, R. Sroka, D. Zaak, M. Siebels and A. Hofstetter, Autofluorescence and 5-aminolevulinic acid induced fluorescence diagnosis of penile carcinoma – new techniques to monitor Nd:YAG laser therapy, *Urol. Res.*, 2002, **30**, 295–300.

20. J.F. Kelly and M.E. Snell, Hematoporphyrin derivative: a possible aid in the diagnosis and therapy of carcinoma of the bladder, *J. Urol.*, 1976, **115**, 150–151.

21. M.A. D'Hallewin and L. Baert, Long-term results of whole bladder wall photo-dynamic therapy for carcinoma *in situ* of the bladder, *Urology*, 1995, **45**, 763–767.

22. U.O. Nseyo, B. Shumaker, E.A. Klein and K. Sutherland, Photodynamic therapy using porfimer sodium as an alternative to cystectomy in patients with refractory transitional cell carcinoma *in situ* of the bladder. Bladder Photofrin Study Group, *J. Urol.*, 1998, **160**, 39–44.

23. J.I. Harty, M. Amin, T.J. Wieman, M.T. Tseng, D. Ackerman and W. Broghamer, Complications of whole bladder dihematoporphyrin ether photodynamic therapy, *J. Urol.*, 1989, **141**, 1341–1346.

24. T.J. Dougherty, M.T. Cooper and T.S. Mang, Cutaneous phototoxic occurrences in patients receiving Photofrin, *Lasers Surg. Med*, 1990, **10**, 485–488.

25. A.P. van der Meijden, R. Sylvester, W. Oosterlinck, E. Solsona, A. Boehle, B. Lobel and E. Rintala, EAU guidelines on the diagnosis and treatment of urothelial carcinoma *in situ*, *Eur. Urol.*, 2005, **48**, 363–371.

26. M. Kriegmair, R. Baumgartner, R. Knuechel, P. Steinbach, A. Ehsan, W. Lumper, F. Hofstadter and A. Hofstetter, Fluorescence photodetection of neoplastic urothe-lial lesions following intravesical instillation of 5-aminolevulinic acid, *Urology*, 1994, **44**, 836–841.

27. M. Kriegmair, R. Baumgartner, R. Knuchel, H. Stepp, F. Hofstadter and A. Hofstetter, Detection of early bladder cancer by 5-aminolevulinic acid induced porphyrin fluorescence, *J. Urol.*, 1996, **155**, 105–109.

28. C.W. Cheng, W.K. Lau, P.H. Tan and M. Olivo, Cystoscopic diagnosis of bladder cancer by intravesical instillation of 5-aminolevulinic acid induced porphyrin fluo-rescence – the Singapore experience, *Ann. Acad. Med Singapore*, 2000, **29**, 153–158.

29. P. Jichlinski, M. Forrer, J. Mizeret, T. Glanzmann, D. Braichotte, G. Wagnieres, G. Zimmer, L. Guillou, F. Schmidlin, P. Graber, B.H. van den and H.J. Leisinger, Clinical evaluation of a method for detecting superficial surgical transitional cell carcinoma of the bladder by light-induced fluorescence of protoporphyrin IX following the topical application of 5-aminolevulinic acid: preliminary results, *Lasers Surg. Med*, 1997, **20**, 402–408.

30. A. Marti, P. Jichlinski, N. Lange, J.P. Ballini, L. Guillou, H.J. Leisinger and P. Kucera, Comparison of aminolevulinic acid and hexylester aminolevulinate induced protoporphyrin IX distribution in human bladder cancer, *J. Urol.*, 2003, **170**, 428–432.

31. J.L. Landry, A. Gelet, R. Bouvier, J.M. Dubernard, X. Martin and M. Colombel, Detection of bladder dysplasia using 5-aminolaevulinic acid-induced porphyrin fluorescence, *BJU. Int.*, 2003, **91**, 623–626.

32. F. Koenig, F.J. McGovern, R. Larne, H. Enquist, K.T. Schomacker and T.F. Deutsch, Diagnosis of bladder carcinoma using protoporphyrin IX fluorescence induced by 5-aminolaevulinic acid, *BJU. Int.*, 1999, **83**, 129–135.

33. P. Jichlinski, L. Guillou, S.J. Karlsen, P.U. Malmstrom, D. Jocham, B. Brennhovd, E. Johansson, T. Gartner, N. Lange, B.H. van den and H.J. Leisinger, Hexyl aminolevulinate fluorescence cystoscopy: new diagnostic tool for photodiagnosis of superficial bladder cancer – a multicenter study, *J. Urol.*, 2003, **170**, 226–229.

34. J. Schmidbauer, F. Witjes, N. Schmeller, R. Donat, M. Susani and M. Marberger, Improved detection of urothelial carcinoma *in situ* with hexaminolevulinate fluo-rescence cystoscopy, *J. Urol.*, 2004, **171**, 135–138.

35. A. Ehsan, F. Sommer, G. Haupt and U. Engelmann, Significance of fluorescence cystoscopy for diagnosis of superficial bladder cancer after intravesical instillation of delta aminolevulinic acid, *Urol. Int.*, 2001, **67**, 298–304.
36. M.A. D'Hallewin, H. Vanherzeele and L. Baert, Fluorescence detection of flat transitional cell carcinoma after intravesical instillation of aminolevulinic acid, *Am. J. Clin. Oncol.*, 1998, **21**, 223–225.
37. C. De Dominicis, M. Liberti, G. Perugia, C. De Nunzio, F. Sciobica, A. Zuccala, A. Sarkozy and F. Iori, Role of 5-aminolevulinic acid in the diagnosis and treatment of superficial bladder cancer: improvement in diagnostic sensitivity, *Urology*, 2001, **57**, 1059–1062.
38. D. Zaak, E. Hungerhuber, P. Schneede, H. Stepp, D. Frimberger, S. Corvin, N. Schmeller, M. Kriegmair, A. Hofstetter and R. Knuechel, Role of 5-amino-levulinic acid in the detection of urothelial premalignant lesions, *Cancer*, 2002, **95**, 1234–1238.
39. A.P. van der Meijden, R. Sylvester, W. Oosterlinck, E. Solsona, A. Boehle, B. Lobel and E. Rintala, EAU guidelines on the diagnosis and treatment of urothelial carcinoma *in situ*, *Eur. Urol.*, 2005, **48**, 363–371.
40. J. Schmidbauer, F. Witjes, N. Schmeller, R. Donat, M. Susani and M. Marberger, Improved detection of urothelial carcinoma in situ with hexaminolevulinate fluorescence cystoscopy, *J. Urol.*, 2004, **171**, 135–138.
41. M.C. Grimbergen, C.F. van Swol, T.G. Jonges, T.A. Boon and R.J. van Moorselaar, Reduced specificity of 5-ALA induced fluorescence in photodynamic diagnosis of transitional cell carcinoma after previous intravesical therapy, *Eur. Urol.*, 2003, **44**, 51–56.
42. M. Kriegmair, D. Zaak, K.H. Rothenberger, J. Rassweiler, D. Jocham, F. Eisenberger, R. Tauber, A. Stenzl and A. Hofstetter, Transurethral resection for bladder cancer using 5-aminolevulinic acid induced fluorescence endoscopy versus white light endoscopy, *J. Urol.*, 2002, **168**, 475–478.
43. C.R. Riedl, D. Daniltchenko, F. Koenig, R. Simak, S.A. Loening and H. Pflueger, Fluorescence endoscopy with 5-aminolevulinic acid reduces early recurrence rate in superficial bladder cancer, *J. Urol.*, 2001, **165**, 1121–1123.
44. T. Filbeck, U. Pichlmeier, R. Knuechel, W.F. Wieland and W. Roessler, Clinically relevant improvement of recurrence-free survival with 5-aminolevulinic acid induced fluorescence diagnosis in patients with superficial bladder tumors, *J. Urol.*, 2002, **168**, 67–71.
45. T. Filbeck, U. Pichlmeier, R. Knuechel, W.F. Wieland and W. Rossler, Reducing the risk of superficial bladder cancer recurrence with 5-aminolevulinic acid-induced fluorescence diagnosis. Results of a 5-year study, *Urologe A*, 2003, **42**, 1366–1373.
46. M. Babjuk, V. Soukup, R. Petrik, M. Jirsa and J. Dvoracek, 5-Aminolevulinic acid induced fluorescence cystoscopy during transurethral resection reduces the risk of recurrence in stage Ta/T1 bladder cancer, *BJU Int.*, 2005, **96**, 798–802.
47. C. Durek, S. Wagner, B. Zeylemaker, R.J. van Moorselaar, J.A. Witjes, M.O. Grimm, R. Muschter, G. Popken, F. König, K.H. Kurth, R. Knüchel and D. Jocham, The impact of Hexvix based fluorescence cystoscopy in treatment decisions of bladder cancer – results of a prospective phase III multicenter study, *J Urol.*, 2004, 171/4, 73-Abstract P278.
48. T. Filbeck, W. Roessler, R. Knuechel, M. Straub, H.J. Kiel and W.F. Wieland, 5-aminolevulinic acid-induced fluorescence endoscopy applied at secondary transurethral resection after conventional resection of primary superficial bladder tumors, *Urology*, 1999, **53**, 77–81.

49. A. Hartmann, K. Moser, M. Kriegmair, A. Hofstetter, F. Hofstaedter and R. Knuechel, Frequent genetic alterations in simple urothelial hyperplasias of the bladder in patients with papillary urothelial carcinoma, *Am. J. Pathol.*, 1999, **154**, 721–727.

50. W. Loidl, J. Schmidbauer, M. Susani and M. Marberger, Flexible cystoscopy assisted by hexaminolevulinate induced fluorescence: a new approach for bladder cancer detection and surveillance?, *Eur. Urol.*, 2005, **47**, 323–326.

51. J.A. Witjes, P.M. Moonen and A.G. van der Heijden, Comparison of hexaminolevulinate based flexible and rigid fluorescence cystoscopy with rigid white light cystoscopy in bladder cancer: results of a prospective Phase II study, *Eur. Urol.*, 2005, **47**, 319–322.

52. P. Steinbach, M. Kriegmair, R. Baumgartner, F. Hofstadter and R. Knuchel, Intravesical instillation of 5-aminolevulinic acid: the fluorescent metabolite is limited to urothelial cells, *Urology*, 1994, **44**, 676–681.

53. D. Zaak, D. Frimberger, H. Stepp, S. Wagner, R. Baumgartner, P. Schneede, M. Siebels, R. Knuchel, M. Kriegmair and A. Hofstetter, Quantification of 5-aminolevulinic acid induced fluorescence improves the specificity of bladder cancer detection, *J Urol.*, 2001, **166**, 1665–1668.

54. S. Tauber, B. Liedl, P. Schneede, F. Liessmann, R. Waidelich and A. Hofstetter, Fluorescence cytology of the urinary bladder, *Urologe A*, 2001, **40**, 217–221.

55. A. Pytel and N. Schmeller, New aspect of photodynamic diagnosis of bladder tumors fluorescence cytology, *Urology*, 2002, **59**, 216–219.

56. L. Holtl, I.E. Eder, H. Klocker, A. Hobisch, G. Bartsch and A. Stenzl, Photodynamic diagnosis with 5-aminolevulinic acid in the treatment of secondary urethral tumors: first in vitro and in vivo results, *Eur. Urol.*, 2001, **39**, 178–182.

57. P. Schneede, P. Munch, S. Wagner, T. Meyer, E. Stockfleth and A. Hofstetter, Fluorescence urethroscopy following instillation of 5-aminolevulinic acid: a new procedure for detecting clinical and subclinical HPV lesions of the urethra, *J. Eur. Acad. Dermatol. Venereol.*, 2000, **15**, 121–125.

58. D. Zaak, A. Hofstetter, D. Frimberger and P. Schneede, Recurrence of condylomata acuminata of the urethra after conventional and fluorescence-controlled Nd:YAG laser treatment, *Urology*, 2003, **61**, 1011–1015.

59. D. Frimberger, P. Schneede, E. Hungerhuber, R. Sroka, D. Zaak, M. Siebels and A. Hofstetter, Autofluorescence and 5-aminolevulinic acid induced fluorescence diagnosis of penile carcinoma – new techniques to monitor Nd:YAG laser therapy, *Urol. Res.*, 2002, **30**, 295–300.

60. D. Frimberger, R. Linke, H. Meissner, H. Stepp, D. Zaak, E. Hungerhuber, R. Waidelich, N. Schmeller, A. Hofstetter and P. Schneede, Fluorescence diagnosis: a novel method to guide radical inguinal lymph node dissection in penile cancer, *World J. Urol.*, 2004, **22**, 150–154.

61. G. Popken, U. Wetterauer and W. Schultze-Seemann, Kidney-preserving tumour resection in renal cell carcinoma with photodynamic detection by 5-aminolaevulinic acid: preclinical and preliminary clinical results, *BJU. Int.*, 1999, **83**, 578–582.

62. R. Waidelich, H. Stepp, R. Baumgartner, E. Weninger, A. Hofstetter and M. Kriegmair, Clinical experience with 5-aminolevulinic acid and photodynamic therapy for refractory superficial bladder cancer, *J. Urol.*, 2001, **165**, 1904–1907.

63. A.P. Berger, H. Steiner, A. Stenzl, T. Akkad, G. Bartsch and L. Holtl, Photodynamic therapy with intravesical instillation of 5-aminolevulinic acid for patients with recurrent superficial bladder cancer: a single-center study, *Urology*, 2003, **61**, 338–341.

64. R.J. Skyrme, A.J. French, S.N. Datta, R. Allman, M.D. Mason and P.N. Matthews, A phase-1 study of sequential mitomycin C and 5-aminolaevulinic acid-mediated photodynamic therapy in recurrent superficial bladder carcinoma, *BJU. Int.*, 2005, **95**, 1206–1210.

65. R. Waidelich, A. Hofstetter, H. Stepp, R. Baumgartner, E. Weninger and M. Kriegmair, Early clinical experience with 5-aminolevulinic acid for the photodynamic therapy of upper tract urothelial tumors, *J. Urol.*, 1998, **159**, 401–404.

66. R. Waidelich, W. Beyer, R. Knuchel, H. Stepp, R. Baumgartner, J. Schroder, A. Hofstetter and M. Kriegmair, Whole bladder photodynamic therapy with 5-aminolevulinic acid using a white light source, *Urology*, 2003, **61**, 332–337.

67. W. Beyer, R. Knüchel, R. Waidelich, H. Stepp, R. Baumgartner and A. Hofstetter, Technical concepts for white light photodynamic therapy of bladder cancer, *Med. Laser Appl.*, 2002, **17**, 37–40.

68. R.A. Weersink, J. Forbes, S. Bisland, J. Trachtenberg, M. Elhilali, P.H. Brun and B.C. Wilson, Assessment of cutaneous photosensitivity of TOOKAD (WST09) in preclinical animal models and in patients, *Photochem. Photobiol.*, 2005, **81**, 106–113.

69. T.R. Nathan, D.E. Whitelaw, S.C. Chang, W.R. Lees, P.M. Ripley, H. Payne, L. Jones, M.C. Parkinson, M. Emberton, A.R. Gillams, A.R. Mundy and S.G. Bown, Photodynamic therapy for prostate cancer recurrence after radiotherapy: a phase I study, *J. Urol.*, 2002, **168**, 1427–1432.

70. D. Zaak, R. Sroka, M. Hoeppner, W. Khoder, O. Reich, S. Tritschler, R. Muschter, R. Knuchel and A. Hofstetter, Photodynamic Therapy by means of 5-ALA induced PPIX in human prostate cancer – preliminary results, *Med Laser Appl*, 2003, **18**, 91–95.

*Chapter 5*

# ALA-PDT and ALA-FD in the Brain: New Prospects for Treating Malignant Gliomas

## Walter Stummer and Reinhold Baumgartner

**Table of Contents**

# Abstract

Surgical management of malignant gliomas remains a major challenge to neurosurgeons. Prognosis of these locally highly invasive tumors is linked to the completeness of their resection. In the brain, however, highly proliferative, marginal tumor tissue is difficult to distinguish, and resections with safety margins are not an option. Five-aminolevulinic acid (ALA) induces specific porphyrin fluorescence within malignant glioma tissue. These porphyrins, being highly fluorescent and photosensitizing, have proven useful for fluorescence-guided resections of malignant gliomas, which has evolved into a practical method for routine use. The potential of ALA as a photosensitizer for adjunctive photodynamic therapy (PDT) is currently under investigation. This article focuses on background and implementation of fluorescence-guided resections of malignant gliomas. It summarizes the rationale for, and present status of PDT using ALA.

## 5.1. Introduction

### 5.1.1. Background

The need for novel, effective, and safe treatments for malignant glioma is obvious. Despite surgery, radiotherapy, and chemotherapy, the median survival time of patients suffering glioblastoma multiforme (WHO Grade IV) still does not greatly exceed 1 year. Patients with anaplastic astrocytoma (WHO Grade III) do little better.

Malignant gliomas are difficult to treat. Modern imaging modalities such as magnetic resonance imaging (MRI) suggest that malignant gliomas are well-circumscribed tumors with regional necrosis surrounded by viable tumor tissue showing marginal contrast enhancement. However, imaging modalities fail to detect the wide margin of infiltrating tumor cells which always accompany these lesions, sometimes extending over a distance of centimeters into functionally important regions of the brain (Figure 1).

Accordingly, it has been argued by some that malignant gliomas are a disease affecting the entire brain rather than a certain brain region and that therapies for local tumor control, such as surgery, are of little help. Nevertheless, in the view of most neurosurgeons surgical cytoreduction of tumor and adjacent tissue containing infiltrating cells of high-density are an accepted treatment modality for these fundamentally incurable tumors. The main aim of surgery is the safe removal of all of the contrast enhancing part of the tumor, which has been shown to be a major factor determining progression and, ultimately, survival prognosis.[1-4] However, this goal is achieved in only less than 30% of cases[1,5,6] because viable, marginal, tumor is difficult to distinguish intraoperatively. In the brain, resections with a safety margin are seldom an option due to the vicinity of functionally important brain regions. In over 80% of cases recurrence is located within 2 cm of initial tumor margins[7] because of residual tumor mass or tumor cells that infiltrate surrounding brain tissue[8,9] (Figure 2).

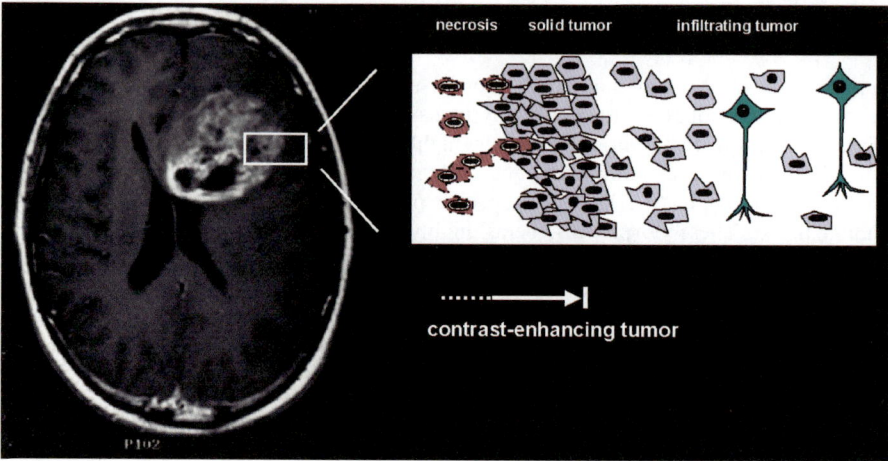

**Figure 1.** Scheme of histological structure characterising malignant gliomas (glioblastoma multiforme). Left: t1-weighted MR image with contrast-enhancement. MRI suggests a circumscribed lesion. Right: Histologically, malignant gliomas are characterised by central necrosis, marginal enhancing, solidly-proliferating tumor and a broad margin of infiltrating tumor cells, which are surgically inaccessible. Only central parts of the tumor lead to contrast-enhancement on MRI.

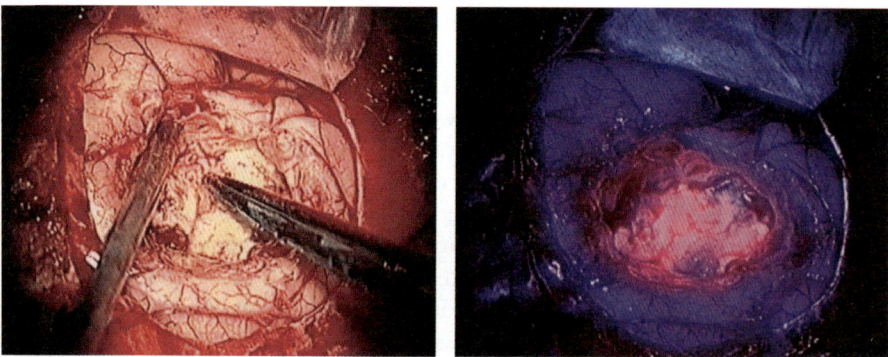

**Figure 2.** Accumulation of porphyrin fluorescence in human malignant gliomas. Left: Resection cavity under white light, showing the opened cerebral cortex with tissue of uncertain dignity at the base of the tumor cavity. Right: Fluorescence image of same cavity and visible red fluorescence identifying tissue as tumor.

Therefore, more aggressive local therapies might serve to prolong survival but it must be remembered that any therapy is essentially palliative because of inevitable tumor recurrence. Consequently, therapy must be safe so as not to impair quality of life for the short period of survival experienced by these patients. Taken together, there are three requirements for local therapies of malignant gliomas:

- therapies should not result in clinical deterioration;
- therapies should enable effective removal of tumor cells; and
- therapies should be highly specific.

Five-aminolevulinic acid (ALA) appears to be a promising new weapon in the fight against malignant gliomas, leading to specific accumulation of porphyrins in malignant glioma cells. The resulting tumor fluorescence is useful for fluorescence-guided resections. Porphyrin-dependent phototoxicity also has shown a significant potential for minimally invasive and specific killing of tumor cells not amenable to surgery. Both modalities are valuable adjuncts for glioma therapy in their own regard, if used for either diagnostic or treatment purposes. Ultimately, however, the best results will be achieved by combining both modalities; safe, but thorough removal of bulk tumor tissue by fluorescence-guided resection, followed by irradiation of the surgical cavity with nonthermal visible light to destroy infiltrating tumor cells.

### 5.1.2. Specificity of Porphyrin-Accumulation in Response to ALA Administration

In terms of the three requirements for local therapies (safety, efficacy, and specificity), specificity is closely linked to the safety of ALA-related therapies for malignant gliomas and this aspect is addressed in greater detail below.

It is informative to examine the reason why ALA-induced accumulation of porphyrins in malignant tissues is more specific and thus safer than other traditional photosensitizers such as hematoporphyrin derivatives or mTHPC. This is because porphyrin synthesis and accumulation occur within the tumor cell itself. Intracellular synthesis and accumulation of porphyrins in response to ALA administration has been amply demonstrated *in vitro* and *in vivo*.[10] Conversely, traditional photosensitizers, which are usually bound to plasma albumin when administered intravenously, depend on the tumor-associated disruption of the blood–brain barrier to enter the tumor, together with edema. They will primarily be located in the extracellular space. They will also dissipate with edema, thus potentially impairing specificity. Furthermore, they are located in the blood vessels outside the tumor and these blood vessels will be susceptible to phototoxic and unspecific damage.[11–13]

But how does ALA enter the brain? ALA is a polar molecule to which the blood–brain barrier is virtually impermeable.[14,15] Thus, a disruption in the blood–brain barrier must be considered a key prerequisite for porphyrin accumulation in brain tumors. In contrast to traditional sensitizers, however, porphyrin synthesis in human malignant glioma has been shown to be strongly linked to tumor proliferation and cell density rather than to the degree of tumor-dependent vascular proliferation.[15] Porphyrin concentration in blood is insignificant, and porphyrin accumulation in endothelial cells has not been observed.[16] Finally, over 300 biopsies taken from the transition zone between fluorescing and non-fluorescing tissue in a pilot series on 50 patients[3] revealed only a single fluorescence-positive biopsy lacking clear evidence of tumor infiltration. This finding was reproduced in the much larger, randomized, phase III drug approval trial for ALA with over 400 patients (see below). All biopsies were taken in patients pretreated with corticosteroids, the patients having received 20 mg ALA kg$^{-1}$ bodyweight administered orally 2.5–3.5 h prior to induction of anesthesia for surgery.

## 5.2. Fluorescence-Guided Resections for Malignant Gliomas Using ALA

### 5.2.1. History and Present Status

The use of ALA-induced tumor fluorescence for delineating residual malignant glioma was initially based on experimental work undertaken at the Laser Research Laboratory and the Department of Neurosurgery at the Klinikum Grosshadern in Munich, Germany. After successful *in vitro* and *in vivo* studies[10] confirming its potential clinical application, ethical committee approval for treating the first patient was accorded to the same team. The first patient harboring a glioblastoma multiforme was operated on in 1995 at the Department of Neurosurgery, Munich, using an ALA dose of 20 mg kg$^{-1}$ body weight, and excitation and detection hardware developed and implemented by the Laser Research Laboratory. This first experience demonstrated convincingly that human malignant gliomas were able to accumulate porphyrin fluorescence in usable amounts. In collaboration with the Carl Zeiss company (Oberkochen, Germany) hardware suitable for fluorescence detection was adapted to a standard neurosurgical operating microscope (OPMI 6) and this was used for operating on a first clinical series of patients.[16,17] This was followed by a larger prospective study of 50 glioblastoma patients, using ALA in combination with the modified Zeiss microscope, which demonstrated feasibility, safety, and potential benefit to patients. In this series, contrast-enhancing tumor in an early post-operative MRI was devoid of residual tumor in 70% of patients, compared to less than 30% in a historic, surgical series. In the remaining patients, the surgeon was aware of residual tumor in all but one patient; that is, he was aware of leaving residual tumor behind which he was reluctant to remove for safety reasons. ALA-derived porphyrins were demonstrated to be highly specific for malignant glioma tissue. This investigator-initiated study was supported by the "Medac Gesellschaft für medizinische Spezialpräparate mbH", Wedel, Germany. The encouraging results of the study provided the incentive for a more structured development, with consecutive Phase I/II (MC-ALS.8-I/GLI) and II (MC-ALS.24/GLI) studies focusing on optimal dose and safety, and finally, a pivotal randomized phase III study (MC-ALS.3/GLI). The sponsor of these clinical trials was the medac company.

*5.2.2. Phase III Drug Approval Trial*

In all, 15 neurosurgical centers* participated in this pivotal, two-armed, randomized, group-sequential, blinded-observer trial on fluorescence-guided resection of malignant gliomas with ALA. The study featured central neuropathological (Institut für Neuropathologie, Universitätsklinikum Bonn) and neuroradiological (Institut für Neuroradiologie, Klinikum der J.-W. Goethe Universität Frankfurt) review. In this study patients with suspected malignant gliomas bearing the potential for complete resection of contrast-enhancing tumor were randomized to receive either 20 mg ALA kg$^{-1}$ body weight for fluorescence-guided resection (ALA-group) or nothing, for conventional white light tumor resection (white light group). Surgery was performed using identical microscopes (Zeiss NC4 OPMI Neuro Fl) available to each participating center. Primary aims of the study were to determine the number of patients without contrast-enhancing tumor on early post-operative MRI and the number with progression-free survival at 6 months. The study included a scheduled interim analysis, which would conclude the study in case of significant differences between the two study arms with regard to the two primary aims. The interim analysis of the first 270 eligible patients fulfilled all expectations. Early post-operative MRI was devoid of residual, contrast-enhancing tumor in 65% of patients in the ALA group compared to only 36% in the white light group. Kaplan–Meier analyses revealed significantly prolonged progression-free survival in ALA patients ($p < 0.01$) compared to white light alone with cumulative 6-months progression-free survival rates of 41% and 21%, respectively ($p < 0.01$). Importantly, neither post-operative neurological status nor Karnofsky Index differed between patients with or without fluorescence-guided resections. The latter analysis was of utmost importance, demonstrating that more radical resections did not impose a higher risk of post-operative deficits.

At present, drug approval for fluorescence-guided resections is pending.

*5.2.3. Practical Aspects of Fluorescence-Guided Resections Using ALA*

*5.2.3.1. Dosage and Timing of ALA Administration*
Structured clinical development of fluorescence-guided resections has concentrated on an oral dose of 20 mg ALA kg$^{-1}$ bodyweight in combination with dexamethason pretreatment (3 × 4 mg per day for 2 days). In the double-blinded, prospectively randomized medac trial MC-ALS.8-I/GLI, lower doses of ALA (2 and 0.2 mg kg$^{-1}$) were tested for efficacy and safety. However, fluorescence was clearly strongest with 20 mg kg$^{-1}$ compared to 2 mg kg$^{-1}$, whereas no useful fluorescence was found with 0.2 mg kg$^{-1}$. There were no differences in the frequency of adverse and severe adverse events among the three doses. Thus 20 mg kg$^{-1}$ seems to be a safe dose from a pharmacological point of view. Higher doses have not been tested on a stringent basis for fluorescence-guided resections of malignant gliomas of the brain. However, systemic side effects with higher doses of ALA, as used for other indications, have been reported. Doses of 40 mg kg$^{-1}$ body weight or more have been

observed to induce hypotension, nausea, vomiting, and increases in plasma liver enzymes.[18-21] On the other hand, since PpIX fluorescence using 20 mg $kg^{-1}$ body weight has proven to be adequate for efficient detection and/or resection of malignant gliomas, there is no apparent need to increase the fluorescence intensity by augmenting the ALA dosage. To the contrary, augmenting the ALA dosage, apart from possibly inducing systemic side effects, may jeopardize the resection process. Possible neurological sequelae of tumor resections using a dose of 20 mg $kg^{-1}$ have now been studied extensively in the medac phase III study MC-ALS.3/GLI, a study with a distinct emphasis on safety: however, it must be kept in mind that higher doses of ALA might lead to more extensive tissue fluorescence, encouraging more voluminous resections and thus evoking a greater risk of neurological deterioration. Moreover, in the animal glioma model,[10] doses of 100 mg $kg^{-1}$ induced significant unspecific fluorescence in perifocal, edematous brain tissue. These observations have now been reproduced (Rachinger *et al.*, submitted for publication), whereas a dose of 20 mg $kg^{-1}$ combined with the usual steroid pretreatment failed to demonstrate the unwanted effect in the same model.

In practice ALA is given orally to patients 2.5–3.5 h prior to induction of anesthesia. The time of administration was derived from initial *in vivo* experiments using the C6-Glioma model in the rat. In these experiments maximal porphyrin fluorescence was observed approximately 6 h after administration. Fluorescence was less after 3 and 9 h.[10] For surgery on patients it was calculated that induction of anesthesia, positioning of the patient, draping, and craniotomy would take approximately 1.5–2 h. Allowing another hour for removal of the clearly discernable tumor core, the surgeons would be 3–3.5 h into the operation before fluorescence-guidance became necessary. Hence, administration should take place approximately 3 h prior to surgery. Prior to application, the compound is easily dissolved in tap water and should be freshly prepared to rule out degradation. In our practice, for a dose of 20 mg $kg^{-1}$, 1.5 g (the contents of a vial of lyophilized compound) are dissolved in 50 mL of tap water and given to the patients. If additional ALA is necessary, the contents of a second vial are dissolved. The solution tastes mildly sour but is well tolerated by patients.

It is common practice to maintain complete restriction of food and drink for at least 6 h prior to elective surgery. Administration of ALA leads to occasional discussions with anesthesiologists due to the volume of fluid required. Pharmacokinetic assessments performed in the medac phase I/II study (MC-ALS.8-I/GLI) have demonstrated complete absorption of the total ALA dose within 1 h of oral administration, so that the solution can be expected to have reached the intestine at the time of anesthesia induction. Complications arising from anesthesia induction after recent ALA administration have not been recorded as yet. However, these observations were made with aqueous solutions of ALA and may not pertain to solutions of ALA in fruit juices, which are more acidic and may not leave the stomach as rapidly.

In normal clinical practice the time limit of 2.5–3.5 h for administration cannot always be exactly maintained. A typical neurosurgical problem is having to postpone a planned procedure, due to unforeseen emergency surgery. For example, the patient with malignant glioma scheduled for surgery is given ALA at 4.30 a.m., prior to anesthesia induction at 7:30 a.m. The emergency arises after administration of ALA. A similar situation might arise if the tumor operation were scheduled subsequent to other operations, and these operations happened to be prolonged because of unforeseeable complications. In these situations it would be unwise to cancel the operation and to reschedule it for another day. How long porphyrin fluorescence remains in human malignant gliomas, and whether fluorescence remains specific over prolonged periods of time are unanswered questions. Also unknown is what side effects would be elicited by a short-term re-administration of ALA.

In our center, operations with ALA are always scheduled for the first position in the morning. Prior to administration of ALA, the person administering the ALA first confirms that no emergency operation or other problems are foreseeable that might interrupt the planned schedule. If emergency operations do have to be performed, the operation of the patient having received the ALA has priority over any other procedures scheduled for the day. Fortunately, fluorescence in human malignant gliomas appears to be very stable. In our experience with postponed cases we have observed strong and useful fluorescence even 12–15 h after ALA administration. In these cases, the fluorescence did not appear to be inferior to that obtained in tumors operated on within the usual time range.

### 5.2.3.2. Light Protection

Sensitization of the skin has been reported after systemic application of ALA.[19–21] Fortunately, in contrast to traditional sensitizers, skin sensitization is short-lived, with a duration of approximately 24 h. Consequently, when using ALA in clinical practice, this temporary side effect must be considered. In our practice and as previously recommended for patients in the medac study MC-ALS.3/GLI, direct exposure of patients to sunlight or strong room light must be avoided. Between the time of ALA administration and induction of anesthesia, low levels of ambient light are permissible. In our opinion the most vulnerable period is during the induction of anesthesia and the positioning of the patient for surgery, prior to draping. During this phase of bustling activity in the neurosurgical operating room, direct illumination by overhead lighting is often unavoidable. We try at least to prevent illumination of the patient's skin by the operating lights. Low levels of ambient light are permissible during the post-operative period. Restrictions regarding light exposure are withdrawn 24 h after ALA administration. Observing these precautions, no severe skin reactions in terms of significant sunburn have been observed, although several patients have had slight rubour of the skin lasting 2–3 days.

### 5.2.3.3. Intraoperative Photosensitization

During the initial application of ALA there was some concern over possible photosensitization of normal brain elicited by the surgical microscope light source. However, spectrographic studies have since confirmed that with 20 mg kg$^{-1}$ in the steroid-pretreated patient, no ALA-dependent porphyrin fluorescence is detectable within normal or edematous, perifocal human brain[3] (Medac study MC-ALS.8-I/GLI). Experimentally, 100 mg of ALA kg$^{-1}$ in the rat did not result in sensitization of normal brain tissue. In the rat edema model, on the other hand, slight sensitization was observed with 100 mg ALA kg$^{-1}$ body weight, with damage detectable to a depth of less than 1 mm using a 635 nm diode laser and 200 J cm$^{-2}$ illumination energy. Conversely, the traditional photosensitizer Photofrin II was found to sensitize normal cortex unspecifically even in the absence of edema,[13] leading to greater damage than ALA pretreatment in the face of brain edema. Illumination densities obtained with the Zeiss microscope are typically in the range of 40–80 mW cm$^{-2}$ and strongly depend on microscope distance from the resection cavity. Even in this setting it would take more than 30 min to reach energy levels consistent with those used for phototherapy of malignant gliomas. In combination, our experience and experimental results fail to demonstrate any risk of damage to normal brain during fluorescence-guided neurosurgery by phototoxic mechanisms.

### 5.2.3.4. Equipment

Adequate equipment for the visualization of porphyrin fluorescence is of utmost importance for the successful implementation of fluorescence-guided resections. In modern neurosurgical practice the surgical microscope is indispensable. Therefore, fluorescence imaging hardware has to be adapted to the microscope. In its simplest form, fluorescence can be visualized by introducing a short pass filter into the excitation light path, in order to filter out the proper excitation wavelength which is shorter than the fluorescence emission wavelength. A long pass filter placed into the observer light path could block out excitation light and allow only red porphyrin-induced fluorescence to pass. However, "monochrome" imaging conducted in this fashion does not allow recognition of tissue outside of fluorescing tumor and therefore is not suitable for neurosurgical applications. For this reason the Zeiss microscope features a refined combination of excitation and emission filters with slightly overlapping transmission,[17] a combination developed by researchers at the Laser Research Laboratory, Klinikum Grosshadern, Munich. Because of the overlap, a small fraction of the excitation light is remitted from tissue, imparting to normal brain a blue color in contrast to bright red porphyrin fluorescence (cf. Chapter 1.6.3.). The degree of filter overlap is crucial for successful imaging. If remitted light is too strong, porphyrin fluorescence is no longer recognizable. If remitted light is too weak, surgery becomes difficult because of the lack of tissue detail. Also, faint autofluorescence in the red spectral region becomes visible and can be mistaken for porphyrin fluorescence. In the first case, sensitivity of detection is impaired: in the second case, specificity may be reduced, a worrisome effect with possible danger to the patient as a consequence of unwanted resections.

In the Medac study MC-ALS.3/GLI, only the Zeiss system with its particular combination of excitation light source, excitation filter, microscope optics and emission filters was tested for safety and efficacy. All participating centers were equipped with identical microscopes. Whether the results from this study can be extended to equipment supplied by alternate microscope manufacturers is unknown. If alternate equipment were more sensitive, it might pick up tissue autofluorescence or diffuse porphyrin fluorescence not seen with the Zeiss system, leading to greater resection volumes, thereby jeopardizing safety. If alternate equipment were less sensitive, this might result in less radical resections. Thus, caution is warranted in this regard.

Finally, in order to visualize fluorescence on a video screen for documentation purposes, specially modified video cameras are necessary. One such camera, which was used in the phase III study, was a 3-chip color CCD-camera from Karl Storz, Tuttlingen, Germany (Tricam SL PDD PAL). This camera was optimized for red porphyrin fluorescence detection by enhanced sensitivity in the wavelength range beyond 600 nm. The camera circuitry enabled automatic or manually adjustable target integration of the porphyrin signal with exposure periods ranging from 1/10,000th of a second to 2 s. Typical exposure times are 1/15th to 1/30th of a second. When switched to the fluorescence mode, the gain of the red channel was automatically increased relative to the other channels (blue and green), in order to increase red sensitivity.

### 5.2.4. Practical Implementation

#### 5.2.4.1. General Aspects
Apart from necessary precautions regarding light exposure, the course of the normal operation does not differ greatly from conventional microneurosurgical operations. Anesthesia induction, patient positioning, draping, and craniotomy are performed in the usual fashion. As in other operations, neuronavigation can be a useful adjunct for planning craniotomy or locating tumors which do not reach the cortical surface. Alternately, for initial localization of tumor, sonography may be used. At times, even when the cortical surface appears inconspicuous, switching to blue excitation light may allow discrimination of subcortical tumor extensions, providing a valuable guide for initial corticotomy. For surgery in eloquent brain regions we have repeatedly used awaken craniotomy for language mapping or for surgery in the vicinity of the motor cortex, and this has never interacted negatively with fluorescence-guided resections. Apart from initial localization, neither neuronavigation nor sonography have provided additional help in defining tumor borders. However, sonography has been useful in defining gross anatomy, for instance when operating on temporal tumors extending toward the basal ganglia. In this situation, sonography is useful for delineating the sylvian fissure and insula in order to define the plane for termination of resection before entering the basal ganglia. It must be borne in mind that tissue fluorescence gives only two-dimensional information on tumor extensions and does not prevent following tumor into eloquent brain. While intact neurological function does not seem likely within fluorescing tissue, surgical manipulations may lead to remote damage away from the immediate resection site, for instance by damaging blood vessels traversing the tumor. Therefore, ALA is a tool that helps

discriminate tumor, but it is at the surgeon's discretion as to whether all of the fluorescing tissue encountered should be removed.

Of course there are different methods for operating on a malignant glioma. The surgeon may prefer to remove necrotic and easily distinguishable solid tumor regions first, predominantly under white light, and then to remove marginal, residual tumor using fluorescence-guided resection. Alternately, he might choose to remain in fluorescing tumor margins at the resection plane. We use simple suction or an ultrasound aspirator for removing tumor. The monopolar loop has also been used but is limiting in ALA-assisted surgery, since it destroys tissue fluorescence. By sucking away tissue debris, fluorescing tissue is re-exposed and it may then be resected. Blood in the resection cavity quenches the fluorescence signal but can be removed easily by suction to give an impression of the fluorescence quality of the tissue. In this respect, fluorescence-guided resection is a dynamic process. Conditions do not have to be optimized for demonstrating fluorescence. Rather, switching from normal to blue excitation light is performed repeatedly during the course of the operation, removing blood oozing onto the tumor surface by suction as necessary. Toward the end, longer periods of the operation can be performed using blue excitation light alone. If unspecific oozing becomes too strong and impairs fluorescence detection, white light illumination with its greater detail is used for coagulating vessels.

### 5.2.4.2. Photobleaching

There has been some concern about photobleaching of porphyrins by microscope illumination, which might destroy tissue fluorescence and impair sensitivity. We have examined this issue in detail in earlier work[16] and found photobleaching to be much slower than anticipated. Under operating light conditions, fluorescence decayed to 36% in 25 min for violet-blue and 87 min for white light excitation. Moreover, during surgery microscope light is usually directed at a small part of the resection cavity, whereas other parts are often covered by coagulated blood or cotton patties. Still, mild fluorescence may be bleached in exposed regions of the tumor which are not removed immediately, and these may be missed. In this situation fluorescence may be refreshed by suction and removal of superficial cell layers.

### 5.2.4.3. Fluorescence Qualities

Tumor fluorescence is not homogenous. Typically, within the normal glioblastoma, three tissue regions can be discriminated with regard to fluorescence. Necrotic tumor accumulates little or no fluorescent porphyrins. Tissue adjacent to necrotic tissue shows strong fluorescence. In this zone, two qualities of fluorescence are regularly observed (Figure 3). Tissue bordering on necrotic tumor usually displays strong, deeply red fluorescence, whereas more marginal tissue displays lighter, pink fluorescence. Analyzing this phenomenon histologically, we found strong fluorescence predominantly to characterize solidly proliferating tumor; that is, tumor cells without intervening, normal brain tissue (Figure 4a). On the other hand, pinkish, marginal fluorescence was found

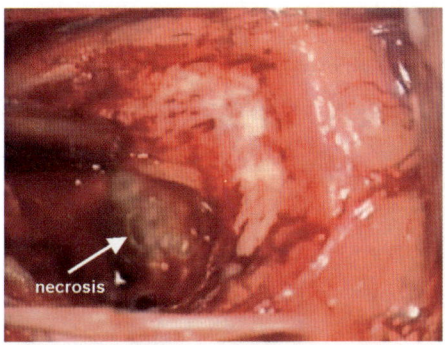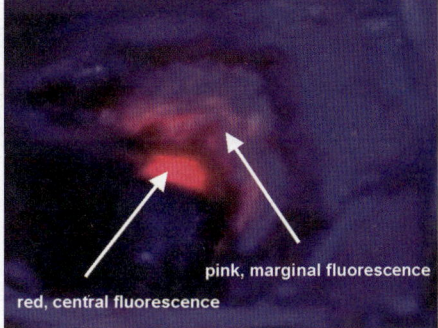

**Figure 3.** Left: Tumor cavity under conventional microscope illumination displaying macroscopic appearance of tumor. Right: Fluorescence image of same cavity showing different fluorescence qualities. Necrotic tumor shows no fluorescence accumulation. Adjacent tumor shows strong red fluorescence. This is surrounded by a margin of tissue with pink fluorescence.

predominantly to represent infiltrating tumor with a moderate to high cell density (Figure 4b).[3] Leaving residual, pink-fluorescing, marginal tissue unresected for fear of ensuing neurological deficits did not necessarily imply residual tumor on post-operative MRI. Our previous analysis showed that in nearly half of the patients without residual tumor on post-operative MRI, parts of marginal tissue with pink fluorescence had been left unresected. This observation also implied that ALA-derived porphyrin fluorescence is more sensitive than MRI in identifying tumor, as are any resection aids that rely on MR imaging, for instance neuronavigation or intraoperative MRI. On the other hand, survival analysis suggested that prognosis was better when fluorescing tissue, including tissue with pink porphyrin fluorescence, was removed completely. This observation was confirmed by the phase III Medac study.

### 5.2.4.4. Optimizing Fluorescence Visualization Intraoperatively
There are a number of factors that interfere with fluorescence observation intraoperatively. Xenon light sources have a limited life span and lose significant intensity after prolonged usage. Loss of intensity has less impact on white light resections. However, any reduction of intensity of violet-blue excitation light is more readily perceived. With the Storz D-Light for instance, when adapted to the Zeiss microscope, burners had to be changed every 250 h to ensure useable excitation light intensities.

During surgery care must be taken to reduce ambient room light. Normal room lighting, especially neon tubes, has a strong red component. Red wavelengths are selectively amplified by the detection equipment and thus lead to red discoloration of non-tumor tissue normally perceived as being blue. This problem is usually easily recognized, when the normal cortical surface displays a general reddish tone. Standard surgical lights are less of a problem in this

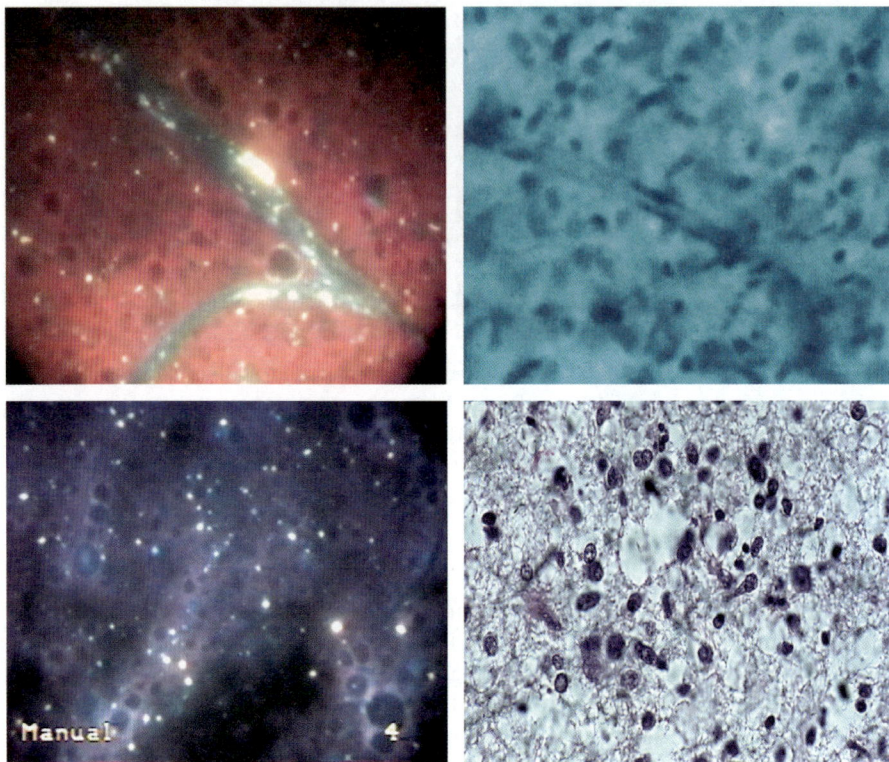

**Figure 4.** (a) Left: Fluorescence microscopic image of smear preparation taken from tumor tissue with strong red fluorescence. Note strong red porphyrin fluorescence with negatively contrasted nuclei. Brightly fluorescing lipid droplets are dispersed throughout. Capillary that traverses specimen shows green autofluorescence but no red porphyrin fluorescence. Right: Aspect of same specimen after fixation and staining, displaying solidly-proliferating tumor without intervening brain tissue. (b) Left: Fluorescence microscopic image of smear preparation taken from marginal tumor tissue with pink fluorescence when examined macroscopically. Red porphyrin is located in the cytosol of tumor cells. Brightly fluorescing droplets of lipids indicate degraded myelin sheathes of white matter tracts. Right: Aspect of same specimen after fixation and staining, displaying infiltrating tumor with intervening brain tissue.

regard, because their red and infrared wavelengths are blocked more effectively. In our practice we turn off all light sources except the surgical lights, which are diverted away from the surgical cavity toward the instrument trays. Of course the operating rooms must be darkened with respect to daylight, which is usually possible in designated neurosurgical operating rooms.

We have found it helpful to reduce light intensity under white light to aid fast adaptation to violet-blue excitation light. Modern neurosurgical microscopes such as the Zeiss NC4 or Pentero systems have excessive white light in any case, and reducing this by 50–70% does little to reduce image quality under white light. However, the system should automatically switch to full intensity when using violet-blue excitation light.

Fluorescence is directly related to the amount of excitation light reaching the tumor. Since excitation light leaves the microscope as an expanding cone of light, intensity will rapidly diminish with distance from the resection cavity. Thus, this distance should be kept as short as possible without interfering with the procedure. Furthermore, apart from distance, shading and angle of illumination are factors influencing excitation light and resulting fluorescence. For instance, excitation light is partially blocked (shaded) by adequately small openings in the cortex used for resections of deep-seated tumors. Furthermore, placement of illumination and observer axes perpendicular to the resection surface will lead to stronger fluorescence than when observation is at an angle to the resection surface. The three factors influencing fluorescence visualization are distance, shading, and angle. All of these factors should be kept in mind when positioning the microscope.

## 5.3. Photodynamic Therapy of Malignant Gliomas Using ALA

Using ALA for fluorescence-guided resections of malignant gliomas results in a larger fraction of patients with complete removal of contrast-enhancing tumor on early post-operative MRI. However, a significant number of patients will have residual, contrast-enhancing tumor, which was left unresected for safety reasons. On the other hand, a significant fraction of patients without residual enhancement on early post-operative MRI will have residual fluorescing tissue left unresected intraoperatively for the same reason. Finally, there is a fourth group of patients not operated on in the first place because of the inaccessibility of their tumors and the risk of post-operative neurological deficits (Figure 5). All four groups of patients would probably benefit from safe and selective methods of removing larger parts of their tumors or residual, infiltrating cells. The photosensitizing properties of porphyrins induced by ALA may well provide such a method, especially for patients undergoing surgery using ALA who could be treated using PDT immediately subsequent to surgery.

Theoretically, three different application formats of PDT might be used to treat patients with malignant gliomas. Intraoperatively, residual tumor adjacent to eloquent brain regions could be irradiated with focused laser light, whereas deep-seated lesions might be treated by interstitial application of laser light. Finally, residual, macroscopically invisible and unresectable tumor cells could be treated by integral forms of PDT, with light illumination of the entire resection cavity (Figure 6).

At present, PDT of malignant gliomas using ALA has passed the experimental stage and is being tested in clinical protocols. Experimentally, C6 gliomas in the rat have been treated successfully and specifically,[13] using ALA and a 635 nm laser, and an energy of 100 J cm$^{-2}$. Using the same model, edema as a possible consequence of ALA-PDT in the brain has been quantified, and its treatment potential using steroids has been defined.[22]

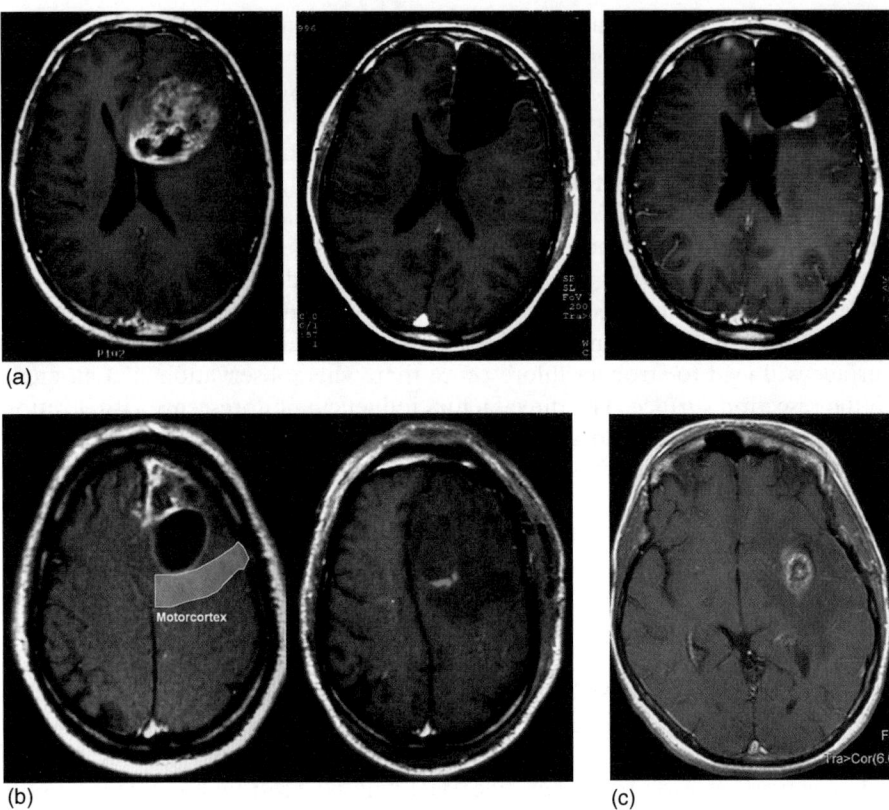

**Figure 5.** (a) MR images of patient operated on for left frontal glioblastoma. Left: Pre-operative image. Center: Early post-operative MRI devoid of residual contrast-enhancing tumor. Right: 6 weeks after surgery tumor recurs at the resection margin. This patient might benefit from PDT of individual infiltrating tumor cells. (b) MR image of patient operated on for left frontal glioblastoma adjacent to motor region. To rule out neurological deficits part of the tumor was left unresected. This patient might benefit from focal and selective PDT of residual tumor. (c) MR image of patient with glioblastoma in the left insula. Meaningful cytoreductive surgery of this tumor carries the risk of significant post-operative deficits (dysphasia, hemiparesis). This patient would benefit from selective PDT of his tumor.

At present, two protocols are being implemented regarding PDT at our institutions (Department of Neurosurgery, Klinikum Grosshadern, Laser Research Institute, Department of Neurosurgery, University of Düsseldorf). The first study is a dose-escalation study (100, 150 and 200 J cm$^{-2}$) using 20 mg ALA kg$^{-1}$ body weight given orally. This study investigates focal PDT in patients with recurrent malignant gliomas with planned resection and anticipated residual, unresectable tumor. Residual tumor is irradiated focally using a 3 W diode laser (Ceramoptec 635 nm) and a fiber/microlens system. Study aims include the assessment of possible side-effects, for example, edema or neurological sequelae, as well as preliminary data on efficacy. Three doses have now

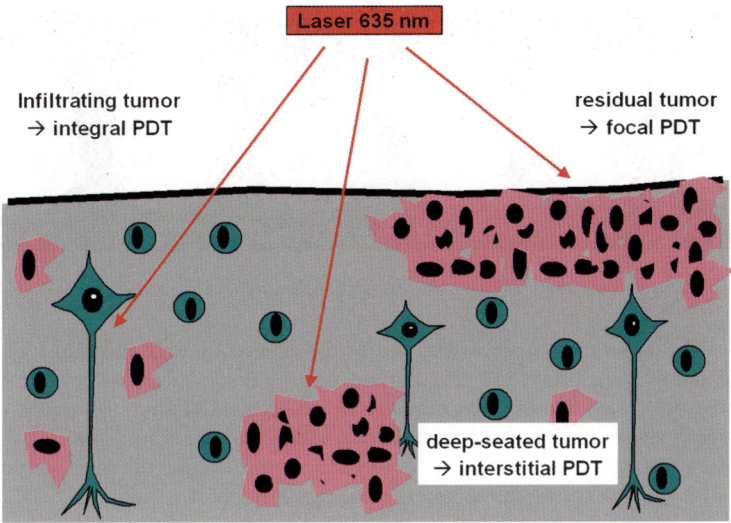

**Figure 6.** Scheme of theoretical application forms for PDT using ALA to treat malignant gliomas. Integral PDT treats residual, infiltrating tumor cells; interstitial PDT treats deep-seated, surgically inaccessible lesions; and focal PDT treats residual tumor adjacent to, or extending into, eloquent brain regions.

been applied (100, 150, and 200 J cm$^{-2}$) and no PDT-related side-effects have been observed.

The second study investigates the safety and possible efficacy of interstitial PDT in patients with unresectable, deep-seated malignant gliomas. In this study, laser diffusers are stereotactically implanted in deep-seated tumors. Placement of diffusers is based on exact pre-operative calculation of three-dimensional light distribution throughout the tumor. The ALA dose used in this study is also 20 mg kg$^{-1}$ body weight given orally.

Preliminary experience gained from this protocol demonstrates the potential of ALA PDT for selectively and effectively killing malignant gliomas (Figure 7).

An alternative form of PDT, metronomic PDT, is being investigated as a possible treatment mode. Theoretically, extended periods of light irradiation will increase the depth of phototoxic cell destruction, enabling larger volumes of tumor kill. For metronomic PDT, light is delivered via chronically implanted devices, for example, light diodes, enabling prolonged irradiation in the awake and mobile patient. This mode of therapy is presently being investigated experimentally.[23]

At present, PDT of malignant gliomas is not restricted to 5-ALA as sensitizer: mTHPC (Foscan) is another compound that is attracting attention. Foscan may be of value because of its absorption peak in the infrared range (652 nm) with the theoretical advantage of greater treatment depths due to greater penetration of light into tissue. Because of its fluorescent properties and

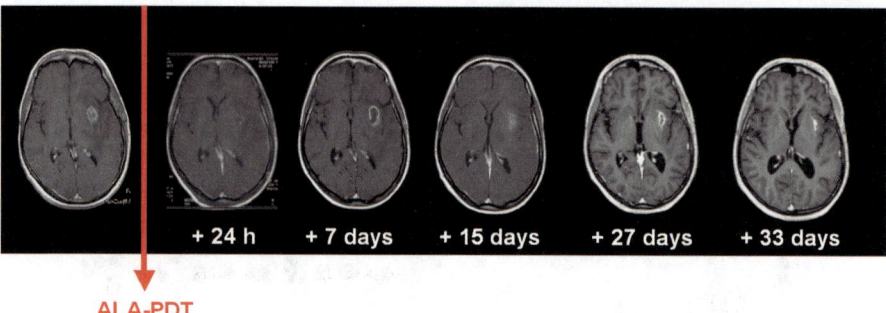

**ALA-PDT**

**Figure 7.** Example of a patient with surgically inaccessible glioblastoma of the left insula treated by interstitial PDT. The patient was treated by stereotactic placement of 4 light diffusors coupled to a diode laser with 635 nm light. MRI obtained on day one after the procedure shows complete attenuation of contrast enhancement. MRI on day seven shows reactive enhancement adjacent to prior tumor. Follow-up MRI demonstrates shrinkage of enhancing region, signifying scar formation.

apparently highly specific accumulation in malignant glioma tissue, Foscan is also under investigation as a substance which enables fluorescence-guided resections followed by intraoperative PDT.[24] In a different approach, 5-ALA is being tested in combination with hematoporphyrin derivative; 5-ALA for fluorescence-guided resections and sensitization of tumor cells, and hematoporphyrin derivative for sensitization of tumor interstitial tissue and tumor blood vessels. First reports on this particular combination of therapies in patients suffering malignant gliomas of the brain and the spinal cord have been presented.[25]

## 5.4. The Future

The goal of enhancing resections using ALA-induced tumor fluorescence has been achieved and this may well define a new standard of surgical care. New adjuvant therapies based on immunology, new chemotherapy strategies and gene therapy will all rely on the possible highest degree of cytoreduction prior to their implementation. In order to test novel adjuvant therapies clinically, conditions regarding resection should be as homogenous as possible, in order to minimize the confounding influence of varying volumes of residual tumor on study results. Thus, fluorescence-guided resections should be implemented in protocols testing novel treatment strategies for malignant gliomas.

Technically, improvements remain to be made. While the present technique of fluorescence visualization is efficacious and safe, the neurosurgeon would prefer a greater amount of excitation light within the tumor cavity, and this is difficult to accomplish using traditional techniques for light transmission within the microscope. An ultimate goal is to incorporate electronic detection methods such as spectrography to detect fluorescence, and to inject the outlines of the fluorescing tissue into the heads-up display available on modern microscopes.

Whether derivatives of ALA are useful for fluorescence-guided resections remains to be established. Possibly, greater blood–brain barrier permeability derived from using more lipophilic derivatives of ALA such as methyl- or hexyl-ALA might allow greater tumor fluorescence and improved visualization or more effective sensitization for PDT. On the other hand, these substance might also cross the cell barrier of normal glia cells and might be metabolized, resulting in unwanted fluorescence of normal brain. In this context, clearly, careful research is still necessary. Furthermore, the toxicological properties of existing derivatives limit their use.

To date little information is available regarding alternate uses of ALA in brain tumor therapy. In our experience meningiomas have been found to synthesize useful amounts of porphyrins to enable their detection, and with metastases results have been ambiguous. However, meningeomas and metastases are easily distinguished using conventional broad-band light sources. Pituitary adenomas may be an entity in which differential accumulation of porphyrins in the pituitary gland and pituitary adenoma may help in discriminating between tumor and normal tissue. However, no published information is available to date concerning uptake of porphyrins in this context.

# References

1. F.K. Albert, M. Forsting, K. Sartor, H.P. Adams and S. Kunze, Early postoperative magnetic resonance imaging after resection of malignant glioma: objective evaluation of residual tumor and its influence on regrowth and prognosis, *Neurosurgery*, 1994, **34**, 45–60.
2. B.C. Devaux, J.R. O'Fallon and P.J. Kelly, Resection, biopsy, and survival in malignant glial neoplasms. A retrospective study of clinical parameters, therapy and outcome, *J. Neurosurg.*, 1993, **78**, 767–775.
3. W. Stummer, A. Novotny, H. Stepp, C. Goetz, K. Bise and H.J. Reulen, Fluorescence-guided resection of glioblastoma multiforme utilizing 5-ALA-induced porphyrins. A prospective study in 52 consecutive patients, *J. Neurosurg.*, 2000, **93**, 1003–1013.
4. M. Lacroix, D. Abi-Said and D.R. Fourney, A multivariate analysis of 416 patients with glioblastoma multiforme: prognosis, extent of resection, and survival, *J. Neurosurg.*, 2001, **95**, 190–198.
5. A. Kowalczuk, R.L. Macdonald, C. Amidei, F.D. III, R.K. Erickson, J. Hekmatpanah, S. Krauss, S. Krishnasamy, G. Masters, S.F. Mullan, A.J. Mundt, P. Sweeney, E.E. Vokes, B.K.A. Weir and R.L. Wollman, Quantitative imaging study of extent of surgical resection and prognosis of malignant astrocytomas, *Neurosurgery*, 1997, **41**, 1028–1038.
6. D.C. Shrieve, E.A. III, P.B. McL, P.Y. Wen, H.A. Fine, H.M. Kooy and J.S. Loeffler, Treatment of patients with primary glioblastoma multiforme with standard postoperative radiotherapy and radiosurgical boost, prognostic factors and long-term outcome, *J. Neurosurg.*, 1999, **90**, 72–77.

7. K.E. Wallner, J.H. Galicich, G. Krol, E. Arbit and M.G. Malkin, Patterns of failure following treatment for glioblastoma-multiforme and anaplastic astrocytoma, *Int. J. Rad. Oncol. Biol. Phys.*, 1989, **16**, 1405–1409.

8. W. Stummer, H.J. Reulen, A. Novotny, H. Stepp and J.C. Tonn, Fluorescence-guided resections of malignant gliomas – an overview, *Acta Neurochir. Suppl.*, 2003, **88**, 9–12.

9. T.C. Origitano and O.H. Reichman, Photodynamic therapy for intracranial neo-plasms: development of an image-based computer-assisted protocol for photo-dynamic therapy of intracranial neoplasms, *Neurosurgery*, 1993, **32**, 587–595.

10. W. Stummer, S. Stocker, A. Novotny, A. Heimann, O. Sauer, O. Kempski, N. Plesnila, J. Wietsorrek and H.J. Reulen, *In vitro* and *in vivo* porphyrin accumu-lation by C6 glioma cells after exposure to 5-aminolevulinic acid, *J. Photochem. Photobiol. B*, 1998, **45**, 160–169.

11. W. Stummer, C. Dautermann, A. Heimann, A. Hassan and O. Kempski, Kinetics of photofrin II in perifocal brain edema, *Neurosurgery*, 1993, **33**, 1075–1081.

12. C. Götz, A. Hasan, W. Stummer, A. Heimann and O. Kempski, Photodynamic effects in perifocal, oedematous brain tissue, *Acta Neurochir. (Wien.)*, 2002, **144**, 173–179.

13. B. Olzowy, C.S. Hundt, S. Stocker, K. Bise, H.J. Reulen and W. Stummer, Photoirradiation therapy of experimental malignant glioma with 5-aminolevulinic acid, *J. Neurosurg.*, 2002, **97**, 26–32.

14. S.R. Ennis, A. Novotny, J. Xiang, P. Shakui, T. Masada, W. Stummer, D.E. Smith and R.F. Keep, Transport of 5-aminolevulinic acid between blood and brain, *Brain Res.*, 2003, **959**, 226–234.

15. A. Novotny and W. Stummer, 5-Aminolevulinic acid and the blood-brain barrier – a review, *Med. Laser Appl.*, 2003, **18**, 36–40.

16. W. Stummer, S. Stocker, S. Wagner, H. Stepp, C. Fritsch, C. Goetz, A.E. Goetz, R. Kiefmann and H.J. Reulen, Intraoperative detection of malignant glioma by 5-ALA-induced porphyrin fluorescence, *Neurosurgery*, 1998, **42**, 518–526.

17. W. Stummer, H. Stepp, G. Möller, A. Ehrhardt, M. Leonhard and H.J. Reulen, Technical principles for protoporphyrin-IX-fluorescence guided microsurgical re-section of malignant glioma tissue, *Acta Neurochirurgica.*, 1998, **140**, 995–1000.

18. M.A. Herman, J. Webber, D. Fromm and D. Kessel, Hemodynamic effects of 5-aminolevulinic acid in humans, *J. Photochem. Photobiol. B*, 1998, **43**, 61–65.

19. P. Mlkvy, H. Messmann, J. Regula, M. Conio, M. Pauer, C.E. Millson, A.J. MacRobert and S.G. Bown, Sensitization and photodynamic therapy (PDT) of gastrointestinal tumors with 5-aminolevulinic acid (5-ALA) induced proto-porphyrin IX (PPIX) – a pilot study, *Neoplasma*, 1995, **42**, 109–113.

20. J. Regula, A.J. MacRobert, A. Gorchein, G.A. Buonaccorsi, S.M. Thorpe, G.M. Spencer, A.R. Hatfield and S.G. Bown, Photosensitisation and photodynamic therapy of oesophageal, duodenal and colorectal tumours using 5-aminolevulinic acid induced protoporphyrin IX – a pilot study, *Gut*, 1995, **36**, 67–75.

21. K.D.M. Fan, C. Hopper, P.M. Speight, G.A. Buonaccorsi, A.J. MacRobert and S.G. Bown, Photodynamic therapy using 5-aminolevulinic acid for premalignant and malignant lesions of the oral cavity, *Cancer*, 1996, **78**, 1374–1383.

22. I. Seiro, W. Rachinger, H. Stepp, H.J. Reulen and W. Stummer, Oedema formation in experimental photoirradiation therapy of brain tumours using 5-ALA, *Acta Neurochirurgica*, 2005, **147**, 57–65.

23. S.K. Bisland, L. Lilge, A. Lin, R. Rusnov and B.C. Wilson, Metronomic photo-dynamic therapy as a new paradigm for photodynamic therapy: rationale and

preclinical evaluation. Technical feasibility for treating malignant brain tumors, *Photochem. Photobiol.*, 2004, **80**, 22–23.

24. H. Kostron, T. Fiegele, A. Zimmermann and A. Obwegeser, Foscan mediated photodynamic diagnosis in combination with simultaneous photodynamic therapy as new treatment concept for malignant brain tumors, *10th World Congress of the International Photodynamic Association*, Munich, 2005, abstract S03. 01.

25. S. Kaneko, K. Tokuda, T. Yoshimoto, T. Yamauchi, S. Fujimoto, Yoshizuma, T. Shirasaka, T. Kashiwaba, M. Nishimura and N. Miyashi, PDT following PDD for gliomas patients in the eloquent brain areas and spinal cord using HPE and ALA, *10th World Congress of the International Photodynamic Association*, Munich, 2005, abstract S03. 03.

*Chapter 6*

# Otorhinolaryngology

## C.S. Betz and A. Leunig

**Table of Contents**

## Abstract

In the recent past, fluorescence diagnosis (FD) and photodynamic therapy (PDT) using 5-aminolevulinic acid (5-ALA) induced Protoporphyrin IX (PPIX) has attracted increasing scientific and clinical attention as a new modality for the management of (pre-)cancerous conditions of the upper aerodigestive tract (UADT) and of the outer head and neck skin. Following topical application of 5-ALA to the UADT via rinsing or inhalation, fluorescence imaging helps with both the identification and the superficial delimitation against healthy tissue of malignant epithelial neoplasms and precancerous recurrent respiratory papillomatosis. Yet, fluorescence diagnostic methods using 5-ALA still lack international approvals. At this point, systemic application of 5-ALA does not seem to be recommendable for FD in otorhinolaryngology. Limitations of the method lie in the inability to assess deeper tissue layers, a considerable high number of false positives and photobleaching effects. Internationally approved therapeutic applications of PDT using topically applied 5-ALA or its esters in the head and neck region are limited to actinic keratoses of the face and scalp. Promising results are also reported for the management of other pre- and early malignant, unpigmented facial lesions, and the treatment of precancerous mucosal lesions of the UADT. To date, no general statement can be given as to the value of 5-ALA PDT in UADT cancer. Drawbacks are again the superficiality of the photodynamic effects as well as local and sometimes also systemic side effects. In conclusion, the use of 5-ALA for FD and PDT of head and neck conditions has advanced considerably over the last two decades and shows many potentials for further developments.

## 6.1. Introduction/Motivation

Malignant neoplasms of the upper aero digestive tract (UADT) present a significant problem in modern health care. Mucosal cancer of the head and neck is the fifth most common cancer worldwide and the most common neoplasm in central Asia.[1] As in most other industrialized countries, both incidence and mortality rates of squamous cell carcinoma (SCC) of the UADT show an upward tendency within the German population (Figure 1). Nonmelanomatous skin cancer of the head and neck region account for another large group of malignant neoplasms in humans, comprising mostly basal and SCC. Whereas the incidence rate of this disease has shown a dramatic increase in recent decades and lies way above the one for UADT carcinomas, the cancer-induced deaths are well below those for mucosal malignancies (Figure 2).

In terms of management, patients with carcinomas of the UADT require a multidisciplinary approach. Most patients with advanced carcinomas are treated with surgery, radiation, or a combination of the two. The choice of modality depends upon the site, stage, and resectability of the disease. The role of chemotherapy as adjunctive therapy has not yet been clearly defined. Unlike distant metastases, lymph-node metastases are common and require surgical treatment (neck dissection). However, despite advances in diagnosis and

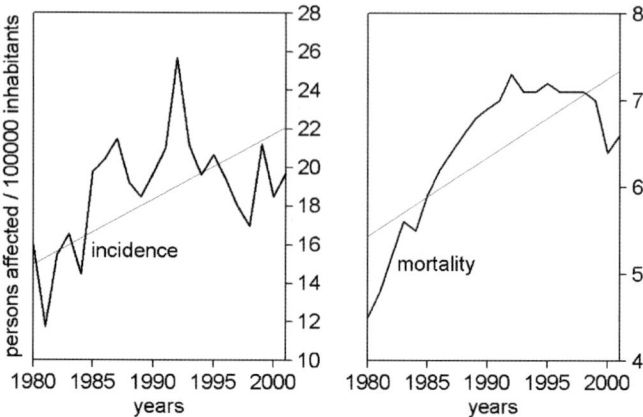

**Figure 1.** Graphical illustration of the development of both, incidence (a) and mortality rates (b) for UADT cancer within the German population over the last two decades.

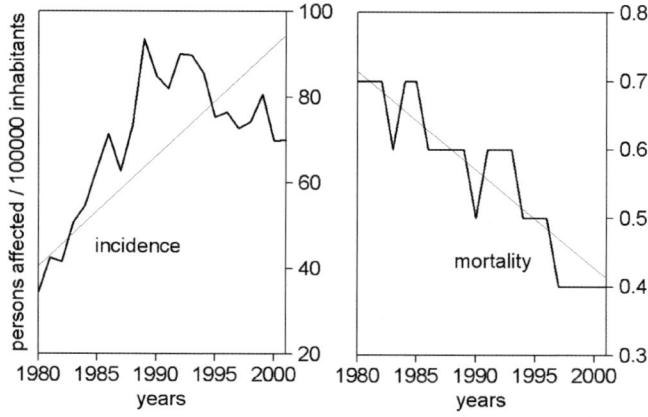

**Figure 2.** Graphical illustration of the development of both incidence (a) and mortality rates (b) for nonmelatomatous skin cancer of the head and neck region within the German population over the last two decades.

treatment in the past 3 decades, the numbers of new cases and deaths from UADT cancer are stagnating or even rising, and death and 5-year survival rates of this disease have remained largely unchanged.[2] Nonmelanomatous skin cancers (tumor precursors or early lesions) are regularly treated by local excision or other means of limited tissue destruction (cryotherapy, laser ablation, curettage). Only in larger tumor stages are the same strategies applicable. Even though the mortality rates of this group of tumors is rather small and even declining, the treatment often leaves the patient with functionally or cosmetically unacceptable results.

In view of the above limitations, research into new and promising diagnostic and therapeutic procedures is still required. One of these newly introduced methods is fluorescence diagnosis (FD) and photodynamic therapy (PDT) using 5-aminolevulinic acid (ALA)-induced Protoporphyrin IX (PpIX). In the present chapter, the most obvious diagnostic and therapeutic deficits in the management of cancer of the UADT and head and neck skin that could potentially be overcome by FD or PDT are presented.

### 6.1.1. Diagnosis

#### 6.1.1.1. Early Detection of (Pre)malignant Lesions

Improved early detection, followed by an immediate and radical surgical resection of the (pre-)malignant mucosal lesions is known to improve the prognosis of patients. A larger epidemiological study demonstrated that the earliness of first diagnosis has a significant positive correlation with the cure rates of these diseases[3] (Figure 3). As a clinical reality, the goal of improved early detection is not easily realizable, since tumor precursor lesions such as dysplasias and carcinoma *in situ*, as well as early carcinomas, are sometimes present in widespread chronically irritated or macroscopically inconspicuous areas, and therefore escape both visual examination and imaging techniques. A novel technique for an improved demarcation of (pre)malignant mucosal lesions would therefore be of a great value.

In principle, the same limitations apply for nonmelanomatous skin cancers and their precursors. A more detailed description is presented in the Dermatology chapter of this book.

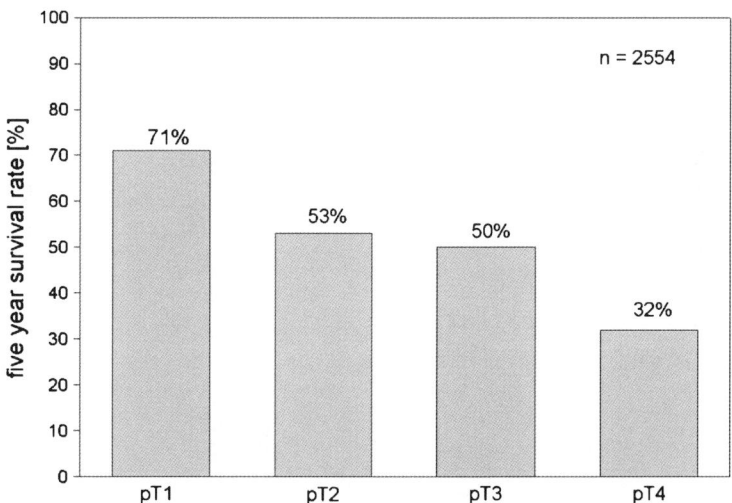

**Figure 3.**   Graphical illustration of the influence of the tumor size (pT-stage) on the 5-year survival rate. (Data taken from the "Munich Tumor Registry 1998".)

*6.1.1.2. Exact Delimitation of Tumor Borders*
A complete excision of all cancerous tissue (preferably within the first operative session) seems to be of utmost importance for the prognosis of patients with carcinomas of the UADT, since it is known that premalignant or even malignant remnants after resection correlate with an increased rate of tumor recurrence[4,5] (Figure 4). However, especially when tongue-like, submucosal spreadings of malignancy or diffuse infiltration into surrounding tissue layers is present, the intraoperative definition of the exact tumor borders may pose a serious problem to the operating surgeon. Such problems can result in prolonged operation times or, in the worst case, can prevent a successful resection in terms of tumor-free borders, as well as contributing to increased rates of local recurrence.

Again, the treatment of dermatological tumors of the face and scalp would also benefit from a better pre- and intraoperative detection of the exact tumor borders just like the mucosal neoplasms (refer to the Dermatology chapter).

*6.1.1.3. Management of Recurrent Respiratory Papillomatosis (RRP)*
Papillomatous, HPV (human papilloma virus)-associated mucosal lesions of the upper (and partly even, the lower) aero digestive tract pose a heavy and sometimes life-threatening burden for the patients concerned because of the exceptionally high rate of recurrences. In adults, these mucosal papillomas show a malignant degeneration in some cases. To date, the treatment is mostly restricted to a symptomatic removal of macroscopically visible lesions, which in

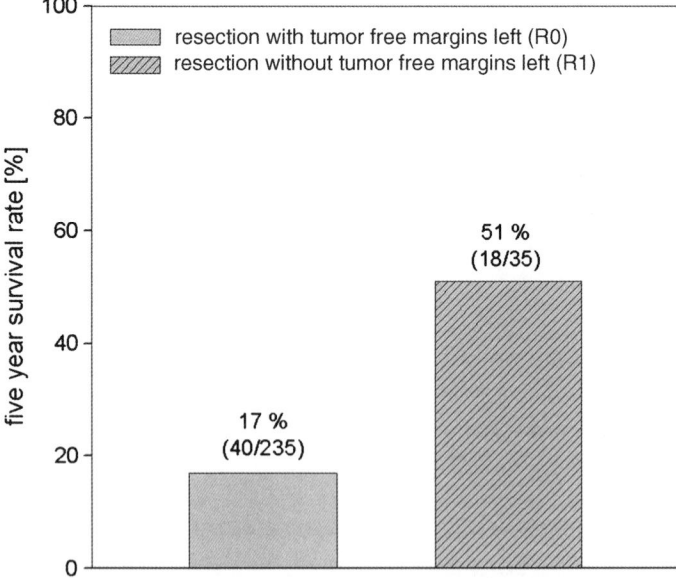

**Figure 4.** Graphical illustration of the influence of microscopically tumor-free resection margins on the 5-year survival rates of head and neck cancer. (Data adapted from Chen *et al.*[4])

most cases is performed by $CO_2$–laser ablation. Even after complete resection of the visible lesions, however, studies have shown that HPV-DNA remains within clinically inconspicuous mucosa.[6] Being more radical in surgical removal does not seem to pose an alternative, since (especially in the laryngeal region) this would lead to heavy functional deficits as laryngeal synechias.[7] It would therefore be very useful for the treating physician to have a diagnostic tool at hand that allows the detection of all infected mucosal areas.

As shown above, there currently exists a diagnostic deficit that may possibly be targeted by novel diagnostic approaches. As such FD could serve both in the early detection and delimitation of malignant tissue toward healthy mucosa and in the demarcation of virus-infected mucosal regions in RRP.

### 6.1.2. Therapy

#### 6.1.2.1. Recurrent Respiratory Papillomatosis
The dilemma of sufficient treatment (but not excessive treatment) of the mucosal area in RRP is always of major concern. Apart from the diagnostic aspects, an optimized therapeutic regimen that includes a gentle but still complete removal of all mucosal areas affected by viral infection may greatly improve the patients' prognosis.

#### 6.1.2.2. Precancerous Mucosal Lesions
Leukoplakias and the much rarer occurring erythroplakias are chronic mucosal lesions of varying size, which show a considerable tendency for malignant degeneration (15.6–39.2% in leukoplakia and >90% for erythroplakia).[8] Upon detection of these lesions, it is generally recommended that representative biopsies be taken, followed by regular follow-up visits. However, with this strategy, it seems to be possible that areas of malignant degeneration within the lesion may be overlooked. A complete resection (and thus a more definite diagnosis) of the lesions, which are sometimes quite elaborate in size, may go along with prolonged wound healing and functional deficits, and even after complete resection these precancerous lesions tend to recur.[9] A more gentle but complete removal of the lesions (as potentially possible by PDT) therefore seems desirable.

#### 6.1.2.3. Field Cancerization
Because of chronic mucosal damage via exogenously applied noxae ("condemned mucosa"), patients with neoplastic lesions of the UADT may develop the so-called field cancerization, that is, malignancies which are widespread, or which occur in different areas synchronously.[10] A complete resection is complicated, however, because the superficial tumor borders and early secondary carcinomas are often not distinguishable from healthy surrounding mucosa. A tumor-selective, gentle ablation of malignantly degenerated areas could therefore be helpful.

### 6.1.2.4. Tumors not Curatively Resectable

Just as clinically important are the large group of extensively growing, but not curatively resectable, tumors. Even after exhaustive surgical and radio-/ chemotherapeutical treatment modalities, a mere reduction of the tumor bulk in a gentle manner would be desirable for palliative reasons.

### 6.1.2.5. Pre- and Early Malignant Lesions of Facial Skin

Last but not least, actinic keratoses are the most common premalignant facial skin lesions, showing a yearly rate of conversion into malignant, invasive SCCs of 0.25–1%, so that once this diagnosis has been established, a therapeutical measure is desirable.[11] Treatment options include surgical excision, cryotherapy, curettage, electric cauterization, and local chemotherapy. As with the resection of early basaliomas, the cosmetic outcome is often dissatisfying, so that gentler, alternative therapeutic strategies seem desirable.

In conclusion, from a therapeutic standpoint special challenges in head and neck cancer that could potentially be overcome by PDT are pre- and early malignant mucosal- and facial skin lesions, recurrent respiratory papillomatosis, and the group of not curatively treatable tumors of the UADT.

## 6.2. Approval Status

### 6.2.1. Diagnosis

Currently, all methods of FD in the UADT are still at the stage of experimental and clinical studies. So far there has been no approved indication or procedure for this field. However, there are two commercially available systems for fluorescence-guided examinations of the UADT: one is tailored for FD of head and neck tumors with and without using ALA-induced PpIX (Storz-PDD-System, see the appendix), whereas the other (Xillix LIFE ENT System, Xillix Technologies Corp., Richmond, BC, Canada) uses autofluorescence only for the discrimination of tissue status.

### 6.2.2. Therapy

To our knowledge, at the time of publication of this book the use of ALA-based PDT for the treatment of recurrent respiratory papillomas of the UADT has been neither published in the medical literature, nor granted approval. As for PDT of SCC's of the UADT and its precursors using ALA-induced PpIX, the method of treatment has been described in individual publications, but here too there have been no official approvals for its usage to date. Apart from ALA-induced PpIX, only one photosensitizer, meta-tetra-hydroxyphenyl-chlorin (mTHPC) (Foscan, BioLitec AG, Jena, Germany) has achieved Europe-wide approval with the European Agency for the Evaluation of Medical Products (EMEA) for palliative treatment of patients with extensive SCCs of the head and neck region. It applies after failure of earlier performed therapies and for cases where radiation therapy, surgical resection or systemic chemotherapy are not suitable.[12]

Two methods for PDT using precursors of PpIX have now been approved by various authorities for the treatment of dermatological diseases in the head and neck area as described below.

(i) The so-called Levulan Kerastick (DUSA Pharmaceuticals Inc., see the appendix) contains ALA in a liquid solution as the (pre-)active compound. It has been approved by the Food and Drug Administration of the United States of America for therapy of nonhyperkeratotic actinic keratoses of the face and scalp.[13]

(ii) An alternative compound, Metvix (PhotoCure ASA, see the appendix) consists of a cream preparation containing ALA-methylester. The current approvals for Germany, the Scandinavian Countries, England, and Australia/New Zealand involve the treatment of thin or nonhyperkeratotic and nonpigmented keratoses of the face and scalp as well as of superficial and/or nodular basal cell carcinoma (BCC) unsuitable for other available therapies because of possible treatment related morbidity and poor cosmetic outcome (such as lesions on the mid-face or ears, lesions on severely sun damaged skin, large lesions or recurrent lesions).

## 6.3. Procedure

### 6.3.1. Fluorescence Diagnosis

Even though FD of neoplastic UADT lesions using ALA-induced PpIX has not yet gained approval as either a diagnostic procedure or a medical device, this promising approach has found its way into larger experimental studies at various hospital sites. Consequently, a fairly standardized set of required equipment and recommended procedures has been developed over the last few years. (For FD of dermatologic head and neck lesions, please refer to the Dermatology chapter of this book.)

#### 6.3.1.1. Substance, Application, Dosage, and Incubation
5-aminolevulinic acid hydrochloride may be used in a sterile packing from various suppliers (see the appendix). The substance should be diluted just prior to application (<5 min) in order to minimize early degradation.

For topical application in the oral cavity and the anterior oropharynx, the combination of a 0.4% rinsing solution (200 mg ALA per 50 mL tap water) and an application time of 15 min have provided the best results according to our own experience, and verified by other research groups.[14,15] Topical application to the larynx can be achieved using an inhaler, which should ideally produce

**Table 1.** ALA application parameters for FD of cancers in the oral cavity

| | |
|---|---|
| Substance | 5-aminolevulinic acid hydrochloride |
| Application | ⇒ Rinsing: 15 min; 0.4% (200 mg ALA per 50 mL H$_2$O) |
| | ⇒ Inhalation: 15 min; 0.6% (30 mg ALA per 5 mL 0.9% NaCl) |
| | ⇒ Systemic application: 5–25 mg ALA kg$^{-1}$ BW in 50 mL H$_2$O (via NGT) |
| Incubation | Topical application 1–2 h; systemic application 3 h |

droplets of a larger diameter than for pulmonary inhalation. At our institution, the ultrasound inhaler used (T/A Vernebler-S, Medanz, Starnberg, Germany), produces a median droplet diameter of 9 µm which should allow for a good deposition of the substance on the vocal folds. In this case a 0.6% ALA solution (30 mg ALA per 5 mL 0.9% NaCl) and an inhalation time of 15 min has proven to be the most effective combination. While inhaling, intermittent vocalization seems to enhance the deposition of the ALA solution on the vocal chord level.[16]

For systemic application of ALA, the optimum dosage has not yet been well defined but according to our studies, it seems to lie between 2.5 and 25 mg ALA kg$^{-1}$ body weight. Other groups have reported similar optimal dosages.[17–19] The formulation, which is diluted in approximately 50 mL of mineral water, should be applied via a temporary nasogastric tube in order to prevent additional topical effects on the mucosa.

Spectroscopic examinations in a larger group of patients ($n > 150$) at our facility have shown an optimal incubation time (hours after application of ALA) of 1–2 h following topical rinsing/inhalation and 3 h after systemic application. Our findings are not in total agreement with Svanberg et al., who have proposed an incubation time of 5–8 h following systemic application.[17] In our experience, such long incubation times lead to accumulated PpIX in the innocuous mucosa to a larger extent than for a 3 h application. ALA application parameters are summarized in Table 1.

### 6.3.1.2. Fluorescence Imaging

Following application of the ALA formulation and incubation, fluorescence imaging may be performed either with or without general anesthesia.

The 405-nm blue-violet excitation light may be generated by a filtered lamp system, novel LED sources, or lasers (e.g., krypton-ion-laser, dye laser). Today, commercially available excitation light sources for PpIX are usually filtered xenon short arc lamps, which provide blue-violet excitation light at wavelengths around 405 nm and can alternatively be used as regular white light sources. The output port of the lamps allows one to connect conventional fiber- and fluid optical cables.

For fluorescence detection and visualization in the oral cavity and the oropharynx, the observer or the detection unit, respectively, should be equipped with a longpass filter to totally exclude the reflected excitation light (e.g., OG515, Schott KG, Mainz, Germany): the resulting fluorescence image shows a red to green contrast from tumor and normal tissue. As the only commercially available system for FD using ALA-induced PpIX in the head and neck region to date, the Storz PDD System uses an endoscope with integrated filters coupled to an electronically modified and highly red-sensitive color CCD camera (see Table 2). Alternatively, regular endoscopes with a detection filter attached to the proximal end of the ocular may be used. Because of accessibility of the oral cavity from outside, however, the use of an endoscope is not mandatory, and an optimized detection unit with improved capturing capacity of the weak fluorescence signals is currently being developed at our facility. The fluorescence images that have been captured and processed by the camera are then passed to an ordinary color monitor.

Fluorescence imaging in the endolarynx is very similar to that used in the oral cavity, with some minor differences. The fluorescence examinations are usually better performed under general anesthesia and on an intubated patient since, according to our own experience, the captured fluorescence light intensity in indirect laryngoscopy is quite low for the attainment of high-quality fluorescence images. Excitation light sources as well as detection units are almost the same as those used for fluorescence imaging in the oral cavity. Because of the lower overall lighting conditions as compared to the mouth, the use of other detection filters which allow for a minor part of the remitted blue-violet light to pass through (*e.g.* GG455, Schott KG, Mainz, Germany) can considerably enhance image quality. Accordingly, the normal tissue then appears light blue as compared to the reddish appearance of the tumor tissue. Relevant parameter values of the technical equipment used in our clinical studies are summarized in Table 2 below.

### 6.3.2. Photodynamic Therapy

#### 6.3.2.1. Precancerous Mucosal Lesions of the UADT

Even though several attempts have been made to apply the photodynamic treatment of premalignant or malignant lesion of the UADT using topically or systemically applied ALA, both the necessary equipment and the procedure itself have not yet been as well standardized as for FD. However, some general statements can be made at this point.

First, for precancerous (and maybe even early cancerous) lesions of the oral cavity, for example, leukoplakias, a 10–20% ALA ointment preparation is being recommended with an application time of up to 2 h.[20–22] There will obviously be (depending on the production of saliva and the compliance of the patient) a high percentage of substance washout, so the amount of ALA uptake into the lesions will vary greatly. While some attempts are being made to

**Table 2.** Parameter values of the technical equipment used in our clinical studies on FD of cancers in the oral cavity

| | |
|---|---|
| Light source | $\Lambda \approx 405$ nm (*e.g.*, D-LIGHT C SCB; Karl Storz) |
| Light cable | Conventional fiber cable/fluid optical cable (adapter type Storz) |
| Detection unit | Endoscope with integrated detection filter (*e.g.*, Hopkins II type endoscope (PDD), Karl Storz) |
| Camera | Red sensitive CCD-camera (*e.g.*, Tricam SL-PDD SCB; Karl Storz) |

**Table 3.**  Drug and light application parameter values for PDT of early mucosal lesions of the UADT

| Substance | 5-aminolevulinic acid hydrochloride |
|---|---|
| Application | ⇒ Oral cavity: 2 h; 10–20% ALA ointment preparation |
|  | ⇒ Larynx: 30 min; 4.0% (400 mg ALAper10 mL NaCl) |
| Incubation: | ≤ 6 h |
| Activation: | 635 nm; 100 J cm$^{-2}$ |

overcome this problem, the result after illumination is hard to predict at this stage. Topical application to the larynx for the treatment of precancerous lesions may be achieved, as in FD, via the inhalation technique. Also for PDT a much higher concentration of the ALA solution is being recommended for treatment purposes as compared to diagnostic purposes (4% instead of 0.6%), and the duration of the application should be at least 30 min.[23]

For both anatomic regions, a total light dose of around 100 J cm$^{-2}$ at 635 nm is suggested, based on the treatment of precancerous lesions[20–23] following an incubation time of up to 6 h (Table 3). It seems advisable to use an argon-ion-pumped dye laser or a diode laser system (producing laser light at the recommended wavelength) in combination with spherically- or cylindrically-formed light applicators (see the appendix).

### 6.3.2.2. Cancerous Lesions of the UADT

The standardization of equipment and procedure for PDT, as for FD, has not yet been achieved. From the few authors who have performed PDT of malignant UADT lesions and published their results,[21,24,25] a systemic administered dose of about 60 mg ALA kg$^{-1}$ BW is being recommended, and the total light dose (at 635 nm) should be between 100 and 200 J cm$^{-2}$ (Table 4). Since the tumor selectivity of ALA-induced PpIX is quite low for this way of application, a careful shading of the healthy tissue surrounding the lesion, which would also be illuminated during photoactivation, is required. For this kind of application, it should be remembered that a generalized photosensitization of the whole integument may be present for up to 48 h post administration of the drug.

### 6.3.2.3. Pre- and Early Malignant Facial Lesions

For the treatment of pre- and early malignant facial lesions, for example, superficial basal and SCC, and solar or actinic keratoses, most authors of the many publications on this topic recommend the use of a 10–20% ALA ointment.[26–28] Following local application to the diseased skin, an occlusive dressing is applied and left on the lesion for an incubation time of approximately 4–8 h. After removal, the lesions should be illuminated with red light centered around 635 nm (just as in the treatment of UADT lesions) with a total dose of 50–150 J cm$^{-2}$ (Table 5).

Two compounds have now been granted international approval for PDT of pre- and early malignant facial lesions. The recommended procedure for their use is summarized below.

**Table 4.**  Drug and light application parameter values for PDT of more advanced cancers of the UADT

| Substance: | 5-aminolevulinic acid hydrochloride |
|---|---|
| Application: | $\Rightarrow$ Systemic Application: 60 mg ALA kg$^{-1}$ BW in 50 mL H$_2$O (via NGT) |
| Incubation: | $\leq$ 6 h |
| Activation: | 635 nm; 100–200 J cm$^{-2}$ |

**Table 5.**  Drug and light application parameter values for PDT of facial lesions

| Substance | 5-aminolevulinic acid hydrochloride |
|---|---|
| Application | $\Rightarrow$ Facial skin/scalp: 4–8 h; 10–20% ALA ointment preparation |
| Incubation | 4–8 h (equal to ALA application) |
| Activation | 635 nm; 50–150 J cm$^{-2}$ |

(i) The pen-like Levulan Kerastick (DUSA Pharmaceuticals Inc., see the appendix) is a two-component system consisting of a plastic tube containing two sealed glass ampoules (one with a solution vehicle and one with ALA-HCl as a dry solid) and an applicator tip. Just prior to the application of the substance, a 20% ALA solution is prepared by breaking the ampoules inside the tube with finger pressure and mixing the contents by shaking the applicator. Immediately following this procedure, the liquid substance is applied to the keratotic lesions (not to the perilesional skin, periorbital areas, or mucosal surfaces) *via* the applicator tip. Once the initial application has dried, the formulation is applied once more in the same manner. A time stretch of 14–18 h should be allowed for photosensitization, so it may be advisable to apply the compound within the afternoon hours in order to be able to perform the illumination of the lesions on the following morning. The actinic keratoses should not be washed during this time, and the patient should be advised to wear adequate protective apparel to shade the treated actinic keratosis lesions from intensive lighting until treatment.

At the second visit (14–18 h after application), the actinic keratoses to be treated are gently rinsed with water and patted dry. Photoactivation is then accomplished with illumination from the so called BLU-U Blue Light Photodynamic Therapy Illuminator at 400–450 nm. The choice of blue excitation light rather than the more deeply penetrating red light at 635 nm has been justified by the superficiality of the lesions to be treated.[29] The use of this light source (provided by the same company as the compound itself), which is comprised of seven horizontally mounted U-shaped fluorescent tubes with a plastic chassis, is therefore mandatory as defined in the protocol of approval. A 1000 s exposure is required to provide the 10 J cm$^{-2}$ light dose needed for an adequate phototoxic reaction (Table 6).

(ii) The second compound, Metvix (PhotoCure ASA, see the appendix) contains 160 mg g$^{-1}$ ALA-methylester as the active ingredient in a cream preparation. In this case the recommendation is to prepare the surface of

**Table 6.**  Drug and light application parameter values for PDT using Levulan Kerastick and BLU-U

| Substance | 5-aminolevulinic acid hydrochloride |
|---|---|
| Application | 20% ALA solution; 2 × topical application via applicator tip |
| Incubation | 14–18 h |
| Activation: | BLU-U (400–450 nm); 1000 s (10 J cm$^{-2}$) |

**Table 7.**  Drug and light application parameter values for PDT using Metvix and
Actilite$^{®}$

| Substance | 5-aminolevulinic acid methyl ester (ALA-ME) |
|---|---|
| Application | 160 mg g$^{-1}$ ALA-ME cream preparation; 1 × topical application (1 mm thick) |
| Incubation | 3 h |
| Activation | 635 nm; 75 J cm$^{-2}$ (*e.g.* Actilite, Photocure ASA) |

actinic keratosis and superficial BCC before applying the Metvix cream
by removing scales and crusts and roughening the surface of the lesions.
In the case of nodular BCC, the intact epidermal keratin layer (if
present) should be removed as well. The cream is then applied to the
lesion itself and to the surrounding 0.5–1.0 cm of normal skin at a
thickness of about 1 mm by using a spatula. This is followed by a 3-h
incubation under an occlusive dressing.

After removal of the dressing and cleansing of the area with normal
saline, the lesion is exposed to red light centered at about 635 nm with a
total light dose of 75 J cm$^{-2}$ (Table 7). The same company that provides
Metvix also offers a light source for photoactivation: the Actilite lighting
system uses red emitting diodes ($\lambda$ = 634 ($\pm$3) nm) arranged in a
rectangle. Its use in conjunction with Metvix, however, is not dictated
by the approval in this case, so other light sources may be used as well.
Even though ,the healthy skin surrounding the lesions has been in
contact with ALA-methylester as well, it does not need to be protected
during illumination. Whereas actinic keratoses are usually adequately
cured by one treatment session only, BCC lesions should undergo two
consecutive treatments one week apart.

## 6.4.  ALA-FD

### 6.4.1. Indications

Even though not approved by national or international authorities, in investigational clinical studies over the last decade several indications with promising
results have been developed for FD in the head and neck region by detection
and visualization of ALA-induced PpIX fluorescence (see Section 6.3 on
"Procedure"). These systems, arranged by importance, are described below.

(For dermatological indications in the head and neck region, please refer to the Dermatology chapter.)

### 6.4.1.1. *Identification of pre- and Early Malignant Mucosal Lesions*

According to the study results of our own as well as other research groups, FD following topical application of ALA as an oral rinse or inhalation, in conjunction with regular clinical inspections, provides for a higher rate of identification of carcinoma precursors and early carcinomas of the UADT than normal inspection or imaging techniques alone.[30–33,14–15] Typical findings are presented in Figure 5.

In our examination of 133 patients, severe dysplasias (D III) and Carcinomata *in situ* (Cis) were subjectively better detectable in fluorescence examinations in comparison with regular white light inspections, which could be verified by histopathological evaluations of a total of 345 biopsies (sensitivity: fluorescence examination 93.1%; white light inspection 89.7%). SCC's (pT1–pT4), however, were correctly identified by their macroscopic appearances in both

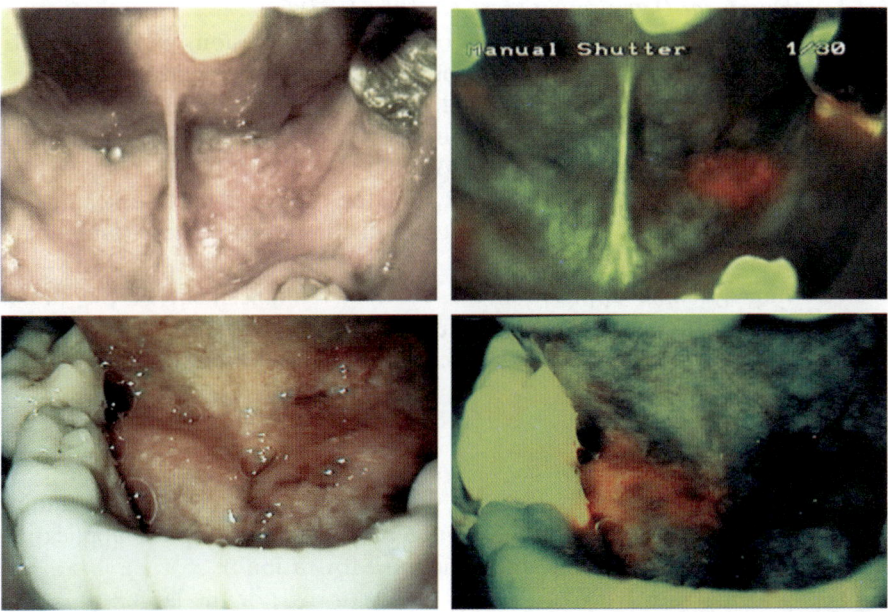

**Figure 5.** Exemplary findings visualizing the value of fluorescence diagnosis using ALA for the identification of UADT tumors. The two upper images show a pT1 SCC of the floor of the mouth in a 57-year-old male patient in regular inspection (left), and following topical ALA application in fluorescence mode (right). The two lower images show the finding of an early secondary carcinoma of the right floor of the mouth in a 49-year-old male patient (left: regular inspection; right: fluorescence image) with a known SCC of the left upper alveolar ridge, which was detected only via fluorescence diagnosis.

fluorescence examinations and normal inspection in nearly 100% in our studies. In these cases, fluorescence diagnosis may nevertheless serve as a screening tool for the detection of early secondary carcinomas, which show a significantly increased frequency of occurrence in the UADT and were identified in our study in three cases only via fluorescence diagnosis. Other groups have reported the sporadic appearance of false negative findings in the case of large, centrally necrotic tumors,[14] which could not be verified by our examinations. The quality of fluorescence imaging following application of ALA seems to be independent of the tumor grading; yet a high grade of cornification of the malignant lesions seems to have a negative effect on the identification of the tumors.

Following systemic administration of ALA (2.5–25 mg kg$^{-1}$ BW), our own examinations of 60 patients, as well as other studies, have shown a large percentage of false-negative findings (*i.e.*, malignant areas missed in fluorescence examinations), even though other groups have reported an occasional identification of carcinoma precursors or additional findings via fluorescence diagnosis.[17–19] For the identification of pre- and early malignant mucosal lesions of the UADT, topical application of ALA is therefore preferable to the systemic application.

### 6.4.1.2 Superficial Delimitation of SCC of the UADT

The improved superficial delimitation of precancerous lesions and SCC (pT1–pT4) of the UADT from healthy mucosa offers another advantage of fluorescence diagnosis using topically applied ALA (Figure 6). In 133 patients that were examined at our institution, tumor borders were subjectively better distinguishable than with regular inspection in 17.8% of the cases, which makes fluorescence diagnosis a useful tool for the head and neck surgeon, especially in the case of operative resection of the tumors. These results have been independently verified.[15]

Following systemic application of ALA, an enhancement of the tumor demarcation does not seem to be achievable. Systemic administration of ALA is thus not a recommended procedure for the delimitation of mucosal neoplasias at this time.

### 6.4.1.3. Screening Method for the Identification of UADT Papillomas

Apart from SCCs and their precursors, the RRP seems to be especially suitable for fluorescence staining with ALA-induced PpIX after topical application (Figure 7) (see Section 6.1 on "Introduction/Motivation").

In 12 patients with RRP of the oral cavity ($n = 2$) or the larynx ($n = 10$), fluorescence diagnosis following topical application of ALA led to a reliable identification and delimitation of these benign neoplasias.[34] In four out of the twelve patients examined in this only study using this method for the detection of RRP lesions so far, additional suspicious red fluorescent regions were found in macroscopically completely inconspicuous mucosa, which proved to be satellite-papillomas in the histopathological evaluation. This may finally lead to an improvement of the long-term results: Prospective randomized

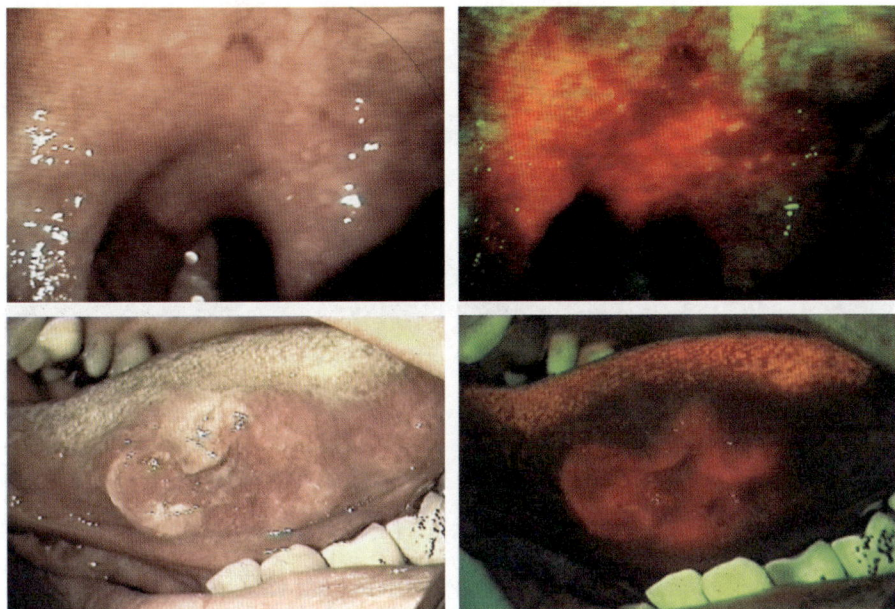

**Figure 6.** Demonstration of tumor delimitation abilities of ALA fluorescence diagnosis. On the upper left is shown a superficial spreading SCC of the right soft palate in a 49-year-old patient with macroscopically indistinct tumor borders. The delimitation becomes much clearer in fluorescence diagnostic mode using ALA induced PpIX (right). On the lower images, a pT2 SCC of the left rim of the tongue is shown in both regular white light imaging (left) and fluorescence diagnostic mode (right), where the tumor borders again become much clearer. As always, the dorsum of the tongue shows a false positive red fluorescence.

comparative trials are needed to provide information concerning the recurrence rates of the disease following treatment with and without the help of fluorescence diagnosis.

Because of the lack of clinical experience, no statement can be made at this point concerning the value of fluorescence diagnosis of RRP of the UADT after systemic administration of ALA.

### 6.4.2. Limitations

Fluorescence diagnosis using ALA-induced PpIX in otolaryngology also has remaining limitations, which are discussed below.

#### 6.4.2.1. Superficial Staining Following Topical ALA Application

Fluorescence microscopic studies on tumor tissue specimens of the UADT have shown a depth of ALA-induced PpIX staining of 0.9 mm on an average.[35] With

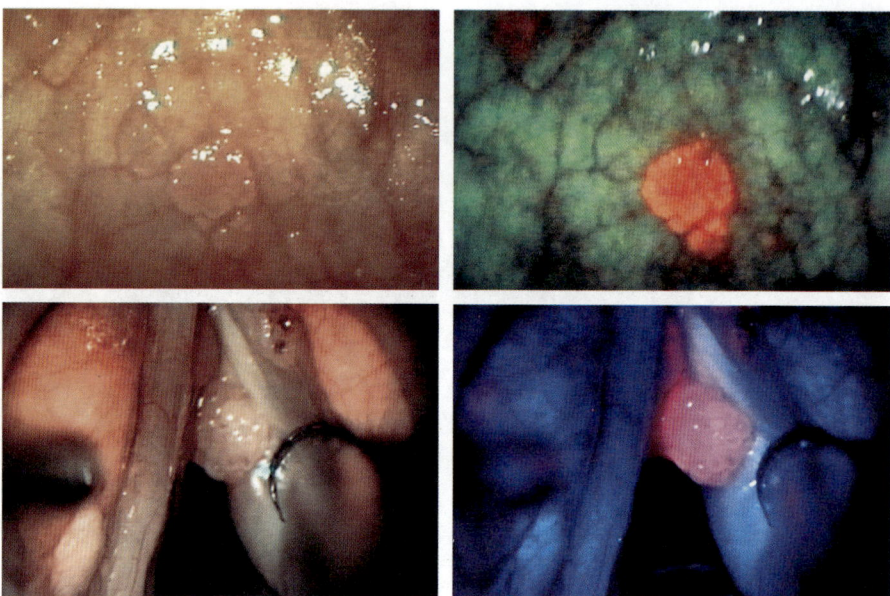

**Figure 7.** Demonstration of the improved visualization of mucosal papillomas of the UADT via ALA fluorescence diagnosis. The image on the upper left shows a papillomatous lesion of the rim of the tongue in a 37-year-old patient under white light inspection: however, the same mucosal alterations (upper right) are much emphasized in the fluorescence image. In the lower pictures, a laryngeal papillomatosis is being depicted. Again, the fluorescence image (lower right) gives a much clearer and more definite impression of the lesion than normal inspection (lower left).

a known limitation of the optical penetration of the light used for fluorescence excitation of only 0.3 mm, this fact is irrelevant for the detection and superficial delimitation of malignant mucosal lesions. An intraoperative identification of neoplastic spreading in the depth of the tissue seems to be possible only with a systemic application of ALA, which, at the moment does not provide reliable results.

*6.4.2.2. General Limitations Following Systemic Administration of ALA*
Even though some promising results have been published on systemic administration of ALA for fluorescence diagnosis of oral and laryngeal malignomas,[17–19] this method seems to be inferior to fluorescence staining using topically applied ALA in terms of quality of tumor identification and superficial delimitation. Additionally, side effects occur much more frequently, and are usually more severe, when choosing this method of administration (see Section 6.5.3 on "Unwanted effects"). A possible indication would be in-depth tumor staining, a requirement for fluorescence-guided resections.

### 6.4.2.3. Occurrence of False Positives

For all indications and routes of administration of ALA, our own investigations have shown a considerable number of false-positives, *that is*, benign tissue falsely rated suspicious owing to red fluorescence. However, these false-positives always numbered fewer than for regular inspection. In general, chronically inflamed tissue shows the highest rate of false red fluorescence. As a rule, the dorsums of tongues are also always (falsely-positive) red-fluorescing (Figure 6): malignant disease, however, is very seldom found in this location. Therefore, all chronically inflammatory processes of the UADT (*e.g.* chronic, radiogenic mucositis after radiation therapy) as well as the dorsums of tongues have to be seen as limitations to fluorescence diagnosis with ALA.

### 6.4.2.4. Photobleaching

The phenomenon of "photobleaching", a degradation of red-fluorescing PpIX molecules by light irradiation into non or weakly fluorescing photoproducts, leads to a time limitation for fluorescence examinations. A power density of approx. 50–200 mW cm$^{-2}$ from the excitation light source is usually incident onto the observed tissue surface. In the examinations that we performed, good subjective results concerning the quality of macroscopic fluorescence imaging could be obtained for the first 5 min of each examination only. After only 10 min, the subjectively visible red fluorescence had completely vanished due to photobleaching effects and a fluorescence contrast between tumor and surrounding normal tissue was no longer detectable. It is therefore important to minimize photobleaching beforehand so that the best tumor detection conditions prevail during the fluorescence measurement period.

### 6.4.2.5. Bacterial Coatings

Whitish to yellowish appearing bacterial coatings on necrotic tumor surfaces, and on gingival and periodontal tissue in patients with poor oral hygiene show a strong, unspecific red fluorescence after (and sometimes even before) application of ALA.[14–15,31,32] *Prevotella oralis*, *Porphyromonas gingivalis*, and *Staphylococcus aureus* proved to be the most dominant colonies in a microbiological analysis of the bacteria responsible for this unspecific red fluorescence.[14] The authors recommended a mechanical removal of bacterial coatings from tumor surfaces and necks of teeth before administration of ALA in order to minimize such false positive findings. While such fluorescence is sometimes without doubt of the "false-positive" type, there may be some merit in using such fluorescence as a guide since these coatings are usually found exclusively in those typical locations mentioned above. In case of doubt, the surface can be wiped off to reveal any PpIX fluorescence that is produced in the tumor tissue itself.

### 6.4.3. Unwanted Effects

#### 6.4.3.1. Local Symptoms

The occurrence of local symptoms (*i.e.*, within or around the area examined) may be caused either by mucosal irritations due to the acidity of the ALA solution administered, or by the potentially phototoxic effects of PpIX. In 193 patients who have undergone fluorescence diagnosis at our institution to date, only two have experienced stronger coughing during ALA inhalation; another two have reported a weak burning sensation during illumination of the tumor, following systemic application of ALA. According to these observations and to those of other groups,[14,15] severe local symptoms are quite rare and not generally anticipated.

#### 6.4.3.2. Systemic Symptoms

Because of the systemic resorption of ALA following oral, or, to a much smaller extent, topical application as an oral rinse or inhalation,[36] the occurrence of porphyria-like neurovisceral and photocutaneous side effects cannot be ruled out. The spectrum of neurovisceral symptoms after a high-dose, oral application runs from nausea, vomiting, and transient liver dysfunction with increased transaminases to severe drop-offs of the blood pressure due to a reduced vascular resistance.

In patients, topical application of ALA ($n = 133$) as oral rinse or inhalation led to no systemic side effects; minimally increased levels for PpIX in plasma and skin could be determined only following inhalative administration.[36] Systemic application of 2.5–25 mg ALA per kg body weight ($n = 60$) resulted in a transient, slight increase of liver function parameters in one case, in a short drop-off of the blood pressure during anaesthetization in another case, and vomiting 8 h after intake of ALA in a third patient. Additionally, four patients showed a general cutaneous sensitization associated with a slight reddening of the skin, which disappeared completely after 48 h. Accordingly, increased levels for PpIX in both skin and blood could be determined following oral administration of ALA.[36]

In conclusion, one should be prepared for possible neurovisceral and photocutaneous side effects after systemic administration of ALA, especially when applying higher dosages ($>20$ mg ALA kg$^{-1}$ body weight). For topical ALA applications, systemic unwanted effects are unlikely.

## 6.5. ALA-PDT

### 6.5.1. Indications

ALA-based PDT in the head and neck region has now been approved for the treatment of actinic keratoses and early BCC of the face and scalp (see Section 6.2 on "Approval Status"). Apart from that, there are also several other indications for the application of PDT using ALA-induced PpIX without national or international approvals but with considerable clinical experience and promising results. The following section will deal briefly with both approved and nonapproved aspects of the method.

### 6.5.1.1. Treatment of pre- and Early Malignant, Unpigmented Facial Lesions

For superficial precancerous lesions, as well as early malignomas (actinic/solar keratoses, early SCC/BCC), complete remissions following ALA PDT are being reported in 71–100% of cases.[26–28] Complete remission rates of actinic keratoses with approved Levulan Kerastick have been reported to be within the same range (72–88%)[29,37,38] (Figure 8).

Because of an enhanced lipophilicity and an accelerated tissue penetration, esterification of ALA leads to shorter incubation times, and the applied dose may be reduced.[39,40] Numerous scientific publications on the use of this prodrug of ALA (marketed as "Metvix") within recent years confirm the growing interest in this compound[41–43] (Figure 9). The rates of complete remission achieved by the substance in the treatment of actinic keratoses are comparable to those for nonesterified ALA being used (69–91%).

Whereas the above photodynamic methods for treatment of nonpigmented facial lesions following topical ALA application do not seem to exhibit significant differences from standard procedures (surgery, cryotherapy, chemotherapy using topical 5-fluorouracil) with respect to the overall cure rates,[11,44] it has been reported repeatedly that better cosmetic results are usually attainable. Therefore, the acceptance by patients may be improved.

### 6.5.1.2. Treatment of Precancerous Mucosal Lesions of the UADT

To date, all of the scientific publications on the use of ALA PDT for precancerous mucosal lesions deal with leukoplakic lesions.[20–22,45] For oral, pharyngeal, and laryngeal leukoplakias, the authors report complete remission rates of 42–100%. The high variations in results obtained by the different groups may be due to significant differences in the treatment protocols concerning application and dosage. In one study that initially found a complete remission in 10 out of 12 cases with widespread oral leukoplakic lesions after ALA PDT (10% ALA ointment, 100 J cm$^{-2}$ lighting), a longer term follow-up showed a recurrence in two out of ten patients within 6 months.[22]

It must be stressed that even though larger clinical studies are necessary to evaluate the effectiveness of ALA PDT for the treatment of leukoplakia in the UADT as compared to standard surgical measures, the method might be preferable to others because of the excellent cosmetic results, the possibility of treating multifocal changes and the good patient tolerance.[22]

### 6.5.1.3. Treatment of Cancerous Lesions of the UADT

In comparison with the broad use of PDT in the head and neck region following topical application of ALA for the treatment of precancerous and

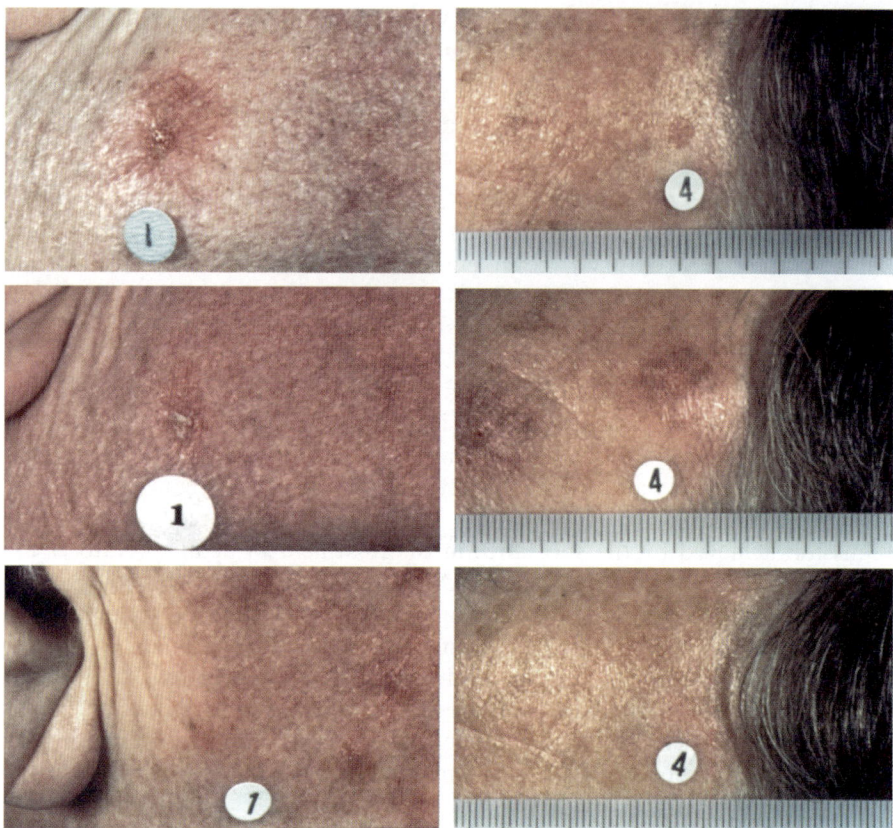

**Figure 8.** Clinical examples of the FDA-approved use of the Levulan Kerastick for the treatment of nonhyperkeratotic actinic keratoses. Both the left and the right column show two such lesions (left: right cheek; right: left front). Upper row: before PDT. Middle row: 24 h after; bottom row: 8 weeks post PDT (Pictures are published with authorization of DUSA Pharmaceuticals Incorporated).

superficially growing cancerous lesions, its use following systemic substance administration for treating more advanced tumors in both a curative and a palliative intent has so far been reported in very few publications.[21,24,25] This may be because of a markedly decreased selectivity for cancerous tissue as compared to local application and a comparatively low photodynamic activity (see "Fluorescence Diagnosis"). For this indication, purified hemotoporphyrin derivative (*e.g.* Photofrin®, Axcan Pharma, Inc., Quebec, Canada) as a classical first-generation photosensitizer with broad clinical use and mTHPC, with an especially high absorption rate of the excitation light, have gained markedly more attention than ALA-induced PpIX. According to the authors using this method, however, tumor-selective necroses at a thickness of 0.1–1.3 mm can be obtained following administration of 60 mg kg$^{-1}$ body weight ALA. It might therefore be a useful concept for both gentle and effective therapy of smaller- to

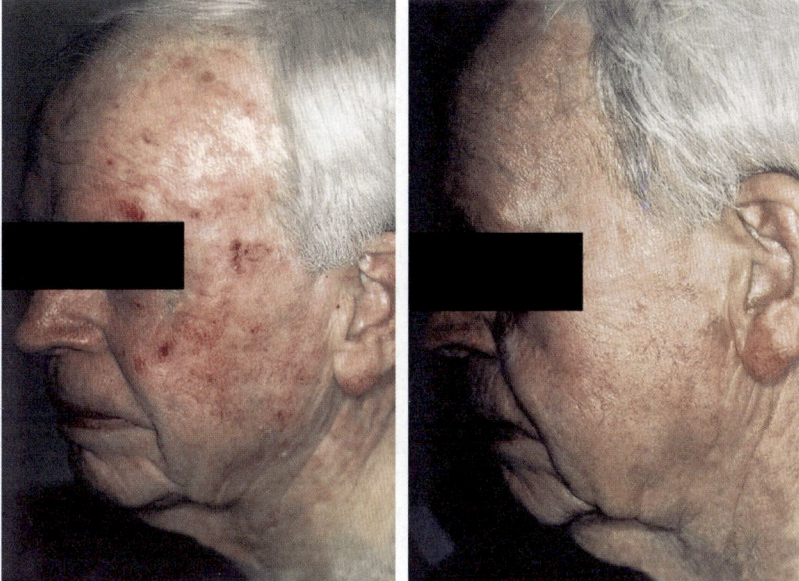

**Figure 9.** Clinical example of the internationally approved application of Metvix for treatment of facial actinic keratoses. The left picture shows a left facial side covered with actinic keratoses. After PDT (right), the skin lesions have completely resolved (Publication of these pictures has been authorized by PhotoCure ASA).

medium-sized malignant lesions of the UADT. When repeated after certain time intervals, this method may offer a palliative, or even curative, therapeutic option for larger head and neck tumors.

### 6.5.2. Limitations

As for fluorescence diagnosis, the following limitations apply to PDT in the head and neck region using ALA.

#### 6.5.2.1. Superficial Photodynamic Effects

Because of limitations concerning the penetration depth of the photoactivating light into the tissue (approximately 0.75 mm at $\lambda = 635$ nm), effective PDT when using incident lighting is restricted to (epi-)dermal and mucosal lesions that do not infiltrate into deeper tissue layers. To that depth, the photosensitizer PpIX is usually found in amounts adequate for PDT, even if the substance has been applied topically (see "Fluorescence Diagnosis").

For this reason, high rates of recurrence have been reported after treatment of BCC[46] and SCC[25] in the head and neck region. The method, even with systemic administration of ALA, should therefore be performed only for the treatment of pre- and very early and only superficially infiltrating malignant lesions.

For the treatment of larger tumors, possible ways to overcome this problem include the application of repeated superficial PDTs, interstitial PDT, and intraoperative PDT of the tissue defect left from palliative tumor debulking (all following systemic administration of ALA). To date, however, no studies have been published on this topic.

### 6.5.2.2. The Absence of Tissue Specimens for Histopathological Diagnoses

In typical precancerous lesions such as leukoplakias of the UADT mucosa and in solar/actinic keratoses of the face and scalp, a malignant conversion (*i.e.*, invasive growth of at least part of the lesion) is usually determined by histopathological diagnosis of resected tissue specimens. In the case of ALA-PDT, the superficial lesion becomes necrotic after treatment, and the healing process requires days to weeks following treatment. Where invasive growth into deeper tissue layers has been missed by clinical inspection and random biopsies preceding PDT, malignant cells may survive the treatment and continue to grow beneath the newly forming skin layer. Therefore, the clinical evaluation of the lesions should be made beforehand with great care, and the sites of ALA-PDT should be inspected in regular follow-ups after the treatment.

### 6.5.2.3. Topical Administration of ALA in the UADT

Topical administration of ALA to the UADT as ointment or inhalation does not provide for predictable and constant PpIX concentration over the entire time of application to penetrate the mucosa. Thus, the final concentration of PpIX within the lesions, as well as the photodynamic effects, may vary greatly. One option is to use systemic administration of ALA, but this would result in lower tumor selectivity and a higher potential for side effects (see Section "6.5.3. Unwanted effects").

### 6.5.2.4. Light Application in the UADT

Similar to the problems that arise with topical application, a uniformly even light application (and thus a reproducible dosimetry) is not easily accomplished in a complex geometric setting like the UADT. Therefore, more advanced light applicators are being used for the illumination of hollow organs like the oral cavity, pharynx, larynx, or the cavities formed by tumor resection. Ideally, their outer shapes should correspond to the form of the cavity to be illuminated, and a suitable scattering medium within the applicators should be included for isotropic light intensities on the tissue surfaces.[47,48] Commonly used examples are spherically or cylindrically formed applicators, which may be obtained either in preset dimensions or be blown into shape *in situ* by either liquid or air pressure.[48,49] Despite these advances in light application techniques and equipment within recent years, an even light distribution may often be hindered by variable geometric settings such as crevices and folds, leading to somewhat unpredictable therapeutic effects.

### 6.5.3. Unwanted Effects

#### 6.5.3.1. Local Symptoms
Most patients experience local discomfort of varying intensity during ALA-PDT. Typically it is described as a stinging or burning sensation within the irradiated tissue, which is highest at the beginning of the procedure and tends to gradually decrease towards the end.[21,25] In some cases, it has taken up to 24 h for complete resolution.[13,38] For both approved compounds, local symptoms have been analyzed statistically and are presented in a more structured form in published data sheets.[13,50] Recommendations for analgesic treatment of the painful symptoms during and directly following illumination run from local anesthesia and intensive cooling[51] to the administration of systemic pain-relieving substances.[21]

Erythema and edema are usually encountered at the treated sites,[37,38] but these usually resolve within 1–4 weeks after PDT. Additionally, scaling and crusting in combination with itching sensations is often noted as a local side effect of treatment.[13,50,52]

The most common unwanted long-term local side-effects of ALA-PDT are hypo- and hyperpigmented areas within the treated area.[13,50] More serious local adverse events are rarely reported. In one published case, a patient developed a malignant melanoma at the site of the scalp (Clark level II) that was previously (4 years to 8 months prior) treated four times for a solar keratosis by ALA-PDT.[53] A direct tumor initiation could not be demonstrated in this case report, nor could it clearly and obviously be ruled out. The authors therefore recommend regular follow-up examinations, particularly when the treatment is administered repeatedly to chronically sun-damaged skin.

#### 6.5.3.2. Systemic Symptoms
Concerning the possible systemic symptoms following administration of ALA, the same applies as for fluorescence diagnosis, except that the systemically applied dose of ALA is usually much higher in this case, naturally leading to a higher frequency and severity of neurovisceral and cutaneous side effects. In a study of 18 patients receiving palliative ALA-PDT for advanced head and neck cancer, the administration of 60 mg kg$^{-1}$ BW ALA led to nausea and vomiting in one third of cases, and to a prolonged general cutaneous photosensitivity, which lasted up to 48 h.[25]

No systemic symptoms are to be expected from topically applied ALA. As well, light irradiation during PDT treatment generally does not lead to adverse side effects.

## 6.6. Conclusions

In recent decades, a growing list of clinical indications has developed for the application of fluorescence diagnosis and PDT using ALA-induced PpIX in the head and neck region.

For fluorescence diagnosis, they comprise the detection and delimitation of pre- and early malignant mucosal lesions of the UADT and the detection of mucosal papillomas following topical application of ALA (Dermatologic indications are discussed and summarized in the corresponding chapter of this book). The value of the method is so far limited by the superficiality of the staining following topical ALA application, the generally reduced tumor specificity after systemic substance administration and the short time span for each examination due to photobleaching effects. Furthermore, the method produces a fairly high rate of false positive findings in chronically inflamed/dysplastic tissue, on the dorsums of tongues and at locations with bacterial coatings (*e.g.*, necrotic tumor surface, gingival mucosa). To date there has been no official approval for the use of this method in the areas of the head and neck.

Indications for ALA-PDT are primarily the treatment of superficially growing, non-pigmented pre- and early malignant lesions of the face and scalp following topical substance application. For this method, two products have so far achieved approval status by international authorities. Even though the clinical experience is still low, other useful applications might be in the treatment of premalignant mucosal lesions of the UADT following topical substance application or even the palliative reduction of tumor mass in advanced head and neck malignancies after systemic administration of ALA. The methods are again limited by the superficiality of the photodynamic effect, due largely to the penetration depth of the excitation light. As for other photosensitizers (*e.g.*, purified hematoporphyrin derivative, mTHPC), this drawback might be manageable by either an interstitial placement of fibers for the application of excitation light or a repeated therapy within short time intervals. Additionally, the absence of histopathological diagnosis in the case of PDT treatment may cause important features of the lesion's infiltrative behavior to be missed. Further, the complex anatomical structure of the UADT makes it difficult to achieve an even topical deposition of the substance and a reproducible and constant illumination of the lesions to be treated, so the photodynamic effects are sometimes difficult to observe. Finally, both the local and the systemic side effects (only after systemic administration of ALA) are to be taken into consideration when considering the application of this method.

In conclusion, both fluorescence diagnosis and PDT using ALA-induced PpIX offer promising possibilities in both diagnosis and treatment of head and neck cancer with only minor restrictions.

## 6.7. Final Note

For both fluorescence diagnosis and PDT using ALA, interesting and promising indications have been developed, and some are now in routine use in the head and neck region. This development, however, is not complete: novel protocols will be described and tested, and existing protocols will be methodologically improved and evaluated as to their value in larger, more objective, future randomized trials.

Future perspectives in the diagnostic field include the detection and differentiation of infectious agents within the head and neck region as well as the intraoperative detection of parathyroid glands in adenomatous hyperparathyroidism. Furthermore, image-processing procedures may enhance the significance of fluorescence findings. Therapeutically, further developments may still be anticipated in the treatment of cutaneous head and neck (pre)malignancies and of tumor precursors in the upper aero-digestive tract.

# References

1. R. Sankaranarayanan, E. Masuyer, R. Swaminathan, J. Ferlay and S. Whelan, Head and neck cancer: a global perspective on epidemiology and prognosis, *Anticancer Res.*, 1998, **18**, 4779–4786.
2. S.J. McQuone and D.W. Eisele, *Carcinoma of the oral cavity and pharynx*, Essential *otolaryngology/head and neck surgery*, K.J. Lee (ed), Appleton & Lange, Stamford, 1999, 527–547.
3. S.J. Silverman, Early diagnosis of oral cancer, *Cancer*, 1988, **62**, 1796–1799.
4. T.Y. Chen, L.J. Emrich and D.L. Driscoll, The clinical significance of pathological findings in surgically resected margins of the primary tumor in head and neck carcinoma, *Int. J. Radiat. Oncol. Biol. Phys.*, 1987, **13**, 833–837.
5. A. Dunsche and F. Harle, Precancer stages of the oral mucosa: a review, *Laryngorhinootologie*, 2000, **79**, 423–427.
6. B.M. Steinberg, W.C. Topp, P.S. Schneider and A.L. Abramson, Laryngeal papillomavirus infection during clinical remission, *N. Engl. J. Med.*, 1983, **308**, 1261–1264.
7. V. Bonkowsky, *Erkrankungen des Larynx und Trachea: Gutartige Tumoren*, In: *Praxis der HNO-Heilkunde, Kopf- und Halschirurgie*, J. Strutz and W. Mann (eds), Georg Thieme Verlag KG, Stuttgart, 2001, 501–507.
8. B.W. Neville and T.A. Day, Oral cancer and precancerous lesions, *CA Cancer J. Clin.*, 2002, **52**, 195–215.
9. G. Lodi, A. Sardella, C. Bez, F. Demarosi and A. Carrassi, Systematic review of randomized trials for the treatment of oral leukoplakia, *J. Dent. Educ.*, 2002, **66**, 896–902.
10. D. Slaughter, H. Southwick and W. Smejkal, Field cancerization in oral stratified squamous epithelium, *Cancer*, 1953, **6**, 963–968.
11. E.W. Jeffes III and E.H. Tang, Actinic keratosis. Current treatment options, *Am. J. Clin. Dermatol.*, 2000, **1**, 167–179.
12. European Agency for the Evaluation of Medical Products (EMEA), Foscan: Summary of product characteristics, (www.emea.eu.int/humandocs/Humans/EPAR/foscan/foscan.htm), 2001.
13. Food and Drug Administration (FDA), Levulan Kerastick, (www.fda.gov/cder/foi/label/1999/20965lbl.pdf), 1999.
14. W. Zenk, W. Dietel, P. Schleier and S. Gunzel, Visualizing carcinomas of the mouth cavity by stimulating synthesis of fluorescent protoporphyrin IX, *Mund Kiefer Gesichtschir.*, 1999, **3**, 205–209.
15. B.M. Lippert, C. Külkens, N. Klahr, B.J. Folz and J.A. Werner, *5-Delta-Aminolävulinsäure induzierte Fluoreszenzdiagnostik bei Karzinomen der oberen Luft- und Speisewege-erste Ergebnisse*, *Fluoreszenzdiagnostik und Photodynamische*

*Therapie, Lippert, BM.*, S. Schmidt and J.A. Werner (eds), Shaker Verlag GmbH, Aachen, 2000, 65–73.

16. H. Kumazawa, M. Asako, T. Yamashita and K.S. Ha, An increase in laryngeal aerosol deposition by ultrasonic nebulizer therapy with intermittent vocalization, *Laryngoscope*, 1997, **107**, 671–674.

17. K. Svanberg and I. Wang, Fluorescence diagnostics of head and neck cancer utilizing oral administration of delta-Aminolevulinic Acid, *SPIE*, 1995, **2371**, 129–141.

18. K. Svanberg, I. Wang, S. Colleen, I. Idvall, C. Ingvar, R. Rydell, D. Jocham, H. Diddens, S. Bown, G. Gregory, S. Montan, S. Andersson-Engels and S. Svanberg, Clinical multi-colour fluorescence imaging of malignant tumours – initial experience, *Acta Radiol.*, 1998, **39**, 2–9.

19. I. Wang, L.P. Clemente, R.M. Pratas, E. Cardoso, M.P. Clemente, S. Montan, S. Svanberg and K. Svanberg, Fluorescence diagnostics and kinetic studies in the head and neck region utilizing low-dose delta-aminolevulinic acid sensitization, *Cancer Lett.*, 1999, **135**, 11–19.

20. A. Kübler, T. Haase, M. Rheinwald, T. Barth and J. Muhling, Treatment of oral leukoplakia by topical application of 5-aminolevulinic acid, *Int. J. Oral Maxillofac. Surg.*, 1998, **27**, 466–469.

21. A. Sieron, G. Namyslowski, M. Misiolek, M. Adamek and A. Kawczyk-Krupka, Photodynamic therapy of premalignant lesions and local recurrence of laryngeal and hypopharyngeal cancers, *Eur. Arch. Otorhinolaryngol.*, 2001, **258**, 349–352.

22. A. Sieron, M. Adamek, A. Kawczyk-Krupka, S. Mazur and L. Ilewicz, Photodynamic therapy (PDT) using topically applied delta-aminolevulinic acid (ALA) for the treatment of oral leukoplakia, *J. Oral Pathol. Med.*, 2003, **32**, 330–336.

23. M. Malczewski, J. Birecki, K. Symonowicz, A. Bronowicz, P. Ziolkowski, J. Rabczynski, K. Baranska and P. Marszalik, Clinical applications of photodynamic therapy in the treatment of laryngeal lesions, *Otolaryngol. Pol.*, 1999, **53**, 671–675.

24. W.E. Grant, C. Hopper, A.J. MacRobert, P.M. Speight and S.G. Bown, Photodynamic therapy of oral cancer: photosensitisation with systemic aminolaevulinic acid, *Lancet*, 1993, **342**, 147–148.

25. K.F. Fan, C. Hopper, P.M. Speight, G. Buonaccorsi, A.J. MacRobert and S.G. Bown, Photodynamic therapy using 5-aminolevulinic acid for premalignant and malignant lesions of the oral cavity, *Cancer*, 1996, **78**, 1374–1383.

26. P. Wolf, E. Rieger and H. Kerl, Topical photodynamic therapy with endogenous porphyrins after application of 5-aminolevulinic acid. An alternative treatment modality for solar keratoses, superficial squamous cell carcinomas, and basal cell carcinomas?, *J. Am. Acad. Dermatol.*, 1993, **28**, 17–21.

27. R.M. Szeimies, S. Karrer, A. Sauerwald and M. Landthaler, Photodynamic therapy with topical application of 5-aminolevulinic acid in the treatment of actinic keratoses: an initial clinical study, *Dermatology*, 1996, **192**, 246–251.

28. R. Fink Puches, A. Hofer, J. Smolle, H. Kerl and P. Wolf, Primary clinical response and long-term follow-up of solar keratoses treated with topically applied 5-aminolevulinic acid and irradiation by different wave bands of light, *J. Photochem. Photobiol. B.*, 1997, **41**, 145–151.

29. E.W. Jeffes, J.L. McCullough, G.D. Weinstein, R. Kaplan, S.D. Glazer and J.R. Taylor, Photodynamic therapy of actinic keratoses with topical aminolevulinic acid hydrochloride and fluorescent blue light, *J. Am. Acad. Dermatol.*, 2001, **45**, 96–104.

30. M. Mehlmann, C.S. Betz, H. Stepp, S. Arbogast, R. Baumgartner, G. Grevers and A. Leunig, Fluorescence staining of laryngeal neoplasms after topical application of 5-aminolevulinic acid: preliminary results, *Lasers Surg. Med.*, 1999, **25**, 414–420.

31. A. Leunig, C.S. Betz, M. Mehlmann, H. Stepp, S. Arbogast, G. Grevers and R. Baumgartner, Detection of squamous cell carcinoma of the oral cavity by imaging 5-aminolevulinic acid-induced protoporphyrin IX fluorescence, *Laryngoscope*, 2000, **110**, 78–83.

32. C.S. Betz, H. Stepp, P. Janda, S. Arbogast, G. Grevers and R. Baumgartnerand A. Leunig, A comparative study of normal inspection, autofluorescence and 5-ALA-induced PPIX fluorescence for oral cancer diagnosis, *Int. J. Cancer*, 2002, **97**, 245–252.

33. W. Zheng, K.C. Soo, R. Sivanandan and M. Olivo, Detection of neoplasms in the oral cavity by digitized endoscopic imaging of 5-aminolevulinic acid-induced protoporphyrin IX fluorescence, *Int. J. Oncol.*, 2002, **21**, 763–768.

34. A. Leunig, C.S. Betz, M. Mehlmann, H. Stepp, S. Arbogast, G. Grevers and R. Baumgartner, A pilot series demonstrating fluorescence staining of laryngeal papilloma using 5-aminolevulinic acid, *Laryngoscope*, 2000, **110**, 1783–1785.

35. A. Leunig, M. Mehlmann, C.S. Betz, H. Stepp, S. Arbogast, G. Grevers and R. Baumgartner, Fluorescence staining of oral cancer using topical application of 5-aminolevulinic acid: fluorescence microscopic studies, *J. Photochem. Photobiol. B.*, 2001, **60**, 44–49.

36. K. Rick, R. Sroka, H. Stepp, M. Kriegmair, R.M. Huber, K. Jacob and R. Baumgartner, Pharmacokinetics of 5-aminolevulinic acid-induced protoporphyrin IX in skin and blood, *J. Photochem. Photobiol. B.*, 1997, **40**, 313–319.

37. D. Ormrod and B. Jarvis, Topical aminolevulinic acid HCl photodynamic therapy, *Am. J. Clin. Dermatol.*, 2000, **1**, 133–139.

38. D.J. Piacquadio, D.M. Chen, H.F. Farber, J.F. Fowler Jr., S.D. Glazer, J.J. Goodman, L.L. Hruza, E.W. Jeffes, M.R. Ling, T.J. Phillips, T.M. Rallis, R.K. Scher, C.R. Taylor and G.D. Weinstein, Photodynamic therapy with aminolevulinic acid topical solution and visible blue light in the treatment of multiple actinic keratoses of the face and scalp: investigator-blinded, phase 3, multicenter trials, *Arch. Dermatol.*, 2004, **140**, 41–46.

39. J.M. Gaullier, K. Berg, Q. Peng, H. Anholt, P.K. Selbo, L.W. Ma and J. Moan, Use of 5-aminolevulinic acid esters to improve photodynamic therapy on cells in culture, *Cancer Res.*, 1997, **57**, 1481–1486.

40. J. Kloek, W. Akkermans and G.M.J.B. van Henegouwen, Derivatives of 5-aminolevulinic acid for photodynamic therapy: enzymatic conversion into protoporphyrin, *Photochem. Photobiol.*, 1998, **67**, 150–154.

41. R.M. Szeimies, S. Karrer, S. Radakovic-Fijan, A. Tanew, P.G. Calzavara-Pinton, C. Zane, A. Sidoroff, M. Hempel, J. Ulrich, T. Proebstle, H. Meffert, M. Mulder, D. Salomon, H.C. Dittmar, J.W. Bauer, K. Kernland and L. Braathen, Photodynamic therapy using topical methyl 5-aminolevulinate compared with cryotherapy for actinic keratosis: a prospective, randomized study, *J. Am. Acad. Dermatol.*, 2002, **47**, 258–262.

42. M. Freeman, C. Vinciullo, D. Francis, L. Spelman, R. Nguyen, P. Fergin, K.E. Thai, D. Murrell, W. Weightman, C. Anderson, C. Reid, A. Watson and P. Foley, A comparison of photodynamic therapy using topical methyl aminolevulinate (Metvix) with single cycle cryotherapy in patients with actinic keratosis: a prospective, randomized study, *J. Dermatolog. Treat.*, 2003, **14**, 99–106.

43. D.M. Pariser, N.J. Lowe, D.M. Stewart, M.T. Jarratt, A.W. Lucky, R.J. Pariser and P.S. Yamauchi, Photodynamic therapy with topical methyl aminolevulinate for actinic keratosis: results of a prospective randomized multicenter trial, *J. Am. Acad. Dermatol.*, 2003, **48**, 227–232.

44. C.A. Morton, The emerging role of 5-ALA-PDT in dermatology: is PDT superior to standard treatments?, *J. Dermatolog. Treat.*, 2002, **13**(Suppl. 1), 25–29.

45. A.C. Kübler, M. Scheer and J.E. Zoller, Photodynamic therapy of head and neck cancer, *Onkologie*, 2001, **24**, 230–237.

46. A. Martin, W.D. Tope, J.M. Grevelink, J.C. Starr, J.L. Fewkes, T.J. Flotte, T.F. Deutsch and R.R. Anderson, Lack of selectivity of protoporphyrin IX fluorescence for basal cell carcinoma after topical application of 5-aminolevulinic acid: implications for photodynamic treatment, *Arch. Dermatol. Res.*, 1995, **287**, 665–674.

47. W. Beyer, Systems for light application and dosimetry in photodynamic therapy, *J. Photochem. Photobiol. B.*, 1996, **36**, 153–156.

48. W. Bäumler, *Lichtquellen*, in *Klinische. Fluoreszenzdiagnostik. und Photodynamische. Therapie, Szeimies*, R. -M., D. Jocham, M. Landthaler (eds), Blackwell Verlag GmbH, Berlin/Wien, 2003, 39–58.

49. J.G. Ofner, B. Bartl, S. Konig and W.F. Thumfart, Photodynamic therapy in selected cases at the ENT Clinic, Innsbruck: case reports, *J. Photochem. Photobiol. B.*, 1996, **36**, 185–187.

50. A.S.A. PhotoCure, Metvix® Data Sheet, 2003.

51. K. Lang, K.W. Schulte, T. Ruzicka and C. Fritsch, Aminolevulinic acid (Levulan) in photodynamic therapy of actinic keratoses, *Skin Therapy Lett.*, 2001, **6**, 1–2.

52. S. Lang, R. Baumgartner, R. Struck, A. Leunig, R. Gutmann and J. Feyh, Photodynamische Diagnostik und Therapie von Neoplasien der Gesichtshaut nach topischer Applikation von 5-Aminolävulinsäure, *Laryngorhinootologie*, 1995, **74**, 85–89.

53. P. Wolf, R. Fink-Puches, A. Reimann-Weber and H. Kerl, Development of malignant melanoma after repeated topical photodynamic therapy with 5-aminolevulinic acid at the exposed site, *Dermatology*, 1997, **194**, 53–54.

*Chapter 7*

# ALA-PDT in Gynecology

## Peter Hillemanns, Mathias Fehr and René Hornung

**Table of Contents**

## Abstract

Reasonable indications for a clinical application of ALA in Gynecology are varied: Most studies have been performed for the treatment of cervical or vulvar intraepithelial neoplasia (CIN and VIN). ALA-PDT provides a treatment option that is considerably less invasive than the standard procedures and after healing shows favorable cosmesis. Care must be taken, however, with regard to efficiency, which may be as low as 50% complete response rate. Another possible field for ALA-PDT is the ablation of endometrium or even removal of endometrial cancer, but is currently only recommendable for patients not eligible for surgery. ALA-based fluorescence diagnosis has been studied for intraoperative detection of endometriosis and peritoneal metastasis of ovarian cancer, where the sensitive detection and precise localization of micrometastases seems to be feasible.

## 7.1. Photodynamic Therapy (PDT) of Cervical Intraepithelial Neoplasia (CIN)

### 7.1.1. Introduction/Motivation

Cervical cancer is the second most common malignancy worldwide with about 500,000 new cancers diagnosed each year. Cervical intraepithelial neoplasia (CIN) is its precancerous condition localized at the squamocolumnar junction of the cervix uteri. The prevalence of CIN has increased during recent decades, especially among younger women. Screening for cervical carcinoma and its precursors relies upon mass screening, including mainly cervical cytology and more recently testing for Human Papillomavirus (HPV) DNA. Because of recently established screening programs in developed countries, the incidence of cervical cancer has decreased by about 50–60%. Various treatment methods already exist for CIN, all of which have a proven satisfactory efficacy: the differences are mainly complication rates and costs. Excision techniques such as cold knife conization or loop electrosurgical procedure have the advantage of providing valuable pathological specimens. However, they are more commonly associated with postoperative bleeding. Destructive techniques like cryotherapy or laser ablation are frequently handicapped by unsatisfactory postoperative colposcopy or, in the latter, increased costs. The major disadvantage common to all methods is the substantial destruction or excision of cervical stroma. Depending on the dimension of the cone or the extent of the destruction, current methods may cause cervical incompetence with premature deliveries and low-birth-weight babies or, adversely, scar stricture with increased risk of infertility and caesarean section.[1,2]

5-ALA-mediated PDT provides an alternative therapy, avoiding complications like postoperative bleeding and cervical stenosis with consequent infertility and cervical incompetence leading to premature deliveries and low birth weight, respectively.

### 7.1.2. Approval Status

Photodynamic therapy of CIN is an investigational procedure and should so far be performed only within clinical studies.

### 7.1.3. Procedure

Five to ten milliliters of a 10% w/v solution of 5-ALA dissolved in either 0.9% sodium chloride or Lutrol F-127 (thermolabile bioadhesive 19% poloxamer 407 gel) are topically applied to the portio uteri using an appropriately sized cervical cap (Cervix Adapter, Wisap, Sauerlach, Germany). After 4 to 8 h, the cervical cap is removed and fluorescence is excited using a filtered short-arc xenon lamp at 375–440 nm (violet light) and a power output of 200 mW (D-Light, Karl Storz GmbH, Tuttlingen, Germany). 5-ALA-induced protoporphyrin IX (PpIX) can be detected by visual inspection or video imaging using a yellow filter to view the emission in the range 470–700 nm (Figure 1). For PDT, laser light at 635 nm wavelength (maximum output power 1.9–2.1 W) is coupled to a 600 μm flexible optical fiber and imaged onto the tissue to produce a homogeneously irradiated field of the whole portio uteri. For the treatment of the endocervical canal, a cylindrical light applicator (diameter 4 mm, length 20 mm) with a backscattering surface for light homogenization can be used. Much better is the use of a specifically designed portio applicator for the simultaneous illumination of the endo- and ectocervical canal. A total irradiation with a dose of 100 J cm$^{-2}$, and with irradiance limited to 150 MW cm$^{-2}$ to avoid thermal effects, is recommended (Figure 2). Usually, there is no local anesthesia necessary. In only a few cases, PDT of the

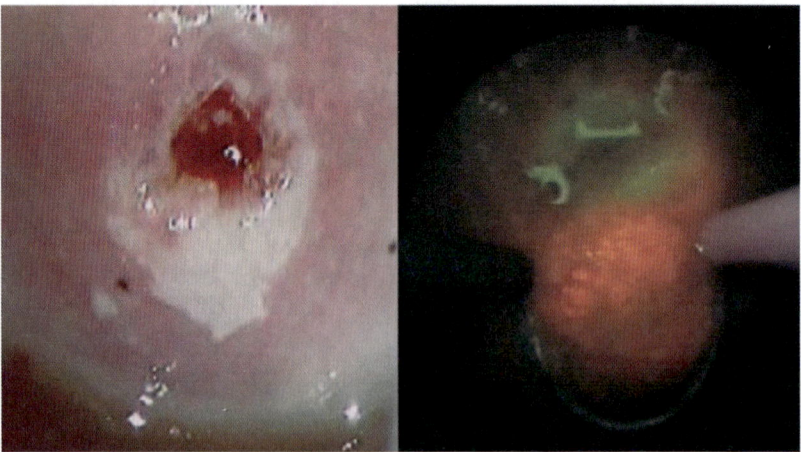

**Figure 1.** Right: CIN 2 lesion visualized by its characteristic red protoporphyrin IX fluorescence at the posterior lip of the portio uteri. Left: white lesion seen in colposcopy after topical application of acetic acid.

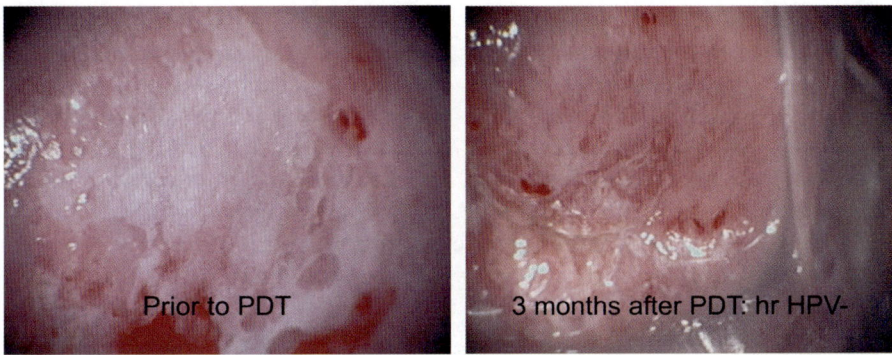

Prior to PDT

3 months after PDT: hr HPV-

**Figure 2.** CIN 3 lesion before and after ALA-induced protoporphyrin IX-mediated PDT.

endocervical canal caused a cramping sensation with the need for intra/para-cervical Xylocain injection.

### 7.1.4. Advantages and Limitations

Most female patients are still in their reproductive years and seek a nonsurgical treatment option for CIN in order to avoid the risk of cervical incompetence and preterm delivery. For CIN lesions, PDT studies have shown equivocal outcomes to date. Although Hillemanns et al. could verify specific protoporphyrin IX fluorescence of CIN after local administration of 1% aqueous solution of 5-ALA,[3,4] photodynamic therapy after topical application of 10–20% ALA showed a regression of CIN lesion size but did not eradicate the CIN lesion entirely. Similar results were presented in two other studies.[5,6] However, PDT with 5-ALA applied to the cervix as a 12% aqueous solution was efficient in improving the grading of the PAP smears in 19 out of 20 CIN patients, and in the eradication of 80% of the cervical HPV infections.[7] The same group demonstrated that ALA-PDT (91%) was nearly as effective as conization (100%) in the clearance of the CIN 2 lesions (with 11 cases in each group) and with respect to the HPV clearance.[8] ALA-derivatives such as hexyl-aminolevulinate are currently being investigated in preclinical and phase I–II studies with promising prospects.

## 7.2. Photodynamic Therapy of Vulvar Intraepithelial Neoplasia

### 7.2.1. Introduction/Motivation

Vulvar intraepithelial neoplasia (VIN) describes dysplasia of the squamous epithelium of the vulva. VIN 3 is thought to be a precursor of squamous cell vulvar cancer and is identical to the term, *carcinoma in situ* or Morbus Bowen.[9] The incidence of VIN 3 is increasing, while the mean age of affected patients is decreasing: 28–52% of patients with VIN 3 are aged 40 years or younger.[10] Reports on the natural history of VIN 3 indicate progression to invasive cancer in at least 25% of untreated and 5% of treated patients. VIN is often multifocal, which makes VIN a difficult disease to treat with a high rate of recurrence. With the appreciation that progression to invasive disease is uncommon in conservatively treated patients and that younger women are increasingly afflicted with this disease, more conservative surgical treatments, such as wide local excision and laser evaporation, have been adopted. Nevertheless, surgical excision in patients with extensive disease is both mutilating and in about 25% of the cases is associated with persistent disease. The preservation of normal vulvar appearance and function is of paramount importance to young women. Although $CO_2$ laser ablation may appear to be effective initially in nonhair-bearing sites, recurrence rates of 79% have been reported after 10 years of follow-up.[11] Of the topical therapies used to manage VIN, 5-fluorouracil (5-FU) is poorly tolerated because it causes severe pain and it is unpredictable in its effectiveness. Imiquimod, an immune-response modifier, has shown conflicting therapeutic results.[12]

### 7.2.2. Approval Status

At present, treatment of cancer precursors in the lower female genital tract with ALA-PDT is an investigational procedure and should be performed only within clinical studies.

### 7.2.3. Procedure

Prior to drug application, the vulvar skin is washed with an aqueous 0.4% chlorhexidine solution to reduce the bacterial flora. Ten grams of a 10% ALA gel is spread over the entire vulva. The vulva is covered by a nonadherent dressing (Tegaderm, Johnson and Johnson Products Inc., Arlington, NJ). Patients are asked to lie in bed for 3–5 h. Fluorescence of the lesions can be evaluated using a Storz D-Light hysteroscope (D-Light; Karl Storz, Inc., Tuttlingen, Germany) with blue light at 380–440 nm. The skin is viewed through an observation long-pass filter to select an emitted wavelength range of 440–700 nm (Figures 3 and 4). For PDT, the dye laser is tuned to emit 635 nm light and is coupled to a 600 μm quartz fiber ending in a front lens for homogenous illumination (Medlight SA, Lausanne, Switzerland). To avoid thermal effects, the irradiance at the surface of the skin is kept below 100 mW $cm^{-2}$ and an optical dose of 100–120 J $cm^{-2}$ is administered.

During irradiation, pain with a burning and stinging sensation may arise. Although 60% of patients can be treated without anesthesia, interruption of illumination for regional anesthesia may be necessary. Pain is not associated with treatment failure but more intense pain may correlate with larger skin areas and subsequently more pronounced swelling.

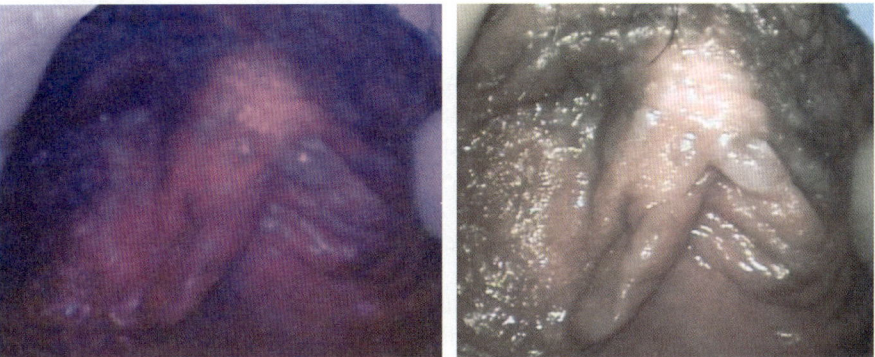

**Figure 3.** Left: *In vivo* fluorescence of VIN 3 on the prepuce of the clitoris and right labium majus 120 min after topical application of 10% ALA-gel. Right: corresponding picture using white light. Fluorescence of the vestibule is still weak.

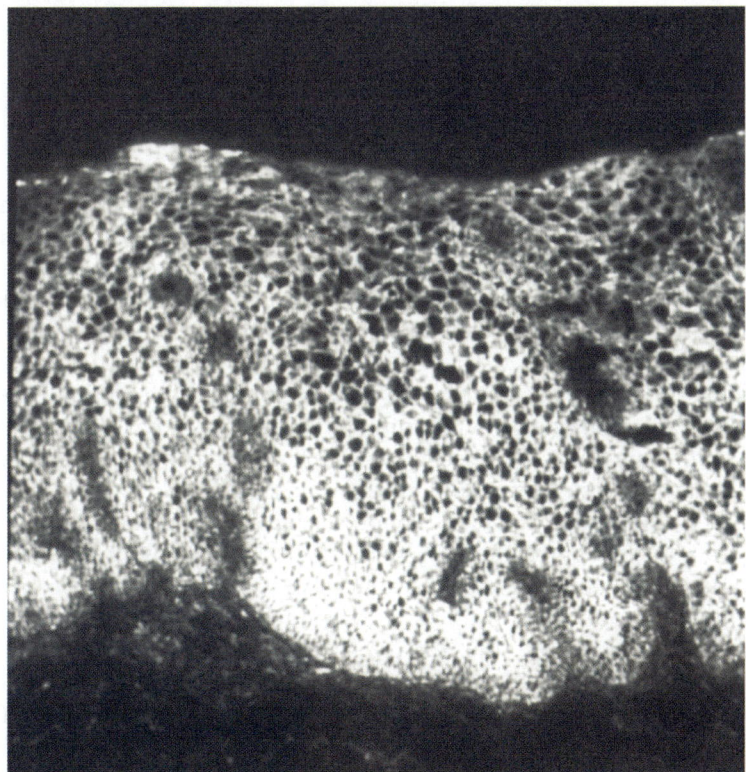

**Figure 4.** Fluorescence micrograph of a poorly keratinized VIN 3. Fluorescence is limited to the epithelial layer.

*7.2.4. Advantages and Limitations*

The clearance rate of VIN after ALA-PDT is about 40–60%. Martin-Hirsch *et al.* observed a clearance rate of 40% in eight patients using 100 J cm$^{-2}$ nonlaser light.[13] Peng *et al.*[14] reported a complete response rate of 50% in 18 patients and Hillemanns *et al.* have reported a complete clearance rate of about 50% in 19 patients with VIN 3.[15] Fehr *et al.* have reported a 60% clearance rate among 21 patients.[16] These results are similar to those of laser evaporation or local excision mostly because of the multifocal nature of VIN disease. The results suggest that PDT of VIN using ALA may have a similar efficacy to those of conventional treatment modalities while involving a shorter healing time. Furthermore, preservation of normal morphology was observed to be excellent with ALA-PDT. Although predictors of response must be defined more clearly and further refinements of the technique are required, photodynamic destruction of these lesions offers unique advantages and deserves further investigation. PDT using topical ALA did not induce ulcers or scar formation and healing times after PDT are far shorter than those published for laser evaporation.

# 7.3. Photodynamic Therapy of Endometrium

*7.3.1. Introduction/Motivation*

Endometrial ablation is an alternative to hysterectomy for the treatment of dysfunctional uterine bleeding. Various techniques have been developed, including hysteroscopic techniques based on thermal coagulation of the endometrial layer by either electrocoagulation[17–19] or laser.[20] Unsatisfactory bleeding control of dysfunctional uterine bleeding, risk of uterine perforation and fluid overload, as well as the need for general anesthesia, led to the development of alternative techniques, such as photodynamic endometrial ablation. The feasibility of selective endometrial ablation by means of PDT has been reported in several animal models.[21–32] Special devices have been developed to provide homogenous light distribution in the human uterine cavity.[29,33,34] A few studies using preformed porphyrin-based photosensitizers in human subjects have shown that the endometrium can successfully be photosensitized and that photodynamic endometrial ablation is feasible.[29,35–37] Diffusion properties of ALA following topical application allow sufficient photosensitization of the endomyometrial junction, the uterine layer underneath the endometrium responsible for endometrial regeneration. Destruction of this layer is critical for a long-lasting PDT effect and bleeding control. Excellent diffusion properties of ALA compensate in part for the limited penetration depth of red excitation light at 635 nm. Photodynamic endometrial ablation has also been suggested for the treatment of endometrial cancer.[35–38] Surgery is currently the mainstay of the treatment of endometrial cancer. However, nonsurgical treatments such as PDT may be an option for patients who are not eligible for surgery.

### 7.3.2. Approval Status

Photodynamic endometrial ablation is an investigational procedure and should so far be performed only within clinical studies.

### 7.3.3. Procedure

Patients can be recruited into an endometrial ablation study only following written informed consent obtained according to local regulations. Malignancy must be ruled out by diagnostic hysteroscopy and curettage. Patients with a history of porphyria or the desire to preserve fertility are not eligible for PDT of the endometrium. Premenopausal patients are treated in the proliferative phase of the menstrual cycle. According to Wyss et al.[29,39,40] 400 mg mL$^{-1}$ ALA (ASAT, Zug, Switzerland) is buffered to pH 5.5 with 10 N NaOH. A volume of 1.5–2.0 mL is injected into the uterine cavity through a hysterosalpingography catheter (Sholkoff Balloon Catheter, COOK, Switzerland) using a sterile 0.22 mm filter unit (Millex-GS, Milipore Products Division, Bedford, MA). Three to six hours following ALA administration an optical dose of 160 J cm$^{-2}$ at a wavelength of 635 nm is applied to the endometrium using an intracavitary-positioned, reflecting balloon-light diffuser (Medlight SA, Lausanne, Switzerland). Alternatively, an intrauterine light probe consisting of three flexible optical fibers, each of which contains a cylindrical light diffuser converging to one bundle resembling the shape of the uterine cavity, may be used.[33] The light dose may be fractionated in sequences of 5 min of illumination followed by nonirradiation gaps of 2 min. ALA-PDT can be performed without anesthesia in most patients. (0.2 mg alfentanil hydrochloride i.v. as analgesia may be applied for cervical dilatation.) Some patients, however, need a spinal-block or even total anesthesia. Antibiotic prophylaxis is administered to all patients.

## 7.4. Fluorescence Diagnosis of Endometriosis

### 7.4.1. Introduction/Motivation

Endometriosis is a condition in which endometrium is found in locations outside the uterus. This misplaced tissue may be found on the ovaries, uterus, bowel, bladder, utero-sacral ligaments, or peritoneum. Endometriosis is a frequent clinical problem for women of reproductive age that can markedly influence both reproductive prognosis and quality of life. Typically, this disorder causes dysmenorrhea, chronic or cyclic pelvic pain and infertility, resulting in prolonged medical treatment and repeated hospitalizations for surgery.

The main diagnostic technique remains laparoscopy. The intraoperative diagnosis of endometriosis by conventional laparoscopy is often difficult and inappropriate. As well as classical dark red or brown lesions, endometriosis has a wide range of manifestations, frequently nonpigmented and presenting as small vesicles, nodular lesions, or plaque-type implants. The diagnosis of endometriosis thus depends on the ability of the surgeon to identify unspecific nonpigmented peritoneal changes as indicative of endometriosis, so that the diagnosis can then be confirmed by biopsy and histological examination. The potential for selective accumulation of a photosensitizing drug may be useful in both the diagnosis and the treatment of endometriosis when it is subsequently exposed to photoactivating light. Earlier studies have shown the destruction of endometrial tissue in animal models by photodynamic therapy with topically and systemically applied photosensitizers.[21–27,41–47,30–31] Malik et al.[48] and Hillemanns et al.[49] simultaneously investigated the diagnostic potential of protoporphyrin IX-based fluorescence after systemic administration of ALA for patients with endometriosis. Both groups found that fluorescence diagnosis of endometriosis was feasible and that white or red lesions of peritoneal endometriosis achieved a significantly higher level of PpIX fluorescence after administration of ALA than did normal peritoneum. Usually, these white lesions are more difficult to detect laparoscopically than are pigmented lesions. Nodular and pigmented peritoneal endometriotic lesions and ovarian endometriomas yielded negative fluorescence results and could not be detected by this fluorescence approach. Rarely, tiny false-positive fluorescent spots were visible on video inspection. However, false positive results were related to conventional histologic criteria for endometriosis. Occasionally, in patients with laparoscopically typical disease, biopsy may yield only histologically negative tissue. Surprisingly, the fimbrial mucosa showed intense fluorescence after oral ALA administration.[49] This is of some concern, because illumination of porphyrin may cause phototoxic damage. Hillemanns et al. concluded that at present the clinical implementation of ALA-induced photosensitization for the detection of endometriosis is rather questionable. Young women of reproductive age with problems of infertility are an important subgroup of patients with endometriosis. Because preservation of the tubal and fimbrial mucosa is crucial for fertility, these patients may have to be excluded from ALA-induced porphyrin fluorescence detection for endometriosis until more data are available to assess phototoxic damage to the fallopian tubes. Data from Malik et al., however, seem not to document any damaging effect of PDT on the fallopian tubes.[48]

### 7.4.2. Approval Status

Fluorescence-based endoscopy for endometriosis is an investigational procedure and should so far be performed only within clinical studies.

### 7.4.3. Procedure

Patients can be enrolled into a fluorescence study following written informed consent as per local regulations. Ten mg per kg body weight dissolved ALA is given orally (*e.g.*, in apple juice) 4–6 h prior to surgery. Patients are required to avoid direct sunlight exposure for 48 h following administration of ALA. Laparoscopy is performed according to standard surgical guidelines. A 300 W xenon short arc lamp (380–440 nm) (D-Light, Karl Storz GmbH, Tuttlingen, Germany) is used for excitation in the blue light mode as well as for the conventional white light laparoscopy. A customized light fiber connects the light source with the laparoscope. A video endoscopy system is used comprising special cameras with built-in color balancing systems provided, for example, by Karl Storz can be employed to enhance the sensitivity in the fluorescence mode. Endometriosis is confirmed by biopsy and histology.

## 7.5. Fluorescence Diagnosis of Ovarian Cancer

### 7.5.1. Introduction/Motivation

Standard treatment of ovarian cancer consists of surgical staging and tumor debulking, followed by chemotherapy. Second-look laparotomy or laparoscopy are used to assess the need for further treatment after first-line chemotherapy. Effective second-line chemotherapy regimes for ovarian cancer have evolved in the last decade. Fifty percent of women die of recurrent disease despite negative second look.[50–54] Thus, there is a strong clinical importance of detecting residual disease after completed first-line chemotherapy. The diagnostic potential of ALA-mediated laser-induced fluorescence is being evaluated as an improvement of laparoscopy for the detection of occult intra-abdominal cancer nodules. Preliminary studies in animals revealed promising results regarding the potential use of fluorescence detection of very small intra-abdominal ovarian cancer spreads.[55–59] Figure 5 shows ALA-mediated protoporphyrin IX fluorescence of ovarian cancer nodules in the chick chorion allantois membrane model (CAM). ALA-mediated protoporphyrin IX fluorescence detection of intra-abdominal micrometastases not visible by the naked eye or by laparoscope may therefore be of clinical interest.

Laparoscopic fluorescence detection of endogenous PpIX after intraperitoneal application of ALA in humans was performed by Loning et al.[60] They showed that the procedure is feasible in humans and that it may provide a higher sensitivity for finding peritoneal metastases of epithelial ovarian carcinoma in comparison to conventional laparoscopy. In an earlier pilot study, Wierrani et al.[61] demonstrated that laparoscopy-guided PDT of ovarian cancer recurrences seems feasible.

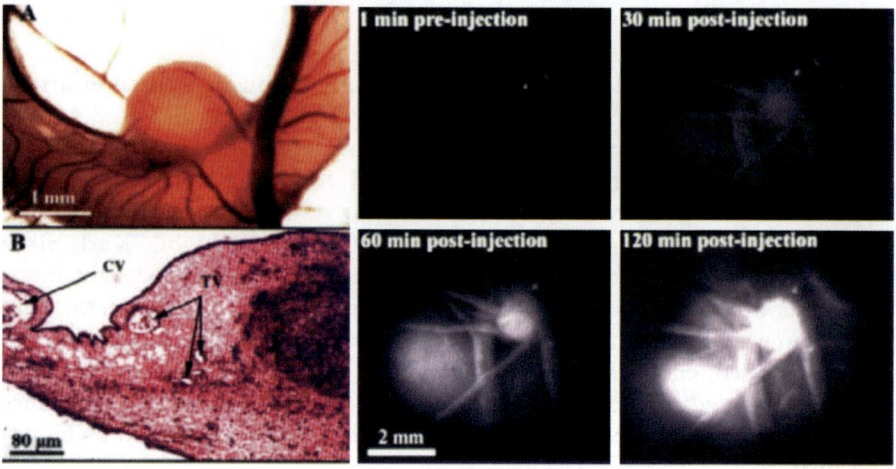

**Figure 5.** Ovarian cancer nodules (Panel A) are grown in the chorion allantois membrane (CAM) model of chicken embryos. Panel B shows the corresponding histology. The black and white micrographs show the *in vivo* fluorescence of ovarian cancer nodules in the CAM model as a function of time following systemic application of ALA. While only background fluorescence can be detected prior to injection, the nodules emit strong fluorescence 1–2 h later. This demonstrates that ALA is absorbed by the peritoneum, is transported by the blood to the tumor nodule, is internalized into the tumor cells, and is metabolized to active Pp IX.

### 7.5.2. Approval Status

There is limited anecdotal data on fluorescence-based endoscopy for ovarian cancer spreads. Therefore, the procedure should be performed only within clinical studies.

### 7.5.3. Procedure

Patients can be enrolled into a fluorescence-based laparoscopy study following written informed consent as per local regulations. Five hours prior to laparoscopy Loning et al.[60] applied intraperitoneally a sterile 1% ALA solution (pH 6.5) (Medac, Wedel, Germany). Thirty mg kg$^{-1}$ body weight was administered via short infusion by a needle placed subcostally on the left side of the body. (A simpler method of application is the oral application of 10 mg ALA kg$^{-1}$ body weight such as for the detection of endometriosis.) After ALA application, the patients were shielded from direct light exposure for 36 h to avoid phototoxic reactions. Five hours after ALA sensitization, fluorescence detection is performed (via laparoscopy or open surgery) using a standard endoscope combined with a blue light source emitting at wavelengths of 350–440 nm to induce PpIX fluorescence (Figure 6). Photobleaching in the white light mode may be avoided by using a special filter restricting the wavelengths to 455–700 nm. Malignancy must be confirmed by histology following biopsy of suspect tissues.

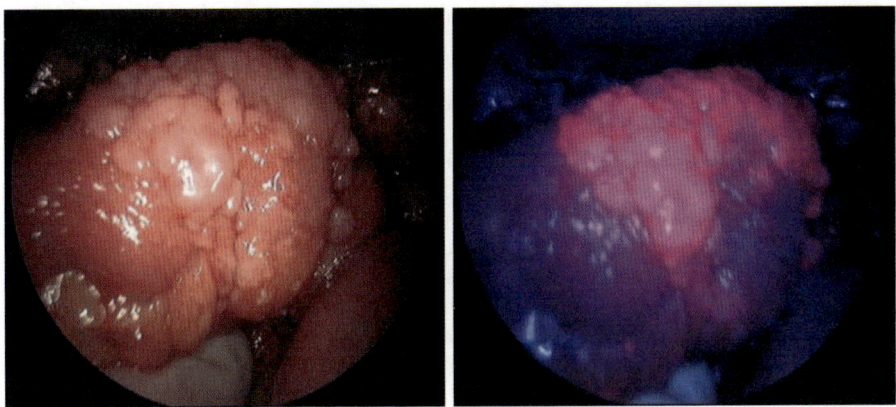

**Figure 6.**  Peritoneal carcinosis on colon visualized by ALA-induced protoporphyrin IX-mediated fluorescence diagnosis.

### 7.6. Advantages/Limitations (PDT of Endometrium, FD of Endometriosis, FD of Ovarian Cancer)

ALA-mediated fluorescence-based diagnosis of benign and malignant diffuse peritoneal disease is an attractive concept from the technical and scientific point of view. However, the use of these complex techniques requires additional training and equipment, restricting its use to major medical centers. In the case of endometriosis, a very common condition, most physicians would initiate treatment based on typical anamnestic and clinical signs even with a negative laparoscopy. On the other hand, experienced surgeons would easily be able to diagnose the disease even if its presentation were limited to some minimal changes. Second-look surgery in the case of ovarian cancers has been abandoned by most centers because of the absence of proven benefit to the patient if chemotherapy is indicated very early in asymptomatic small recurrences. However, innovative approaches in the treatment of recurrent ovarian cancer may, even in the nearer future, result in their early diagnosis with consecutive treatment.

Similarly, ALA-mediated endometrial ablation has some very convincing aspects, especially its low risk of perforation and other surgical complications. The technique, however, must compete with several well-established methods that employ only either a medical or an instrumental intervention. PDT, on the other hand, requires training in both the application of a drug and the use of lasers. Many physicians may shy away from this rather complex but exciting concept.

## References

1. M.A. Bigrigg, B.W. Codling, P. Pearson, M.D. Read and G.R. Swingler, Pregnancy after cervical loop diathermy, *Lancet*, 1991, **337**, 119.

2. P. Mathevet, E. Chemali, M. Roy and D. Dargent, Long-term outcome of a randomized study comparing three techniques of conization: cold knife, laser, and LEEP, *Eur. J. Obstet. Gynecol. Reprod. Biol.*, 2003, **106**, 214–218.

3. P. Hillemanns, M. Korell, M. Schmitt-Sody, R. Baumgartner, W. Beyer, R. Kimmig, M. Untch and H. Hepp, Photodynamic therapy in women with cervical intraepithelial neoplasia using topically applied 5-aminolevulinic acid, *Int. J. Cancer*, 1999, **81**, 34–38.

4. P. Hillemanns, H. Weingandt, R. Baumgartner, J. Diebold, W. Xiang and H. Stepp, Photodetection of cervical intraepithelial neoplasia using 5-aminolevulinic-induced porphyrin fluorescence, *Cancer*, 2000, **88**, 2275–2282.

5. K.A. Keefe, Y. Tadir, B. Tromberg, M. Berns, K. Osann, R. Hashad and B.J. Monk, Photodynamic therapy of high-grade cervical intraepithelial neoplasia with 5-aminolevulinic acid, *Lasers Surg. Med.*, 2002, **31**, 289–293.

6. A.A. Barnett, J.C. Haller, F. Cairnduff, G. Lane, S.B. Brown and D.J. Roberts, A randomised, double-blind, placebo-controlled trial of photodynamic therapy using 5-aminolaevulinic acid for the treatment of cervical intraepithelial neoplasia, *Int. J. Cancer*, 2003, **103**, 829–832.

7. F. Wierrani, A. Kubin, R. Jindra, M. Henry, K. Gharehbaghi, W. Grin, J. Soltz-Szotz, G. Alth and W. Grunberger, 5-Aminolevulinic acid-mediated photo-dynamic therapy of intraepithelial neoplasia and human papillomavirus of the uterine cervix – A new experimental approach [in process citation], *Cancer Detect. Prev.*, 1999, **23**, 351–355.

8. K. Bodner, B. Bodner-Adler, F. Wierrani, A. Kubin, J. Szolts-Szolts, B. Spangler and W. Grunberger, Cold-knife conization versus photodynamic therapy with topical 5-aminolevulinic acid (5-ALA) in cervical intraepithelial neoplasia (CIN) II with associated human papillomavirus infection: a comparison of preliminary results, *Anticancer Res.*, 2003, **23**, 1785–1788.

9. C.M. Ridley, O. Frankman, I.S.C. Jones, S.H. Pincusand and E.J. Wilkinson, (Letter to the editor) New nomenclature for vulvar disease: international society for the study of vulvar disease, *Lancet*, 1989, **20**, 495–496.

10. R.J. Cardosi, J.J. Bomalaski and M.S. Hoffman, Diagnosis and management of vulvar and vaginal intraepithelial neoplasia, *Obstet. Gynecol. Clin. North Am.*, 2001, **28**, 685–702.

11. A. Rodolakis, E. Diakomanolis, G. Vlachos, T. Iconomou, A. Protopappas, C. Stefanidis, H. Elsheikh and S. Michalas, Vulvar intraepithelial neoplasia (VIN) – diagnostic and therapeutic challenges, *Eur. J. Gynaecol. Oncol.*, 2003, **24**, 317–322.

12. C.J. Jayne and R.H. Kaufman, Treatment of vulvar intraepithelial neoplasia 2/3 with imiquimod, *J. Reprod. Med.*, 2002, **47**, 395–398.

13. P.L. Martin-Hirsch, C. Whitehurst, C.H. Buckley, J.V. Moore and H.C. Kitchener, Photodynamic treatment for lower genital tract intraepithelial neoplasia, *Lancet*, 1998, **351**, 1403.

14. Q. Peng, T. Warloe, K. Berg, J. Moan, M. Kongshaug, K.E. Giercksky and J.M. Nesland, 5-Aminolevulinic acid-based photodynamic therapy. Clinical re-search and future challenges, *Cancer*, 1997, **79**, 2282–2308.

15. P. Hillemanns, M. Untch, C. Dannecker, R. Baumgartner, H. Stepp, J. Diebold, H. Weingandt, F. Pröve and M. Korell, Photodynamic therapy of vulvar intraepithelial neoplasia using 5-aminolevulinic acid, *Int. J. Cancer*, 2000, **85**, 649–653.

16. M.K. Fehr, R. Hornung, V.A. Schwarz, R. Simeon, U. Haller and P. Wyss, Photodynamic therapy of vulvar intraepithelial neoplasia III using topically applied 5-aminolevulinic acid, *Gyn. Oncol.*, 2001, **80**, 62–66.

17. J.F. Daniell, B.R. Kurtz and R.W. Ke, Hysteroscopic endometrial ablation using the rollerball electrode, *Obstet. Gynecol.*, 1992, **80**, 329–332.

18. A. DeCherney and M.L. Polan, Hysteroscopic management of intrauterine lesions and intractable uterine bleeding, *Obstet. Gynecol.*, 1983, **61**, 392–397.

19. S.P. Serden and P.G. Brooks, Treatment of abnormal uterine bleeding with the gynecologic resectoscope, *J. Reprod. Med.*, 1991, **36**, 697–699.

20. M.H. Goldrath, T.A. Fuller and S. Segal, Laser photovaporization of endometrium for the treatment of menorrhagia, *Am. J. Obstet. Gynecol.*, 1981, **140**, 14–19.

21. N. Bhatta, R.R. Anderson, T. Flotte, I. Schiff, T. Hasan and N.S. Nishioka, Endometrial ablation by means of photodynamic therapy with photofrin II, *Am. J. Obstet. Gynecol.*, 1992, **167**, 1856–1863.

22. M.K. Fehr, B.J. Tromberg, L.O. Svaasand, P. Ngo, M.W. Berns and Y. Tadir, Structural and functional effects of endometrial photodynamic therapy in a rat model, *Am. J. Obstet. Gynecol.*, 1996, **175**, 115–121.

23. R.A. Steiner, B.J. Tromberg, P. Wyss, T. Krasieva, N. Chandanani, J. McCullough, M.W. Berns and Y. Tadir, Rat reproductive performance following photodynamic therapy with topically administered Photofrin, *Hum. Reprod.*, 1995, **10**, 227–233.

24. R.A. Steiner, Y. Tadir, B.J. Tromberg, T. Krasieva, A.T. Ghazains, P. Wyss and M.W. Berns, Photosensitization of the rat endometrium following 5-aminolevulinic acid induced photodynamic therapy, *Lasers Surg. Med.*, 1996, **18**, 301–308.

25. P. Wyss, B.J. Tromberg, M.T. Wyss, T. Krasieva, M. Schell, M.W. Berns and Y. Tadir, Photodynamic destruction of endometrial tissue with topical 5-amino-levulinic acid in rats and rabbits, *Am. J. Obstet. Gynecol.*, 1994, **171**, 1176–1183.

26. P. Wyss, Y. Tadir, B.J. Tromberg, L. Liaw, T. Krasieva and M.W. Berns, Benzoporphyrin derivative: a potent photosensitizer for photodynamic destruction of rabbit endometrium, *Obstet. Gynecol.*, 1994, **84**, 409–414.

27. P. Wyss, L.O. Svaasand, Y. Tadir, U. Haller, M.W. Berns, M.T. Wyss and B.J. Tromberg, Photomedicine of the endometrium: experimental concepts, *Hum. Reprod.*, 1995, **10**, 221–226.

28. J.Z. Yang, D.A. Van Vugt, J.C. Kennedy and R.L. Reid, Evidence of lasting functional destruction of the rat endometrium after 5-aminolevulinic acid-induced photodynamic ablation: prevention of implantation, *Am. J. Obstet. Gynecol.*, 1993, **168**, 995–1001.

29. P. Wyss, M. Fehr, B.H. Van den and U. Haller, Feasibility of photodynamic endometrial ablation without anesthesia, *Int. J. Gynaecol. Obstet.*, 1998, **60**, 287–288.

30. G.B. Rocklin, H.G. Kelly, S.C. Anderson, L.E. Edwards, R.J. Gimpelson and R.E. Perez, Photodynamic therapy of rat endometrium sensitized with tin ethyl etiopurpurin, *J. Am. Assoc. Gynecol. Laparosc.*, 1996, **3**, 561–570.

31. D.A. Van Vugt, A. Krzemien, B.N. Roy, W.A. Fletcher, W. Foster, S. Lundahl, S.L. Marcus and R.L. Reid, Photodynamic endometrial ablation in the nonhuman primate, *J. Soc. Gynecol. Investig.*, 2000, **7**, 125–130.

32. P. Mhawech, A. Renaud, C. Sene, F. Ludicke, F. Herrmann, I. Szalay-Quinodoz, B.H. Van den and A.L. Major, High efficacy of photodynamic therapy on rat endometrium after systemic administration of benzoporphyrin derivative monoacid ring A, *Hum. Reprod.*, 2003, **18**, 1707–1711.

33. Y. Tadir, R. Hornung, T.H. Pham and B.J. Tromberg, Intrauterine light probe for photodynamic ablation therapy, *Obstet. Gynecol.*, 1999, **93**, 299–303.
34. P.J. Dwyer, W.M. White, R.L. Fabian and R.R. Anderson, Optical integrating balloon device for photodynamic therapy, *Lasers Surg. Med.*, 2000, **26**, 58–66.
35. R. Hornung, M.K. Fehr, B.J. Tromberg, A. Major, T.B. Krasieva, M.W. Berns and Y. Tadir, Uptake of the photosensitizer benzoporphyrin derivative in human endometrium after topical application in vivo, *J. Am. Assoc. Gynecol. Laparosc.*, 1998, **5**, 367–374.
36. J.C. Kennedy, S.L. Marcus and R.H. Pottier, Photodynamic therapy (PDT) and photodiagnosis (PD) using endogenous photosensitization induced by 5-amino-levulinic acid (ALA):mechanisms and clinical results, *J. Clin. Laser Med. Surg.*, 1996, **14**, 289–304.
37. S.L. Marcus, R.S. Sobel, A.L. Golub, R.L. Carroll, S. Lundahl and D.G. Shulman, Photodynamic therapy (PDT) and photodiagnosis (PD) using endogenous photo-sensitization induced by 5-aminolevulinic acid (ALA):current clinical and deve-lopment status, *J. Clin. Laser Med. Surg.*, 1996, **14**, 59–66.
38. H. Koren and G. Alth, Photodynamic therapy in gynaecologic cancer, *J. Photo-chem. Photobiol. B*, 1996, **36**, 189–191.
39. P. Wyss, R. Caduff, Y. Tadir, A. Degen, G. Wagnieres, V. Schwarz, U. Haller and M. Fehr, Photodynamic endometrial ablation: morphological study, *Lasers Surg. Med.*, 2003, **32**, 305–309.
40. A.F. Degen, T. Gabrecht, L. Mosimann, M.K. Fehr, R. Hornung, V.A. Schwarz, Y. Tadir, R.A. Steiner, G. Wagnieres and P. Wyss, Photodynamic endometrial ablation for the treatment of dysfunctional uterine bleeding: a preliminary report, *Lasers Surg. Med.*, 2004, **34**, 1–4.
41. J.A. Chapman, Y. Tadir, B.J. Tromberg, K. Yu, A. Manetta, C.H. Sun and M.W. Berns, Effect of administration route and estrogen manipulation on endome-trial uptake of Photofrin porfimer sodium, *Am. J. Obstet. Gynecol.*, 1993, **168**, 685–692.
42. M. Fortin, M. Lepine, M. Page, K. Osteen, B. Massie, P. Hugo and A.M. Steff, An improved mouse model for endometriosis allows noninvasive assessment of lesion implantation and development, *Fertil. Steril.*, 2003, **80**(Suppl 2), 832–838.
43. E. Malik, A. Meyhofer-Malik, C. Berg, W. Bohm, K. Kunzi-Rapp, K. Diedrich and A. Ruck, Fluorescence diagnosis of endometriosis on the chorioallantoic membrane using 5-aminolaevulinic acid, *Hum. Reprod.*, 2000, **15**, 584–588.
44. B.N. Roy, D.A. Van Vugt, G.E. Weagle, R.H. Pottier and R.L. Reid, Effect of 5-aminolevulinic acid dose and estrogen on protoporphyrin IX concentrations in the rat uterus, *J. Soc. Gynecol. Investig.*, 1997, **4**, 40–46.
45. P. Wyss, R. Steiner, L.H. Liaw, M.T. Wyss, A. Ghazarians, M.W. Berns, B.J. Tromberg and Y. Tadir, Regeneration processes in rabbit endometrium: a photodynamic therapy model, *Hum. Reprod.*, 1996, **11**, 1992–1997.
46. J.Z. Yang, D.A. Van Vugt, J.C. Kennedy and R.L. Reid, Intrauterine 5-amino-levulinic acid induces selective fluorescence and photodynamic ablation of the rat endometrium, *Photochem. Photobiol.*, 1993, **57**, 803–807.
47. J.Z. Yang, D.A. Van Vugt, B.N. Roy, J.C. Kennedy, W.G. Foster and R.L. Reid, Intrauterine 5-aminolevulinic acid induces selective endometrial fluorescence in the rhesus and cynomolgus monkey, *J. Soc. Gynecol. Investig.*, 1996, **3**, 152–157.
48. E. Malik, C. Berg, A. Meyhofer-Malik, O. Buchweitz, P. Moubayed and K. Diedrich, Fluorescence diagnosis of endometriosis using 5-aminolevulinic acid, *Surg. Endosc.*, 2000, **14**, 452–455.

49. P. Hillemanns, H. Weingandt, H. Stepp, R. Baumgartner, W. Xiang and M. Korell, Assessment of 5-aminolevulinic acid-induced porphyrin fluorescence in patients with peritoneal endometriosis, *Am. J. Obstet. Gynecol.*, 2000, **183**, 52–57.
50. L. Muderspach, F.M. Muggia and P.S. Conti, Second-look laparotomy for stage III epithelial ovarian cancer: rationale and current issues, *Cancer Treat. Rev.*, 1996, **21**, 499–511.
51. M.O. Nicoletto and others, Surgical second look in ovarian cancer: a randomized study in patients with laparoscopic complete remission--a Northeastern Oncology Cooperative Group-Ovarian Cancer Cooperative Group Study, *J. Clin. Oncol.*, 1997, **15**, 994–999.
52. N.R. Abu-Rustum, R.R. Barakat, P.L. Siegel, E. Venkatraman, J.P. Curtin and W.J. Hoskins, Second-look operation for epithelial ovarian cancer: laparoscopy or laparotomy?, *Obstet. Gynecol.*, 1996, **88**, 549–553.
53. P.J. Di Saia and W.T. Creasman, in: *Epithelial Ovarian Cancer*, Creasman. WT, (ed), Mosby. -Year Book. Inc., St. Louis:, 1993, pp. 333–425.
54. A.L. Major, G.S. Rose, C.F. Chapman, J.C. Hiserodt, B.J. Tromberg, T.B. Krasieva, Y. Tadir, U. Haller, P.J. DiSaia and M.W. Berns, In vivo fluorescence detection of ovarian cancer in the NuTu-19 epithelial ovarian cancer animal model using 5-aminolevulinic acid (ALA), *Gynecol. Oncol.*, 1997, **66**, 122–132.
55. R. Hornung, A.L. Major, M. McHale, L.H.L. Liaw, L.A. Sabiniano, B.J. Tromberg, M.W. Berns and Y. Tadir, In vivo detection of metastatic ovarian cancer by means of 5- aminolevulinic acid-induced fluorescence in a rat model, *J. Am. Assoc. Gynecol. Laparosc.*, 1998, **5**, 141–148.
56. R. Hornung, M.J. Hammer-Wilson, S. Kimel, L.H. Liaw, Y. Tadir and M.W. Berns, Systemic application of photosensitizers in the chick chorioallantoic membrane (CAM) model: photodynamic response of CAM vessels and 5- aminolevulinic acid uptake kinetics by transplantable tumors, *J. Photochem. Photobiol. B*, 1999, **49**, 41–49.
57. J. Gahlen, J. Stern, H.H. Laubach, M. Pietschmann and C. Herfarth, Improving diagnostic staging laparoscopy using intraperitoneal lavage of delta-aminolevulinic acid (ALA) for laparoscopic fluorescence diagnosis, *Surgery*, 1999, **126**, 469–473.
58. J. Gahlen, R.L. Prosst, M. Pietschmann, M. Rheinwald, T. Haase and C. Herfarth, Spectrometry supports fluorescence staging laparoscopy after intraperitoneal aminolaevulinic acid lavage for gastrointestinal tumours, *J. Photochem. Photobiol. B*, 1999, **52**, 131–135.
59. J.K. Chan, B.J. Monk, D. Cuccia, H. Pham, S. Kimel, M. Gu, M.J. Hammer-Wilson, L.H. Liaw, K. Osann, P.J. DiSaia, M. Berns, B. Tromberg and Y. Tadir, Laparoscopic photodynamic diagnosis of ovarian cancer using 5-aminolevulinic acid in a rat model, *Gynecol. Oncol.*, 2002, **87**, 64–70.
60. M. Loning, H. Diddens, W. Kupker, K. Diedrich and G. Huttmann, Laparoscopic fluorescence detection of ovarian carcinoma metastases using 5-aminolevulinic acid-induced protoporphyrin IX, *Cancer*, 2004, **100**, 1650–1656.
61. F. Wierrani, D. Fiedler, W. Grin, M. Henry, E. Dienes, K. Gharehbaghi, B. Krammer and W. Grunberger, Clinical effect of meso-tetrahydroxyphenyl-chlorine based photodynamic therapy in recurrent carcinoma of the ovary: preliminary results, *Br. J. Obstet. Gynaecol.*, 1997, **104**, 376–378.

*Chapter 8*

# ALA-PDT in Gastroenterology

## Hugh Barr, Helmut Messmann and Esther Endlicher

**Table of Contents**

# Abstract

The use of photodynamic therapy (PDT) for tissue destruction is highly attractive for the endoscopic and minimally invasive therapy of gastro-oesophageal cancer. Protoporphyrin IX (PpIX) diagnosis offers the possibility of fluorescence detection of early dysplastic and neoplastic change, and aminolevulinic acid (ALA)PDT is suitable for the eradication of earlier asymptomatic mucosal disease. There are other competing techniques that can be used both to detect and to ablate the mucosa. The way forward for oesophageal cancer in particular is for optical detection, followed by curative therapy of early cancers or precancerous change. Many patients are unfit or unwilling to be subjected to radical and morbid surgery, chemotherapy or radiotherapy. The use of PDT for the eradication of superficial cancers and dysplasia is becoming an area of great interest. There are now consistent data demonstrating the effectiveness of ALA-PDT for the eradication of early adenocarcinoma arising from Barrett's oesophagus. At present Barrett's oesophagus is treatable, but not more than 7 cm at one session. Treatment can be performed on an outpatient basis and patients must receive profound acid suppression with proton pump inhibitor (PPI) therapy.

## 8.1. Introduction

### 8.1.1. The Clinical Indication

The most important issue to be addressed in clinical practice is related to the reason for using fluorescence photodiagnosis and ALA-PDT for the treatment of gastrointestinal diseases. The clinical problems and the advantages must be clear. There are areas where patients can benefit, and the disease can be detected and destroyed using ALA photodiagnosis and PDT: the techniques have been used mostly in the treatment of oesophageal disease, in particular Barrett's oesophagus, dysplasia, and adenocarcinoma. In these cases the disease is superficial and thus amenable to photodiagnosis and ablation with PDT. Since patients with Barrett's oesophagus carry an increased risk of developing invasive cancer via dysplasia, a minimally invasive removal of the metaplastic Barrett's epithelium, or a reliable early detection and localization of any appearing dysplasia, is mandatory.

Other less widespread indications such as ulcerative colitis are mentioned in the following sections.

### 8.1.2. Advantages of Photodiagnosis

Currently, the gold standard for diagnosis is histopathological assessment after excision biopsy, either at open operation or, increasingly, using an endoscope or laparoscope. Thus suspicious tissue is removed for assessment and then processed and stained in various ways to be rendered visible to the pathologist. Most gastrointestinal biopsies show few or no abnormalities, and the patient is reassured but no further action is required. There are huge patient and service rewards to be had with a real-time, nondestructive system that reduces the number of required biopsies. The aim of fluorescence diagnosis is to image the tissue *in situ* during endoscopic examination.

In addition, pathological assessment is highly subjective. Most cancers can be diagnosed with a high degree of certainty, but the diagnosis of the premalignant change of dysplasia is more challenging. The pathological definition of dysplasia is an unequivocal neoplastic alteration of the gastrointestinal epithelium which has the potential to progress to invasive malignancy that remains confined within the basement membrane of the gland within which it arose.[1] The diagnostic difficulties involved in identifying the type of neoplastic change (1: negative for dysplasia; 2: indefinite for dysplasia; 3: low-grade dysplasia; 4: high-grade dysplasia; or 5: invasive carcinoma) in gastrointestinal mucosa have been demonstrated.[2,3]

Pathologists' agreement was acceptable when the combination of groups 4 and 5 was to be discriminated from the combination of groups 1, 2, and 3. However, a more relevant classification of pathological findings into four groups (1, 2 + 3, 4, 5) showed poorer levels of intra- and inter-observer agreements.[4]

Other problems include sampling error and the time required to process and deal with biopsy specimens. Endoscopic fluorescence diagnosis offers great promise for early detection and targeting of the biopsy site.

### 8.1.3. Advantages of Photodynamic Therapy

PDT is an attractive, predominantly endoscopic technique for the eradication of early neoplastic and preneoplastic lesions. It was first used for the palliation of advanced upper gastrointestinal cancer. Endoscopists were quick to see the potential of PDT for the treatment of cancers accessible by endoscope. At the Tokyo Medical College a patient with a small upper bronchial squamous cell tumor was treated in 1980 using bronchoscopy with PDT, and the tumor was completely eradicated.[5] Large obstructing oesophageal cancers were similarly treated with PDT with good relief of dysphagia, and possible prolongation of patient survival.[6] Photofrin was the first photosensitiser to be licensed for use in the treatment of advanced oesophageal cancer. (This photosensitiser remains the most widely used, and is licensed for the treatment of advanced oesophageal cancer and high-grade dysplasia in Barrett's oesophagus.) ALA remains unlicensed, so regulatory permission for its use must be sought.

It was soon appreciated that PDT offers distinct advantages over thermal and other methods of tumor destruction. In particular, the nature of the biological response allows safe healing with reduced risk of perforation. Also, the exploitation of tissue threshold effects, particularly in the treatment of pancreatic cancer, means that adjacent tissue damage can be minimised.[7] The full potential of these advantages has yet to be realized.

## 8.2. Application of ALA-PDT in Gastroenterology

### 8.2.1. Administration of ALA and Subsequent Tissue Distribution and Localization of Protoporphyrin IX

The crucial event for ALA-PDT and fluorescence diagnosis is the uptake and production of PpIX in gastroenterological tissues. It appears that accumulation is in very specific areas following the oral administration of ALA, whereas the systemic administration of ALA shows a different PpIX distribution, which is different again when ALA is introduced intravenously. This suggests that the metabolic activity of certain cells is important, and accumulation does not depend on the conversion of ALA into PpIX in the liver and subsequent release into the general circulation. PpIX is not easily soluble in water at physiological pH and it shows a very strong affinity for the lipids of cell membranes. Thus it tends to stay within the cells in which it was synthesized.[8]

The most striking feature of ALA-induced PpIX accumulation in the gastrointestinal tract is the very strong localization of PpIX in superficial mucosal surfaces and in the neoplastic cells of cancer and dysplasia (Figure 1a–d).[9] There is no uptake of PpIX in the interstitial compartment of tissue or in the neoplastic stroma as is the case with other photosensitisers, including Photofrin. Surprisingly, there is also very little accumulation of PpIX in the smooth muscle that lines the intestine, even though muscle cells contain large amounts of heme in the form of myoglobin. Similarly, there is very little accumulation of PpIX in the tumor neovasculature, which is often a target site for accumulation of other photosensitisers.

It is apparent that heme biosynthesis is regulated differently in different tissues. The degree of cellular maturation and differentiation is very important (Figure 1c), as are the regulators of iron uptake. At a cellular level ALA-PDT in gastric cancer cells is enhanced by removing the available iron with desferrioxamine. However, there is a decrease in cell phototoxicity at higher doses of desferrioxamine. The explanation for this effect does not seem to be due to the antioxidation properties of desferrioxamine or singlet oxygen trapping.[10] The effectiveness of ALA-induced PDT is increased by the addition of an iron-chelating agent. The amount of damage produced by PDT of the normal colon is three times greater if the iron-chelator hydroxypyridinone is combined with ALA-induced photosensitization. It is therefore possible to increase the amount of tissue damage without increasing the dose of ALA.[11] An alternative approach has used early illumination shortly after ALA administration in order to selectively reduce ferrochelatase activity. This enzyme is more susceptible to physical and chemical damage than other enzymes upstream in the pathway, such as porphobilinogen deaminase.

The details of the pharmacokinetics of ALA and ALA-induced PpIX uptake and synthesis are very important for our understanding of the process of PDT. Such time-dependent drug data are essential for planning optimum treatment intervals and method of administration. There is experimental evidence of selective accumulation of PpIX in metastatic liver tumors in rodents compared with the surrounding liver. After oral administration of 5-aminolevulinic acid administered over 11 days there was significantly more PpIX in the metastases than surrounding liver, and a ratio of 4:1 could be achieved. In comparison, intravenous delivery of Photofrin (2.5 or 5 mg kg$^{-1}$) resulted in higher concentrations in the surrounding liver than in the metastases with a ratio of 3:1.[12] Another experimental study using animal models has shown differential accumulation between normal mucosa: viable colonic tumor of 1:6. In addition this study confirmed that mucosa contained six times more PpIX than submucosa 4 h after administration of ALA, with barely detectable levels of PpIX in the muscle, and a very marked decrease within 6 h. It is important to note that these studies have shown a good correlation between fluorescence distribution and the biological effect of ALA-induced photosensitisation.[13]

The clinical indication studied most extensively with ALA-PDT is Barrett's oesophagus and oesophageal adenocarcinoma. Important clinical information is available following several very elegant and important studies by a group in Rotterdam. There was no selective retention of ALA or PpIX between metaplastic Barrett's oesophagus and the native squamous epithelium (mean time interval 6.7 h).[14,15] In addition, we have confirmed that there is little selective accumulation of PpIX by neoplastic tissue in comparison with adjacent normal mucosa (Figure 1d). However, the PpIX fluorescence in oesophageal carcinoma was 2-3 times that of the surrounding lamina propria. There is also a clear difference in PpIX accumulation between dysplastic Barrett's epithelium and the underlying lamina propria (Figure 1a and Table 1). The lack of tumor selectivity may not be important, since the laser light is administered only to the oesophageal segment with tumor or preneoplastic involvement. As expected, PpIX fluorescence was higher in the tissues of patients who were given 75 mg kg$^{-1}$ of ALA (in five divided doses) than in those receiving 60 mg kg$^{-1}$ in two divided doses (Table 1). However, higher doses of ALA did not increase the differential accumulation of PpIX between normal mucosa and carcinoma.

There have been very detailed studies of the pharmacokinetics of ALA-uptake in patients with Barrett's oesophagus. In one study, the dose regimen was 60 mg kg$^{-1}$ dissolved in 10 mL of orange juice, with approximately 30 mL of water for rinsing the oesophagus. Twenty-six patients were randomized to varying time-intervals between oral administration and endoscopic biopsy of the oesophageal and gastric mucosa. The results showed that the maximum PpIX concentration occurred in Barrett's epithelium 4.6 h after ingestion and in the squamous mucosa at 6.6 h (Figure 1d).[15] Interestingly, 1 h after ingestion the concentrations of ALA in all tissues, including gastric cardia epithelium, was 20 times higher than the concentrations in plasma. These findings suggest that the mechanism of absorption and metabolism is locally through and into the lining cells of the upper intestinal mucosa, possibly facilitated by ALA's acidic nature, which would allow passive diffusion. This property is exploited in the treatment of lesions in the skin and bladder following local administration of ALA and further supports the technique of direct application onto the diseased area of the oesophagus using an endoscopic spray or a gel.[16,17]

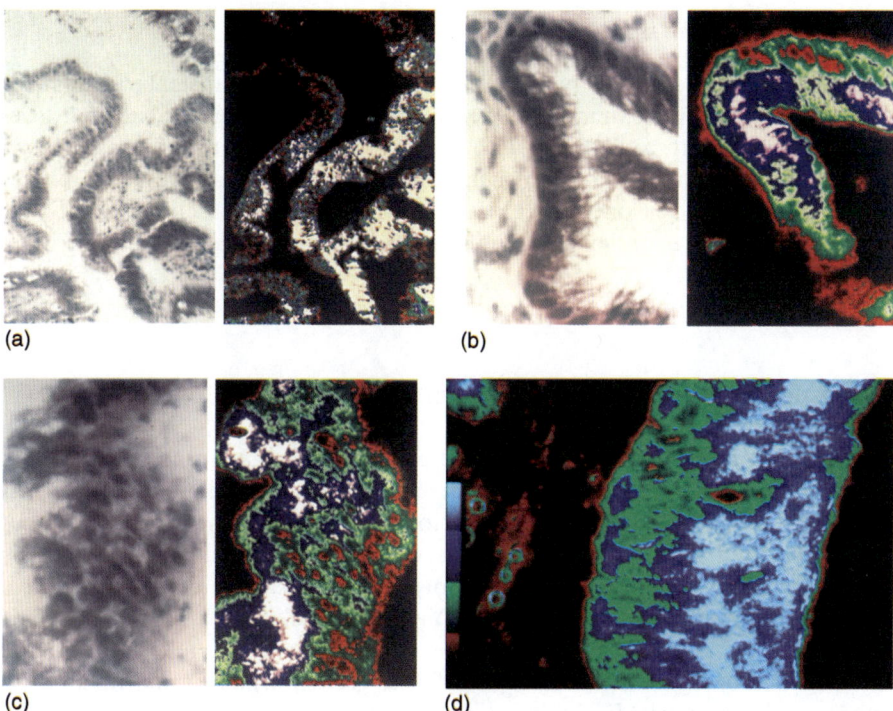

**Figure 1.** False color image and histological section showing the uptake and distribution of PpIX in Barrett's epithelium following the oral administration of ALA (60 mg kg$^{-1}$) 4 h previously.(a) There is very high localization in the mucosal dysplastic Barrett's cells. White indicates high fluorescence (PpIX accumulation) with lower levels through the blue, green to black (scale on picture) indicating lower levels. (b) This image shows detail of fluorescence between high-grade dysplasia and adjacent less dysplastic tissue, and in cells in the stroma. (c) This shows an adenocarcinoma with patchy fluorescence, perhaps indicating differing cell cycle kinetics. There is no accumulation in the tumor stroma: all of the fluorescence is in the neoplastic cellular layer. (d) This image is of normal squamous epithelium indicating that there is still a very high accumulation of PpIX in the adjacent normal non metaplastic or dysplastic tissue.

**Table 1.** Comparison of the normalized ratios of PpIX, measured using fluorescence in oesophageal tissue at two different doses of oral ALA

| ALA dosage | Concentration ratio of PPIX (measured by fluorescence) | |
|---|---|---|
| | 75 mg kg$^{-1}$ | 60 mg kg$^{-1}$ |
| Carcinoma/Barrett's-Epithelium | 1.8 | 1.1 |
| Carcinoma/Squamous-Epithelium | 1.39 | 1.35 |
| Barrett's/Squamous-Epithelium | 0.76 | 1.21 |
| Carcinoma/Barrett's lamina propria | 2 | 0.96 |
| Carcinoma mucosa/lamina propria | 1.94 | 3.70 |

*8.2.2. Pharmacological Effects*

The usual oral dose of 60 mg kg$^{-1}$ of ALA results in transient rise in liver function tests in approximately 50% of patients, returning to a normal level after 3–4 days. At a dose of 75 mg kg$^{-1}$ there is more prolonged liver disturbance and clinical jaundice has occurred, which always resolves after a few days. For this reason the oral dose has not been increased further. Some patients have also experienced nausea and vomiting which can be controlled with antiemetics.

## 8.3. Endoscopic Fluorescence Diagnosis

The identification of many diseases and cancers in the gastrointestinal tract is normally made by visual inspection at endoscopy, the diagnosis being confirmed by pathological analysis after biopsy removal of the suspicious area. However, often by the time a cancer is visible to white-light endoscopy it is relatively advanced, and when visible morphological pathological changes have occurred the disease is often invasive. Yet almost all cancers in the gastrointestinal tract start as a subtle invisible mucosal change (Figure 2) and the interrogation of these surface changes prior to gross morphological transition is vital to early diagnosis. Early detection also allows the possibility of minimally invasive, endoscopic destruction of these preinvasive lesions.

Photodynamic diagnosis is one of the new techniques that allow a better detection of nonvisible or difficult to detect malignant and premalignant lesions.

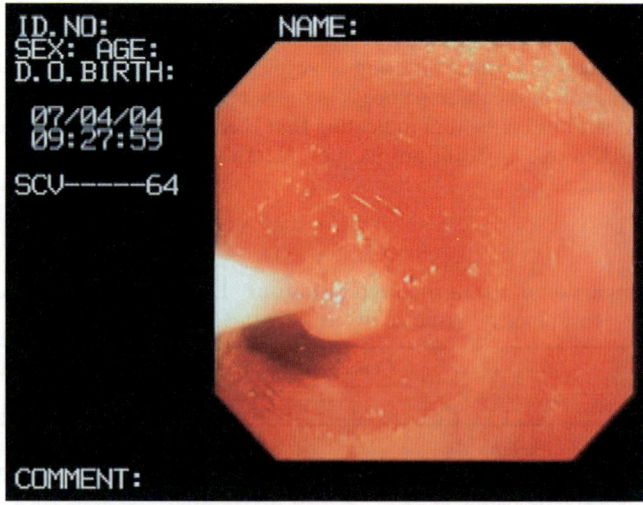

**Figure 2.** A segment of Barrett's oesophagus with a small nodule of high-grade dysplasia prior to treatment.

As in other fields of medicine, there are promising results in gastroenterology, in the detection of early cancers or dysplasia in patients with Barrett's oesophagus or ulcerative colitis.

### 8.3.1. Technical Background

Endoscopy is performed using fiberscopes and the endoscopes are connected to a light source delivering white or blue light (375–440 nm). During endoscopy it is possible to switch between the conventional white-light mode (for visual observation) and the blue-light mode (for fluorescence excitation). Furthermore, the endoscope is adapted to a camera with an image-processing module delivering real-time fluorescence pictures. There are several devices available for the gastrointestinal tract (e.g., D-Light, Storz, Germany and LIFE-GI, Xillix, Canada). While the Xillix laser-induced fluorescence endoscopy gastrointestinal (LIFE-GI) system is based on detecting tissue autofluorescence, the D-Light system is optimized for the detection of 5-ALA induced PpIX fluorescence. Premalignant and malignant lesions will exhibit red fluorescence, while normal tissue will show blue backscatter.

### 8.3.2. Clinical Results

Photodiagnosis has become very important in Barrett's oesophagus for the detection of early cancer and dysplasia.[18] Since there are major problems associated with the differentiation of inflammation from dysplasia, it is usually very important to ensure control of inflammation prior to attempting fluorescence diagnosis. Patients with oesophagitis and inflamed Barrett's oesophagus should receive a full dose of PPI therapy for at least 4 weeks prior to imaging or spectroscopy. The administration of ALA-induced PpIX fluorescence has been more widely used than the autofluorescence technique, since autofluorescence imaging has proven limiting in the latter case.[19]

In 47 patients with Barrett's oesophagus, ten of them with known dysplasia, 58 fluorescence endoscopies were performed after photosensitization with different concentrations of 5-ALA given either orally (5, 10, 20, 30 mg kg$^{-1}$), or locally (500–1000 mg) by spraying the mucosa via a catheter.[20] Fluorescence endoscopy was performed 4–6 h after systemic, and 1–2 h after local ALA application. Photosensitization was performed using a special light source capable of delivering either white or blue light (D-light, Storz, Tuttlingen, Germany). A total of 243 biopsies of red fluorescent ($n = 113$) and nonfluorescent areas ($n = 130$) were taken. In three patients two early cancers and dysplasia, not visible during routine endoscopy, were detected by fluorescence endoscopy. Thirty-three biopsies revealed either low- or high-grade dysplasia. Sensitivity to dysplastic lesions ranged from 60% for local sensitization with 500 mg, to 80–100% after systemic application of 10–30 mg kg$^{-1}$, respectively. However, specificity was best for local ALA application (70%) while systemic administration of ALA produced values of 27% and 56% (Table 2). Using 5

**Table 2.**  Sensitivity and specificity for the detection of dysplasia in Barrett's oesophagus correlated with the ALA concentration and application mode

| 5-ALA application mode | Examinations (biopsies) ($n$) | Sensitivity (%) | Specificity (%) |
|---|---|---|---|
| 500 mg locally | 22 (96) | 60 | 69 |
| 5 mg kg$^{-1}$ orally | 4 (20) | - | 70 |
| 10 mg kg$^{-1}$ orally | 13 (56) | 80 | 56 |
| 20 mg kg$^{-1}$ orally | 13 (44) | 100 | 51 |
| 30 mg kg$^{-1}$ orally | 6 (27) | 100 | 27 |

mg kg$^{-1}$, no red fluorescence in dysplastic lesions was found. No severe side
effects were noted.

Gossner et al. studied 35 patients with upper gastrointestinal dysplasia
and early cancers. Patients were examined 3 h after sensitization with 10
mg kg$^{-1}$ 5-ALA. Neoplasia was detected with a sensitivity and specificity
of 85% and 70%, respectively.[21] Mayinger et al. achieved a sensitivity
and specificity of 85% and 53%, respectively, in the detection of premalignant
and malignant oesophageal lesions: their data were collected from 22
patients, 6–7 h after photosensitisation with 15 mg kg$^{-1}$ 5-ALA administered
orally.[22]

Inflammatory bowel disease, especially ulcerative colitis is another important
possible indication for fluorescence endoscopy. Our experience with 37 patients
with ulcerative colitis showed that local administration of 5-ALA via enema or
spray-catheter is superior to a systemic sensitization with 20 mg kg$^{-1}$ 5-ALA
administered orally.[23] In the latter case, sensitivity was low (43%), although
specificity was quite high (73%.); local ALA administration increased the
sensitivity to 87–100%, while the specificity decreased to only 51% for enema
and 62% for spray-catheter.

The combination of imaging with spectroscopic analysis is very useful when a
time-gated, simultaneous, laser-induced fluorescence endoscope (TSLIFE) is
employed.[17] Forty-two patients (9 women, 33 men; mean age 60) with no
macroscopic abnormality in Barrett's oesophagus had 500 mg of ALA applied
topically after washing with sodium bicarbonate as a mucolytic. A further
imaging/spectroscopic endoscopy was performed 2 h later and a double channel endoscope (GIF 2 T 20, Olympus Co, Japan) was used, with a laser beam
(505 nm) incorporated into the fiber bundle. Of 36 patients who had a negative
white-light endoscopy, four (11%) had malignant lesions detected using the
fluorescence diagnostic system. A total of 475 biopsies were taken, 339 from
fluorescent areas and 136 from nonfluorescent negative areas. Sixty biopsies
diagnostic of malignant change were found in the fluorescent areas; ten of early
cancer, 17 showing high-grade dysplasia, and 33 showing low-grade dysplasia.
However, only six biopsies from the fluorescent negative areas showed malignant change; one of early cancer, two with high-grade dysplasia, and three with
low-grade dysplasia. Spectroscopy had a sensitivity of 76% and specificity of
74%. The TSLIFE system had a sensitivity of 92% and specificity of
only 32%.[16]

Recently, Ortner has demonstrated that time-gated fluorescence spectroscopy can differentiate low-grade dysplasia from nondysplastic Barrett's mucosa.[17] In this study, the patients received an aqueous solution of 0.5 g ALA dissolved in 50 mL of 8.4% sodium bicarbonate and 50 mL of sodium chloride (0.9%) sprayed on the mucosa during endoscopy. Fluorescence-guided endoscopy was performed approximately 60–120 min later. The hypothesis is that PpIX dominates the fluorescence spectra of dysplastic lesions. It was observed that the immediate fluorescence spectra exhibited PpIX fluorescence in the 635 and 699 nm region, as well as a broad auto fluorescence spectrum. Nondysplastic mucosa displayed broad autofluorescence without specific spectral signatures. Since the decay time of autofluorescence (2.4 ns) was considerably shorter than the fluorescence lifetime of PpIX (16 ns), the autofluorescence background could be suppressed by detecting the spectra in a delayed time gate. Thus time-gated spectroscopy is a rational approach to optical diagnosis, and the median normalized fluorescence intensity (*i.e.*, the ratio of delayed PpIX fluorescence intensity to immediate autofluorescence intensity of the tissue) can be measured. For the first time, a significant difference between metaplasia and low-grade dysplasia was demonstrated. The normalized fluorescence ratio was 0.51 (range 0.09–1.92) for metaplasia, and 1.89 (range 0.55–3.92) for low-grade dysplasia ($p < 0.005$). The authors also publish on the interobserver agreement and p53 protein expression in order to try and exclude the subjectivity associated with histopathological diagnosis. Overall the interobserver agreement for pathologists is high, but abnormally low for the distinction of low-grade dysplasia from nondysplastic epithelium. Another important finding of this study was that specialized intestinal metaplasia could be differentiated from junctional or fundic epithelium. Overall time-gated fluorescence appeared to be a highly useful technique. It is clear that randomized trials comparing fluorescence imaging to standard endoscopic surveillance are required.

### 8.3.3. Latest Developments

An exciting alternative approach is the use of multiphoton fluorescence imaging with ALA (Figures 3 and 4). Multiphoton imaging, a process which uses high-intensity excitation light of longer wavelength (more penetrating), has many advantages over other methods of fluorescence imaging. These include improved depth penetration, reduction in background contributions from out-of-focus fluorescence, and reduction in dye bleaching and photodamage to the tissue. In two-photon, microscopy improved in-depth imaging performance is achieved in comparison to conventional and even confocal imaging, since the near-infrared wavelengths can penetrate deeply within the tissue, as there is little electronic transition or molecular vibrational absorption.[24]

The hydrophilic properties of 5-ALA limit the amount of uptake and penetration in tissues, particularly after topical administration. Therefore, chemical modification of ALA into its more lipophilic esters seems to be a promising means of overcoming such limitations. In urology there are promising data using the 5-ALA-hexylester for the photodetection of early human-bladder cancer.[25]

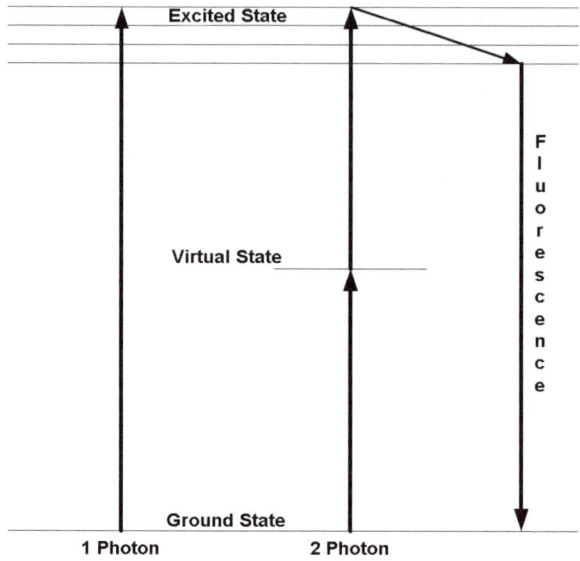

**Figure 3.** One and two photon excitation energy level diagram demonstrating the principle of multiphoton imaging.

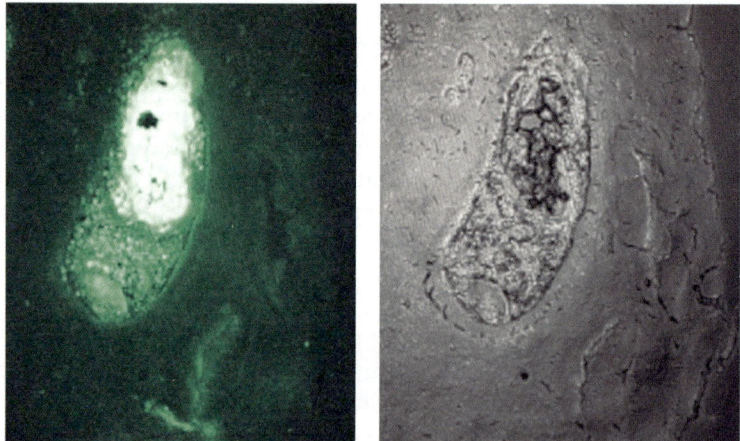

**Figure 4.** A multiphoton image (green) and corresponding phase contrast picture of a nodule of adenocarcinoma in stromal tissue. There is very high fluorescence between the nodule and surrounding area. (The patient had received 60 mg kg$^{-1}$ ALA 4 h previously)[24]

The results of this study indicate that with 5-ALA-hexylester a twofold increase of PpIX fluorescence intensity can be observed using a 20-fold lower concentration as compared to 5-ALA. For the gastrointestinal tract, preliminary clinical data demonstrated selective fluorescence of intraepithelial neoplasia containing adenoma, indicating that hexaminolevulinate-based fluorescence endoscopy has the potential for the detection of premalignant lesions.[26]

Further clinical studies on selectivity of ALA esters in malignant and premalignant lesions are necessary and are in progress. New photosensitizers with higher selectivity and a decreased systemic photosensitivity will hopefully improve photodynamic diagnosis and therapy. Furthermore, the combination of exogenous and endogenous fluorophores will probably improve sensitivity and specificity for the detection of premalignant and malignant lesions.

In future the clinical use of fluorescence endoscopy will depend on new developments with respect to light sources, more sensitive endoscopic camera systems and high-resolution video endoscopes.

## 8.4. Photodynamic Therapy

### 8.4.1. Drug Application

The patient usually receives between 30 and 75 mg kg$^{-1}$ ALA dissolved in orange juice or lemonade: above 75 mg kg$^{-1}$ there are concerns regarding side effects and possible toxicity.[27] The prodrug is administered 3–6 h prior to endoscopic light irradiation. The ALA dose may be fractionated into two aliquots of 30 mg kg$^{-1}$ each, ingested 4 and 3 h prior to PDT. This is our preferred option because it appears to be well tolerated by patients, although there is no clear therapeutic benefit.

There appears to be no real advantage to parenteral administration of ALA.[28,29] I am aware of one incidence of a vaso-vagal attack occurring after parenteral administration, which caused great concern but was entirely self-limiting.

The administration of 5-ALA by direct endoscopic spraying onto dysplastic Barrett's oesophagus is a novel approach for prodrug delivery, allowing for local absorption prior to irradiation. Seven of nine patients had a good PDT response with this approach.[16]

### 8.4.2. Indications and Treatment Procedure

The indications for PDT are for the treatment of dysplastic epithelium with no visible macroscopic lesion. If nodules are present, our group prefers to use endoscopic mucosal resection in order to ensure that invasive cancer is not already present. The major disadvantage of ALA-PDT is that no histology is available following treatment. The length of the segment of dysplasia will limit the amount that can be reasonably treated during one endoscopy session. Lengths of up to 7 cm can be treated with adequate patient safety and comfort.

Precautions are taken to avoid excessive light exposure during transportation of the patient to the endoscopy or operating suite, and exposure to bright operating lights. Treatment is usually performed with topical anesthesia with Xylocaine pharyngeal spray and intravenous sedation of 1–10 mg of midazolam. Also, we have found that analgesia is occasionally administered

(Pethidine 50–100 mg intravenously). Patients photosensitized using ALA often require a prolonged endoscopic light administration and there is often considerable local discomfort and irritation during this time. The patients usually have no recollection of this, but are sometimes very difficult to keep well sedated. The discomfort is manifest, with considerable agitation and restlessness during the procedure. This effect is not seen with other photosensitisers. Throughout treatment, oxygen is delivered via a nasal sponge at a rate of 4–5 L per minute. Repeat sedation is sometimes necessary since the light treatment can last 20–30 min, and considerably longer if a very long segment requires treatment. This is a limitation for ALA-PDT, since other photosensitisers require shorter irradiation times (8–10 min for Photofrin and 2 min for mTHPC), and there appears to be no problem of discomfort. However, the great advantage of ALA is the lack of prolonged systemic photosensitisation and the limitation of damage to the very superficial layer, with little risk of perforation and deeper damage. Perforations have occurred following treatment with other photosensitisers.

It is very important to pay close attention to light dosimetry and to use an appropriate light-centring device (Figure 5).[30,31] The aim is to deliver an even light dose to a defined area of the organ. If only one area is dysplastic or exhibiting early neoplasia, then a bare fiber or a partially shielded device is used for irradiation. To treat long areas, repeated sequential areas are irradiated in 5 cm lengths. There now exist some useful devices, including windowed balloons (Wilson Cook Inc.), which are very easy to use. These inflatable transparent polyurethane balloons can be passed over a guide wire and positioned through the biopsy channel of the endoscope. Inhomogeneous and inconsistent illumination is a problem and occurs because of oesophageal motility, respiratory movements, and patient restlessness. To overcome this problem we often pass a small video endoscope down beside the device to ensure positional stability throughout treatment.

The laser fiber is inserted and the correct wavelength of light chosen (PpIX-635 nm, Photofrin-630 nm, and mTHPC-652 nm). Nonlaser light devices are also highly effective and some include sophisticated monitoring to assess light dose during treatment.[32]

The most appropriate time for light irradiation is 3–4 h after ALA administration, and this time interval appears critical to achieving mucosal destruction

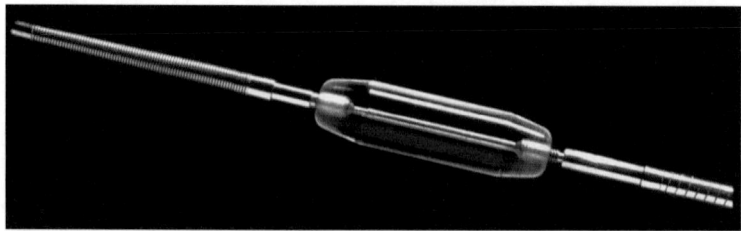

**Figure 5.** An optical centring device for irradiation of the oesophagus. Balloons can also be used.

while sparing oesophageal function.[33] Fractionated illumination has been suggested to improve efficacy, although the correct clinical approach is unclear.[34]

We adjust the position of the irradiating device during the procedure so that the laser treatment is interrupted on occasion. This is important because patient restlessness can cause movement of the irradiation device, as can oesophageal peristalsis. The oesophagus perceives the centring device as a bolus, and has a natural tendency to try to move it through the lumen.

The power density used is between 100 and 400 mW cm$^{-2}$ to provide an energy density of 100 J cm$^{-2}$ from the diffuser to the mucosa. A very important analysis of dosimetry has been published recently, and technical support by trained staff is essential.[31]

There is one comparative study of patients treated with a single laser illumination of 100 J cm$^{-2}$ 4 h after ALA administration, and a fractionated laser illumination of 20 J cm$^{-2}$ 1 h after ALA administration with a further 100 J cm$^{-2}$ 4 h later. Both regimens used a centring balloon in the oesophagus. There is currently insufficient difference in overall clinical outcome to recommend routine fractionation.[35]

Treatment can be performed on an outpatient basis and patients with Barrett's oesophagus must all receive profound acid suppression with PPI therapy. It is recommended for most photosensitisers that direct sunlight and other intense lights be avoided for a period of 4–6 weeks following treatment. However, with ALA this period can be shortened to 2 days, and we have seen no cutaneous photosensitivity after ALA-PDT applications. The major clinical studies of 5-ALA photodynamic therapy for the ablation of high-grade dysplasia have demonstrated eradication of the dysplasia and tumors less than 2 mm in depth.

We now take great care to assess the depth of the lesion using endoscopic ultrasound and careful assessment at endoscopy.

### 8.4.3. Clinical Results

The use of endogenous and exogenous photosensitisers in the endoscopic treatment of superficial cancers and dysplasia is becoming an area of great interest. There are now consistent data demonstrating the effectiveness of ALA and Photofrin for the eradication of early adenocarcinoma arising from Barrett's oesophagus, and mTHPC for destruction of squamous cancer.[32,36]

The main advantage of ALA-PDT is the ease of administration and the absence of side effects. With Photofrin, it is recommended that not more than 7 cm of Barrett's oesophagus is treated at one session in order to avoid stricture formation. Some patients develop small unilateral or bilateral pleural effusions, and atrial fibrillation has been reported. However, the major concern has been the incidence of oesophageal stricture. Approximately 30% of patients developed significant obstruction. These strictures do respond to endoscopic oesophageal dilatation, but occasionally perforation has resulted. These complications do not occur following ALA-PDT. Nevertheless, the results of

Photofrin PDT are very encouraging, with 75–80% of the Barrett's mucosa being converted to neosquamous mucosa. Complete eradication of all metaplastic epithelium occurred in 43 of 100 patients. Dysplasia disappeared in 78 of 100 patients, although 11 developed dysplasia during follow-up and required repeat treatment.[36] Residual areas of abnormal mucosa are best treated by thermal ablation using a laser or an argon plasma coagulator (APC).

The only randomized partially blinded trial for the prevention of cancer in Barrett's oesophagus is still ongoing, with the use of Photofrin as the photosensitiser. The patients, all of whom were confirmed to have high-grade dysplasia in Barrett's oesophagus, were randomized (2:1) such that 138 had PDT and 70 received only omeprazole (PPI). At the 24-month follow-up, ablation of all areas of high-grade dysplasia was noted in 76.8% of patients after PDT versus 38.6% in the control group ($p < 0.0001$). After a mean follow-up of 24.2 months, 13.0% of the PDT patients had disease progression to cancer as compared to 28% after a mean follow-up of 18.6 months ($p = 0.006$).[37]

There has been one preliminary randomized comparison of argon plasma coagulation (APC) with Photofrin PDT. The PDT was performed at one session while APC required an average of three. Dysplasia was eradicated in 10 of 13 (77%) patients treated with PDT and in 11 of 16 patients (69%) after APC. Photosensitivity was seen in 2 (15%) of the PDT patients whereas 3 (19%) of the patients treated with APC experienced dysphagia, pain, and fever.[38]

Following ALA-PDT there is always endoscopic evidence of mucosal necrosis at follow-up endoscopy within 1 week of treatment (Figure 6). The surface is generally covered with a white superficial necrosis. Repeat endoscopies have been performed at 1, 3, 6, and 12 months for the first year. The treated oesophagus is subjected to a standard surveillance biopsy protocol with quadrant jumbo biopsy at 2 cm intervals. The patients are then seen at 6-month intervals. All patients have demonstrated good endoscopic evidence of neosquamous re-epithelialisation by 1–2 months (Figure 7). Table 3 summarizes our results and those of other investigators, and indicates that ALA-PDT is very successful in eradicating dysplastic Barrett's oesophagus.[9,39–41] Areas of metaplastic tissue often remain (Figure 8), but these areas can be eradicated with thermal therapy using a laser or APC. It is difficult to know whether these areas remain because they had not been completely eradicated, or perhaps regeneration had been mixed because of the environmental conditions. The heterogeneous and variable concentration of PpIX after oral administration, and the difficulty of ensuring even light distribution, suggest that some areas may be inadequately treated. Also, the treatment of early carcinoma has clearly demonstrated that tumors must be small in order to be eradicated. A depth of less than 2 mm seems to be critical,[39] since complete destruction at greater depths is difficult.

The most important assessment of the method's success is histopathological. The most striking histological pattern of squamous re-epithelialisation is found following ALA-PDT. We have seen extensive squamous metaplasia deep

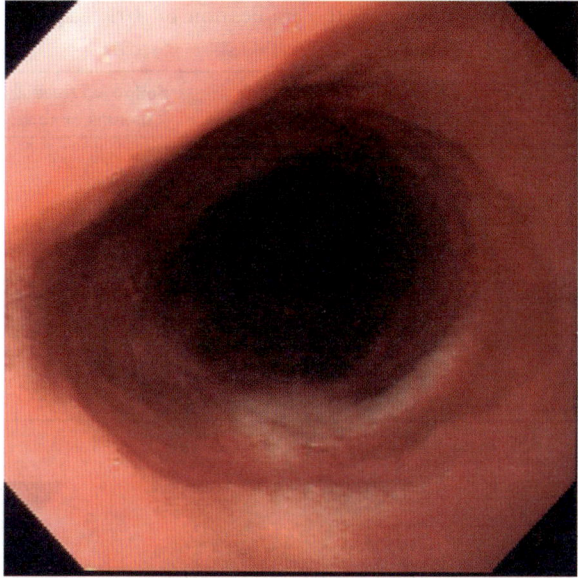

**Figure 6.** The area of abnormality has been treated with ALA-PDT with a bare fiber locally applied. A white necrotic slough is seen 7 days after treatment.

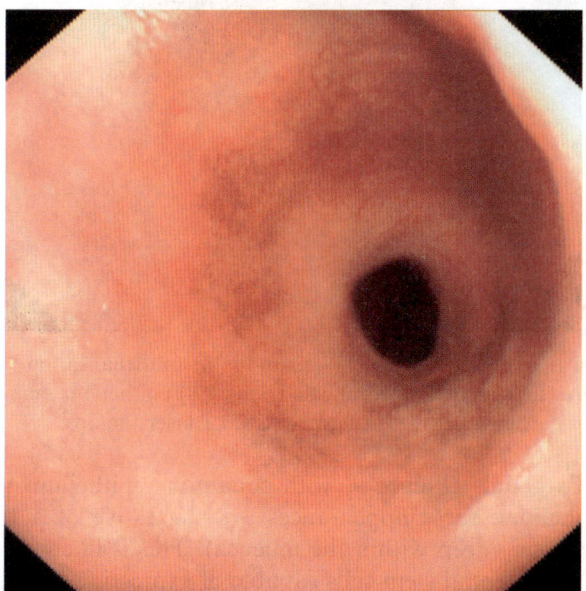

**Figure 7.** An entire segment of Barrett's oesophagus treated circumferentially shows normal-looking neosquamous epithelium post-treatment.

**Table 3.**  Photodynamic therapy for the eradication of early cancer and dysplasia in Barrett's oesophagus[39–41]

| Patient numbers | Stage of cancer (clinical and EUS) | Complete response (%) | Follow-up (median months or range) | Complications (%) |
|---|---|---|---|---|
| 22 AdenoCa | Mucosal | 77 | 9.9 | 0 |
|  | <2 mm: 17 | 100 |  |  |
|  | >2 mm: 5 | 0 |  |  |
| 3 AdenoCa | Tis 1 | 0 |  | 0 |
|  | Tx 2 |  |  |  |
| 12 | All | 16 | (28–36) | 0 |
|  | Tis | 100 |  |  |
|  | T1/2-10 | 0 |  |  |
| 10 | High Grade Dysplasia | 86 | 1–11 |  |
| 5 | High Grade Dysplasia | 80 | 26–120 |  |

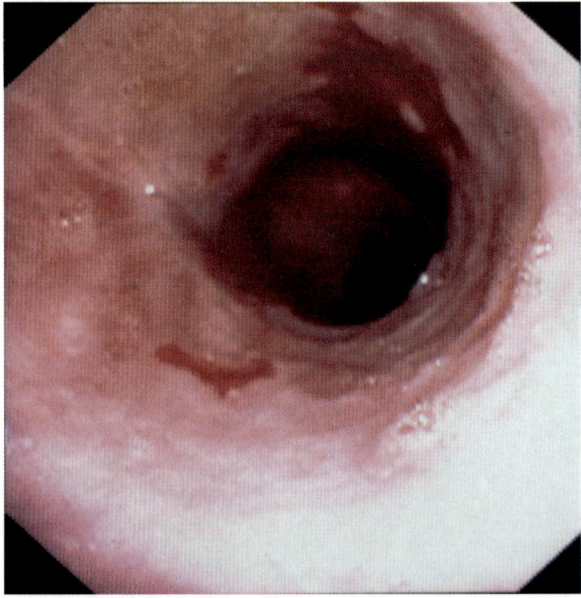

**Figure 8.**   Area treated with extensive squamous re-epithelialisation. A small area of pink fluorescence, indicative of metaplastic (Barrett's) epithelium, remains and can be destroyed with thermal APC or laser therapy.

within Barrett's glands (Figure 9, with squamous epithelium underlying the columnar epithelium: these appearances are suggestive of a true squamous metaplasia occurring deep within the mucosa). This is likely to be due to the existence of pluripotential stem cells capable of expressing a squamous phenotype under certain environmental conditions. The most clinically relevant finding is the discrepancy between the endoscopic assessment and pathological findings. In one study 24 biopsies were clearly identified as coming from a treated segment, yet they showed complete glandular epithelium.

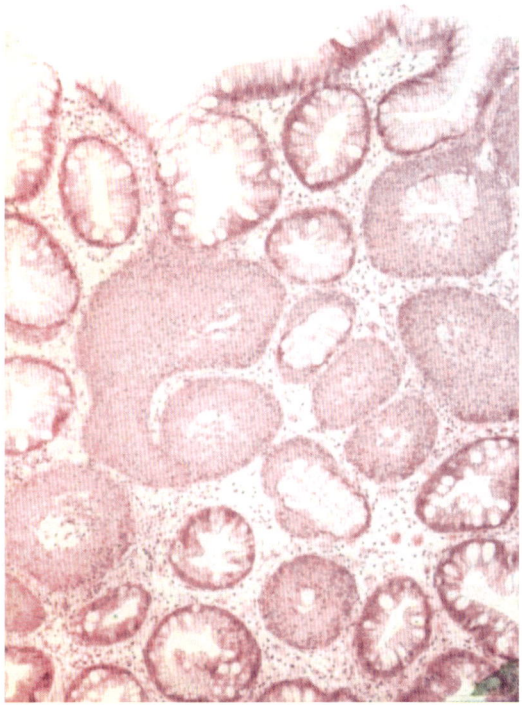

**Figure 9.** Histological picture after ALA PDT showing squamous dysplasia in the Barrett's glands.

This emphasizes the impossibility of identifying cell type at endoscopy in areas of treatment. We have also commonly identified unsuspected buried glands, causing concern that the cancer risk has not been completely removed (Figure 10). It is thus important to take jumbo biopsies in a rigorous manner.

Some comparative trials have been conducted. A recent randomized trial compared ALA-PDT (following continuous light and fractionated irradiation) with APC for the ablation of patients with low-grade dysplasia (8) and no dysplasia (32) in Barrett's oesophagus.[35] The results showed that the mean endoscopic reduction of Barrett's oesophagus at 6 weeks was 51% for ALA with continuous irradiation, 86% following fractionated irradiation, and 93% following APC treatment. A further study has confirmed that the complete ablation of Barrett's epithelium followed APC treatment occurred in 97% of patients compared with only 24% of patients treated with ALA-PDT.[42]

An excellent prospective randomized trial of the treatment of low-grade dysplasia using ALA and irradiation with green light (rather than the usual 630 nm red light) has again confirmed how effective this treatment is in reversing dysplasia/metaplasia. Healing proceeded with the regeneration of neo-squamous epithelium.[43] This study was notable also since there was less

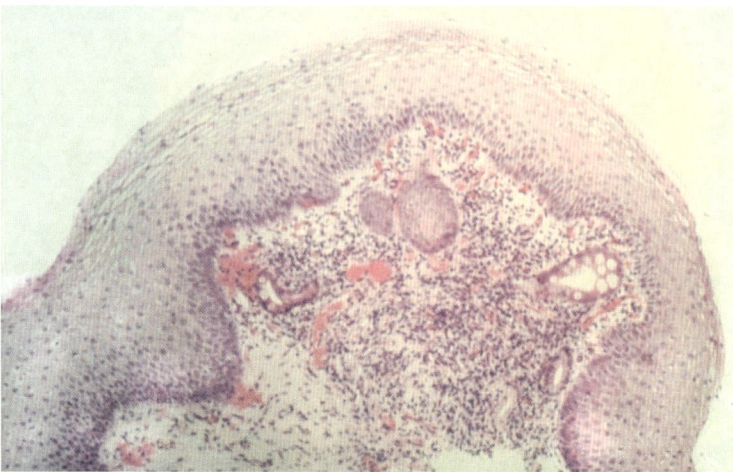

**Figure 10.** Squamous regeneration has occurred in a treated segment, but a gland is seen buried under the neosquamous epithelium. The significance of this is unknown.

evidence of buried Barrett's glands, a most informative and important finding which has yet to be fully explored. The finding of buried glands following mucosal ablation with PDT and other techniques has caused concern of inadequate treatment.

## 8.5. Conclusions

The future of ALA as an agent for gastrointestinal optical diagnosis and PDT is maturing with some clear indications. The main area of intense research and clinical activity is in the detection and eradication of the premalignant changes associated with Barrett's oesophagus. The best results for detection and differentiation of various grades of disease in the mucosa are with ALA fluorescence diagnosis. The eradication of the disease once detected is a subject of much debate: currently for high-grade dysplasia ALA-PDT is a safe and realistic option.

## 8.6. Potential Future Trends of ALA-PDT

ALA-PDT has the advantage of minimal systemic photosensitisation but is limited in the depth of effect. Potential future applications will exploit these two facts. In particular, it is very suitable for treatment of mucosal disease in the upper gastrointestinal tract. Recent data have demonstrated the clear benefit of Photofrin PDT as an effective mucosal ablative method for the eradication of high-grade dysplasia in Barrett's oesophagus and prevention of oesophageal cancer. There is an urgent need for a comparative study with ALA-PDT and this is planned. If ALA is as effective it could become the method of choice for treatment of precancerous mucosal changes.

There are other competing techniques, and this method will have to be compared with endoscopic mucosal resection. The absence of complications and limited depth of damage associated with the technique is clearly beneficial as a minimally invasive method. The delivery of the drug to the diseased tissue with local spray and local application needs to be explored. This may allow ALA photodetection and photodestruction at one session.

The other disease area of great interest, where PDT could possibly make a profound impact, is the bile duct and pancreas. There is a rising incidence of cholangiocarcinoma, and a demonstrated response at least to PDT with external porphyrin.[44]

# References

1. R.H. Riddell, H. Goldman and D. Ransohoff, Dysplasia in inflammatory bowel disease. Standardised classification with provisional clinical application, *Hum. Pathol.*, 1983, **14**, 931–966.

2. B.J. Reid, R.C. Haggitt, C.E. Rubin, G. Roth, C.M. Surawicz, G. Van Belle, K. Lewin, W.M. Weinstein, D.A. Antonioli and H. Goldman, Observer variation in the diagnosis of dysplasia in Barrett's esophagus, *Hum. Pathol.*, 1988, **19**, 166–178.

3. R.C. Haggitt, Barrett's esophagus, dysplasia, and adenocarcinoma, *Hum. Pathol.*, 1994, **25**, 982–993.

4. E. Montgomery, M.P. Bronner, J.R. Goldblum, J.K. Greenson, M.M. Haber, J. Hart, L.W. Lamps, G.Y. Lauwers, A.J. Lazenby, D.N. Lwein, M.E. Robert, A.Y. Toledanao and K. Washington, Reproducibility of the diagnosis of dysplasia in Barrett's oesophagus: A reaffirmation, *Hum. Path.*, 2001, **32**, 268–378.

5. Y. Hayata, H. Kato, C. Konaka, J. Ono and N. Takizawa, Hematoprphyrin derivative and laser photoradiation in the treatment of lung cancer, *Chest*, 1982, **81**, 269–277.

6. J.S. McCaughen, W. Hicks, L. Laufman, E. May and R. Roach, Palliation of esophageal malignancy with photoradiation therapy, *Cancer*, 1984, **54**, 2905–2910.

7. S.G. Bown and C.E. Millson, Photodynamic therapy in Gastroenterology, *Gut*, 1997, **41**, 5–7.

8. J.C. Kennedy and R.H. Pottier, Endogenous protoporphyrin, a clinically useful photosensitizer for photodynamic therapy, *J. Photochem. Photobiol. B: Biol.*, 1992, **14**, 275–278.

9. H. Barr, N.A. Shepherd, A. Dix, D.J.H. Roberts, W.C. Tan and N. Krasner, Eradication of high-grade dysplasia in columnar-lined (Barrett's) oesophagus using photodynamic therapy with endogeneously generated protoporphyrin IX, *Lancet*, 1996, **348**, 584–586.

10. W.C. Tan, N. Krasner and P. O'Toole, Enhancement of photodynamic therapy in gastric cancer cells, *Gut*, 1997, **41**, 14–18.

11. P. Hinnen, R. de and A. Edixhoven, Ferrochelatase inhibition by 5-aminolevulinic acid based photodynamic therapy in human cell lines: clinical importance?, *Gastroenterology*, 1999, **116**, G1848.

12. R. Van Hillegersberg, J.W.O. Van Den Berg, W.J. Kort, O.T. Terpstra and J.H.P. Wilson, Selective accumulation of endogenously produced porphyrins in liver metasasis model in rats, *Gastroenterology*, 1992, **103**, 647–651.

13. J. Bedwell, A.J. MacRobert, D. Phillips and S.G. Bown, Fluorescence distribution and photodynamic effect of ALA-induced PPIX in DMH rat colonic model, *Br. J. Cancer*, 1992, **65**, 818–824.

14. P. Hinnen, F.W.M. de Rooij, E.M. Terlolouw, A. Edixhoven, H. van Dekken, R. Van Hillegersberg, H.W. Tilanus and P.D. Siersema, Porphyrin biosynthesis in human Barrett's oesophagus and adenocarcinoma after ingestion of 5-amino-laevulinic acid, *Br. J. Cancer*, 2000, **83**, 539–543.

15. P. Hinnen, F.W.M. de Rooij, W.C.J. Hop, A. Edixhoven, H. van Dekken, J.H.P. Wilson and P. Siersema, Timing of 5-aminolaevulinic acid-induced photodynamic therapy for the treatment of patients with Barrett's oesophagus, *J. Photochem. Photobiol.*, 2002, **68**, 8–14.

16. M. Ortner, K. Zumbusch, J. Liebetruth, B. Ebert and D. Nolte, Photodynamic therapy in Barrett's esophagus after local administration of 5-aminolaevulinic acid, *Gastroenterology*, 1997, **112**, A633.

17. M.-A.E. Ortner, B. Ebert, E. Hein, K. Zumbusch, D. Nolte, U. Sukowski, J. Weber-Eibel, B. Fleige, M. Dietel, M. Stolte, G. Oberhuber, R. Porschen, B. Klump, H. Hortnagl, H. Lochs and H. Rinneberg, Time gated fluorescence spectroscopy in Barrett's oesophagus, *Gut*, 2003, **52**, 28–33.

18. S.L.W. Wong Kee and K.K. Wang, Optical detection and eradication of dysplastic Barrett's esophagus, *Technol. Cancer Res. Treat.*, 2003, **2**, 51–69.

19. M. Panjehpour, R. Overholt, T. Vo-Dinh, R.C. Haggit, D.H. Edwards and P.F. Buckley, Endoscopic fluorescence detection of high-grade dysplasia in Barrett's esophagus, *Gastroenterology*, 1996, **111**, 93–101.

20. E. Endlicher, R. Knuechel, T. Hauser, R.M. Szeimies, J. Schölmerich and H. Messman, Endoscopic fluorescence detection of low and high grade dysplasia in Barrett's oesophagus using systemic or local 5-aminolaevulinic acid sensitisation, *Gut*, 2001, **48**, 314–319.

21. L. Gossner, H. Stepp, R. Sroka, A. May, M. Stolte and C. Ell, Photodynamic diagnosis (PDD) of high grade dysplasia and early cancer in Barrett's esophagus using 5-aminolevulinic acid (ALA), *Gastroenterology*, 1998, **114**, A136.

22. B. Mayinger, S. Neidhardt, H. Reh, P. Martus and E.G. Hahn, Fluorescence induced with 5-aminolevulinic acid for the endoscopic detection and follow-up of esophageal lesions, *Gastrointest. Endosc.*, 2001, **54**, 572–578.

23. H. Messmann, E. Endlicher, G. Freunek, P. Rummele, J. Scholmerich and R. Knuchel, Fluorescence endoscopy for the detection of low and high grade dysplasia in ulcerative colitis using systemic or local 5-aminolaevulinic acid sensitisation, *Gut*, 2003, **52**, 1003–1007.

24. C. Kendall, A study of Raman spectroscopy for early detection and classification of malignancy in oesophageal tissue, *PhD. Thesis, Cranfield. University* (2003).

25. N. Lange, P. Jichlinski, M. Zellweger, M. Forrer, A. Marti, L. Guillou, P. Kucera, G. Wagnieres and B.H. van den, Photodetection of early human bladder cancer based on the fluorescence of 5-aminolaevulinic acid hexylester-induced pro-toporphyrin IX: a pilot study, *Br. J. Cancer*, 1999, **80**, 185–193.

26. E. Endlicher, C. Gelbmann, R. Knüchel, A. Fürst, R.M. Szeimies, S. Gölder, J. Schölmerich and H. Messmann, Hexaminolevulinate-induced fluorescence endoscopy in patients with rectal adenoma and cancer – a pilot study, *Gastrointest. Endosc.*, 2004, **60**, 449–454.

27. T. Hauser, R.M. Szeimies and J. Scholmerich, Side effects of 5-aminolevulinic acid photosensitization in photodiagnosis of precancerous lesions of the gastrointestinal tract, *Gastroenterology*, 1999, **114**, G3092.

28. J. Van den Boojot, R. Van Hillegersberg, F.W.M. de Rooij, R.W.F. De Bruin, A. Edixhoven-Bosdijk, A.B. Houtsmuller, P.D. Siersema, J.H.P. Wilson and H.W. Tilanus, 5-aminolaevulinic acid-induced protoporphyrin accumulation in tissues: pharmacokinetics after oral or intravenous administration, *J. Photochem. Photobiol. B.*, 1998, **44**, 29–38.

29. W.C. Tan, C. Fulljames, N. Stone, A.J. Dix, N. Shepherd, D.J.H. Roberts, S.B. Brown, N. Krasner and H. Barr, Photodynamic therapy using 5-aminolaevulinic acid for oesophageal adenocarcinoma associated with Barrett's metaplasia, *J. Photochem. Photobiol. B.*, 1999, **53**, 75–80.

30. M. Panjehpour, B.F. Overholt and J.M. Haydek, Light sources and delivery devices for photodynamic therapy in the gastrointestinal tract, *Gastrointest. Endosc. Clin. North America.*, 2000, **10**, 513–532.

31. N. Stone, Standardizing Dosimetry in esophageal PDT. An argument for use of centering devices and removal of misleading units, *Technol. Cancer Res. Treat.*, 2003, **2**, 76–84.

32. A. Radu, G. Wagnieres, B.H. van den and P. Monnier, Photodynamic therapy of early squamous cell cancers of the esophagus, *Gastrointest. Endosc. Clin. North America.*, 2000, **10**, 439–460.

33. J. Van den Boojot, R. Van Hillegersberg, H.J. Van Staveren, R.W.F. De Bruin, P.D. Siersema and H.W. Tilanus, Timing of illumination is essential for effective and safe photodynamic therapy: a study in the normal rat oesophagus, *Br. J. Cancer*, 1999, **79**, 825–830.

34. J. Van den Boojot, H.J. Van Staveren, R.W.F. De Bruin, P.D. Siersema and R. Hillegersberg, Fractionated illumination for oesophageal ALA-PDT: effect on blood flow and PpIX formation, *Las. Med. Sci.*, 2001, **16**, 16–25.

35. M. Hage, P.D. Sieresma, H. van Dekken, E.W. Steyerberg, J. Haringsma, D. van, V. T.E. Grool, R.L.P. van Veen, H.J.C.M. Sterenborg and E.J. Kuipers, 5-Amino-levulinic acid photodynamic therapy versus argon plasma coagulation for ablation of Barrett's oesophagus: a randomised trial, *Gut*, 2004, **53**, 785–790.

36. B.F. Overholt, M. Panjehpour and J.M. Haydek, Photodynamic therapy for Barrett's esophagus: follow-up in 100 patients, *Gastrointest. Endosc.*, 1999, **49**, 1–7.

37. B.F. Overholt, C.J. Lightdale, K. Wang, M. Canto, S. Burdick, H. Barr, N. Macron, R.C. Haggitt, M. Bonner, S.L. Taylor and M. Depot, International Multicenter Partially Blinded Randomised Study of the Efficacy of Photodynamic Therapy (PDT) using Porfimer Sodium (POR) for the ablation of High-Grade Dysplasia (HGD) in Barrett's Esophagus (BE) Results of 24 Month Follow-up, *Gastroenterology*, 2003, **124**(Suppl. 1), A20.

38. K. Ragunath, N. Krasner, V.S. Raman, M.T. Haqqani, C.J. Phillips and W.Y. Cheung, Endoscopic ablation of dysplastic Barrett's oesophagus comparing Argon Plasma Coagulation and Photodynamic therapy: Short term results of a randomized prospective trial, *Br. Med. Laser Annual Conf.*, 2002, Suppl. 1, 5.

39. L. Gossner, M. Stolte, R. Stroke and C. Ell, Photodynamic therapy of high-grade dysplasia and early stage carcinomas by means of 5-aminolaevulinic acid, *Gastroenterology*, 1998, **114**, 447–455.

40. H. Barr, W.C. Tan and N.A. Shepherd, Photodynamic therapy (PDT) using 5-aminolaevulinic acid (5-ALA) for oesophageal adenocarcinoma associated with Barrett's metaplasia, *Laser Med. Sci.*, 1997, **4**, 9.

41. H. Barr, N.A. Shepherd, A. Dix, D.J.H. Roberts, W.C. Tan and N. Krasner, Eradication of high grade dysplasia in columnar-lined (Barrett's) oesophagus using photodynamic therapy with endogenously generated protoporphyrin IX, *Lancet*, 1996, **348**, 584–585.
42. C.J. Kelty, R. Ackroyd, N.J. Brown, T.J. Stephenson, C.J. Stoddard and M.W.R. Reed, Endoscopic ablation of Barrett's oesophagus: a randomised controlled trial of photodynamic therapy vs argon plasma coagulation, *Br. J. Surg.*, 2004, **91**(Suppl. 1), 42.
43. R. Ackroyd, N.J. Brown, M.F. Davis, T.J. Stephenson, S.L. Marcus, C.J. Stoddard, A.G. Johnson and M.W.R. Reed, Photodynamic therapy for dysplastic Barrett's oesophagus: a prospective, double blind, randomized, placebo-controlled trial, *Gut*, 2000, **47**, 613–617.
44. G.C. Harewood, T.H. Baron, A. Rumalla, K.K. Wang, G.J. Gores, L.M. Stadheim and P.C. de Groen, Pilot study to assess patient outcomes following endoscopic application of photodynamic therapy for advanced cholangiocarcinoma, *J Gastroenterol. Hepatol.*, 2005, **20**, 415–420.

*Chapter 9*

# Potential Future Indications

**Alison Curnow**

**Table of Contents**

## Abstract

ALA-PDT can be applied to many clinical situations where cellular destruction or visualisation is required. The clinical applications of this technique are therefore diverse and currently undergoing rapid evolution, as we better understand the mechanism of action of ALA-PDT and its therapeutic potential. As a result new applications are currently being investigated alongside the refinement of the treatment parameters of already established indications. Potential indications of ALA-PDT that are currently being investigated (in either preliminary clinical trials or pre-clinical studies) and that may prove to be of clinical importance in the future are discussed. It is also clear that the clinical outcome of ALA-PDT can be limited in certain circumstances. Several experimental strategies are currently in development, which if successful may improve the future clinical application of ALA-PDT. ALA-PDT is an exciting treatment modality that with careful refinement, application and development will continue to be a useful treatment option in the future.

## 9.1. Introduction

ALA-PDT is a rapidly evolving field. As the previous chapters of this handbook demonstrate the clinical applications of this technique are varied. This diversity of indication with respect to both PDT and photodiagnosis is possible as all nucleated cells have the capacity to generate PpIX from ALA or its esterised conjugates.[1] Furthermore the mechanism of action of ALA-PDT *via* reactive oxygen species produces intracellular oxidative stress,[2] which is capable of damaging cellular components (such as lipids, proteins and nucleotides) universally. As a result this technique is very versatile and can be applied to many clinical situations where cellular destruction or visualisation is required as long as the pathology remains superficial (as red light has only limited penetration into tissue).[3]

Each chapter of this handbook has detailed the present status of ALA-PDT within the main clinical specialities where this technique has been currently established as an appropriate and useful treatment or imaging method. This chapter will focus on potential indications of ALA-PDT that are currently being investigated (in either preliminary clinical trials or pre-clinical studies) and may prove to be of clinical importance in the future. Additionally experimental strategies that may improve the future clinical application of ALA-PDT are also reviewed.

## 9.2. Areas of Future Clinical Potential

### 9.2.1. Infectious Diseases

An area of increasing importance, interest and future potential is the use of ALA-PDT to destroy bacteria, fungi, viruses, yeasts and parasites.[4]

The progressive emergence and spread of microbial drug resistance is a major medical concern. Overuse of antibiotics and poor infection control procedures/policies have been contributing factors to this situation.[5] Alternative treatment and sterilisation strategies are therefore being sought and ALA-PDT (as well as PDT conducted with other photosensitising agents) is currently being investigated for this purpose.

It appears that the reactive oxygen species based mechanism of action of PDT is highly effective at destroying microorganisms of various types with the most likely cellular target being the outer cell wall.[6] Additionally by careful adjustment of the treatment parameters it appears possible to selectively destroy target microorganisms with minimal damage to normal host tissue.[7] This treatment strategy therefore has great potential for the decontamination of extensive wounds and large areas of damaged tissue such as burns and may prove to be a very effective alternative to antibiotic therapy in these circumstances.[8]

Advantages of using PDT in this manner include a broad spectrum of action against gram-positive and gram-negative bacteria, yeasts and mycoplasmas, effective destruction of antibiotic-resistant bacteria and no induction of PDT resistance due to the universal nature of the oxidative damage induced by the photodynamic reaction.[8] Combination of PDT with other treatment options is also feasible.

The potential application of the anti-microbial action of ALA-PDT is therefore wide and diverse. Investigations have also shown that PDT can be a useful treatment of *Heliobacter pylori*, the bacteria associated with stomach ulcers,[9,10] oral bacterial infections (while sparing the normal underlying oral mucosa)[11] as well as generalised wound healing.[12]

Viral infections can also be treated by employing ALA-PDT. Topical ALA-PDT has been demonstrated clinically[13] to be an effective treatment of selected warts caused by the human papilloma virus (HPV). The success of ALA-PDT in gynaecological applications such as cervical intraepithelial neoplasia (CIN) is also attributed in part to the effectiveness of ALA-PDT eliminating the HPV infection that can be associated with this condition.[14] In vulval intraepithelial neoplasia (VIN) if the therapy fails to effectively eradicate HPV, poor clinical results are obtained.[15] ALA-PDT has also been demonstrated experimentally to be useful for the control of herpes genitalis through the destruction of the herpes simplex virus.[16]

Research is also underway on the use of ALA-PDT towards the treatment of parasites,[17] onychomycosis of finger and toenails[18] as well as other dermatological fungal infections[19,20] and more tropical diseases. This area of PDT research is therefore multifaceted in nature and is becoming of increasing importance.

### 9.2.2. Immunological Effects

Besides the anti-microbial action of ALA-PDT, another research area of increasing focus is the immunological effects of ALA-PDT and its potential as a therapy for certain immunological diseases (for instance rheumatoid arthritis) as well as its potential as a technique of bone marrow purging for the treatment of leukaemias with autologous bone marrow transplantation. The full immunological effects of ALA-PDT are complex and are still being elucidated[21] but it has been established that a strong inflammatory response can be induced[22] and that certain aspects of the immune response can be modulated following treatment resulting in a better immunological anti-tumour response.[23] Additionally it has been observed that patients who are immunosuppressed respond poorly to ALA-PDT[24] indicating the importance of the immune system in the overall treatment effect of ALA-PDT in some indications. It is anticipated that this will be an area of further PDT research, which will have a major impact on clinical PDT practice of the future.

### 9.2.3. Pneumology

ALA-PDT is a therapy that is also currently being investigated for various pneumological indications. There are several stages of lung neoplasia where this therapy may be a useful treatment option. Encouraging results have been obtained to date using ALA-induced fluorescence detection *via* bronchoscopy to detect early lung cancers.[25] This disease once located is then irradiated so that it is treated with ALA-PDT.[25] This methodology appears to provide an effective treatment for early central lung cancers and can also be used as an alternative treatment option for unresectable lung cancers or in those patients where conventional surgery is contraindicated.[26]

ALA fluorescence detection has also been shown to be effective in the visualisation of pleural disease.[27] This procedure is conducted *via* thoracoscopy and has been demonstrated to be superior to simple white-light inspection of the pleural cavity, with additional neoplastic disease being detected (with ALA fluorescence detection).[27] Further evaluation of the specificity of this technique for the detection of pleural malignancies is currently being undertaken but this is another clinical application of ALA fluorescence detection with encouraging future clinical potential.

## 9.3. ALA-PDT Enhancement

Besides optimising the treatment parameters of ALA-PDT for the numerous current clinical indications, experimental research is being conducted to investi-gate methods that may enhance the effects of this treatment modality and thus extend its clinical usefulness in a variety of applications.

In dermatology various methods have been used to increase ALA penetra-tion into skin as this has been found to be a major limiting factor when trying to treat thicker skin disease.[28] These methods have been as diverse as tape stripping to remove the scaly stratum corneum prior to topical ALA appli-cation[29] or iontophoresis to drive ALA faster and deeper into the skin.[30]

Some success has also been demonstrated using the solvent dimethylsulfoxide (DMSO) to pre-treat skin before ALA application.[31]

With systemic oral administration it has been found that ALA has a maximal tolerated oral dose of 60 mg kg$^{-1}$ [32] limited by aberrations in liver biochemistry. Some work has been conducted which splits the oral dose into several smaller doses in an attempt to deliver an overall higher dose without excessive adverse effects.[33] Improvement in efficacy has also been achieved in a variety of indications by conducting repeated complete ALA-PDT treatments in relative quick succession.[34,35]

Substantial work has been conducted in order to investigate optimal irradiation parameters. Significant success has been demonstrated with low fluence rates (delivering a set light dose over a longer time period)[36] and fractionated light regimes (interrupting the light dose either for a short period of time (seconds/minutes)[37,38] or alternatively for a long period of time (hours/days)[39]). These methods are hypothesised to improve ALA-PDT outcome by either minimising oxygen depletion, thereby allowing a bigger PDT effect to be produced (as molecular oxygen is essential to the PDT process), or allowing more PpIX to accumulate during a substantial dark period. A related but different technique employs hyperbaric conditions to artificially elevate oxygen levels to improve PDT outcome.[40]

It is also possible to further manipulate the haem biosynthesis pathway on which ALA-PDT treatment is based. This can temporarily further increase the level of PpIX accumulation so that a greater PDT effect is produced on irradiation. This can be done by employing iron chelating agents such as EDTA,[41] desferrioxamine,[42] 1,10-phrenanthroline[43,44] or hydroxypyridinones[45] to slow the conversion of PpIX to haem. Alternatively, the enzyme which catalyses this reaction (ferrochelatase) can be inhibited.[46] It is also possible to increase the level of PpIX accumulation by upregulating its synthesis.[43] Recently experimental research has started to investigate whether ALA-PDT effects can be enhanced by modulating molecular or apoptotic processes that occur during this treatment.[47] By further understanding the mechanism of action of ALA-PDT and the disease processes that one is trying to treat, further advances in this respect will be possible.

An area of increasing interest is the combination of ALA-PDT with other PDT modalities such as Photofrin[48] or ATX-S10 (Na)[49] where longer wavelengths of light can be used to achieve deeper effects where required. Experimentally two-photon PDT is also being pursued.[50,51] ALA-PDT can also be combined with hyperthermia[52] or bioreductive drugs[53] to produce enhanced effects. Additionally PDT in general is also being delivered in combination with other more traditional oncology treatments such as chemotherapy,[54] radiotherapy[55] or surgical techniques.[56] These PDT combination therapies appear to be a very promising avenue of potential advancement that require further investigation and should be relatively simple to translate into clinical practice.

Much, however, can be achieved by simply optimising ALA-PDT treatment parameters for each individual indication. Furthermore, with the development of real-time non-invasive monitoring systems (for PpIX fluorescence, oxygen levels and local blood flow as well as singlet oxygen production) it may be possible for PDT practitioners in the future to monitor clinical ALA-PDT as it is in progress. This should allow treatment parameters to be determined and optimised on a patient-to-patient basis, thus producing the best clinical effects possible in each individual patient.

It is clear that some of these methods of ALA-PDT enhancement, if they prove to be efficacious, will be easier to implement into clinical practice than others. The diversity of strategies does indicate, however, the complexity of the mechanism underlying this treatment modality and the importance of adhering closely to the treatment parameters for the indication being treated. This accumulating body of evidence also indicates that ALA-PDT is still being evolved and perfected and that it is highly likely that new treatment modifications will be implemented into clinical practice for some time to come.

ALA-PDT practice in the future is likely to involve more combinations with existing treatment options with more medical practitioners practising PDT alongside other therapeutic modalities. As real-time monitoring of the PDT process in action becomes more feasible, this should enable treatment parameters to be optimised on an individual basis in indications where a generalised regime is not adequate and maximising the ALA-PDT efficacy is essential for a successful clinical outcome.

## 9.4. Conclusions

The clinical applications of ALA-PDT are diverse and numerous. It has been clearly established that ALA-PDT can provide effective treatment and early diagnosis of a number of malignant and pre-malignant conditions[57] as well as being able to effectively destroy a range of microorganisms associated with a variety of infectious diseases.[4] The adoption of this treatment modality in clinical practice has therefore been rapid. Reducing costs of suitable light sources and increased awareness of the benefits of PDT by medical practitioners and patients alike have accelerated this adaptation.

ALA-PDT has been approved for several indications in many countries around the world, and the low incidence of serious complications and versatility in drug delivery (topical, local or systemic) has facilitated these developments. The ALA-PDT field is rapidly evolving with a variety of new applications being investigated alongside the refinement of the treatment parameters of those indications, which have already been established. Although the effects of ALA-PDT can be limited in certain circumstances, there are a number of methods of enhancement that are currently in development, which may prove to be of clinical use in the future (in applications where bigger effects are required). These techniques may therefore further extend the clinical application of this exciting treatment modality beyond the limits of current ALA-PDT practice.

# References

1. Q. Peng, K. Berg, J. Moan, M. Kongshaug and J.M. Nesland, 5-Aminolevulinic acid-based photodynamic therapy: Principles and experimental research, *Photochem. Photobiol.*, 1997, **65**, 235–251.
2. M.C. Luna, A. Ferrario, S. Wong, A.M. Fisher and C.J. Gomer, Photodynamic therapy-mediated oxidative stress as a molecular switch for the temporal expression of genes ligated to the human heat shock promoter, *Cancer Res.*, 2000, **60**, 1637–1644.
3. Q. Peng, T. Warloe, K. Berg, J. Moan, M. Kongshaug, K.E. Giercksky and J.M. Nesland, 5-Aminolaevulinic-based photodynamic therapy: Clinical research and future challenges, *Am. Cancer Soc.*, 1997, **79**, 2282–2308.
4. G. Jori and S.B. Brown, Photosensitized inactivation of microorganisms, *Photochem. Photobiol. Sci.*, 2004, **3**, 403–405.
5. M. Sharma, H. Bansal and P.K. Gupta, Photodynamic inactivation of antibiotic resistant strain of *Pseudomonasaeruginosa* by porphyrins induced by amino-laevulinic acid, *Indian J. Med. Res.*, 2002, **116**, 99–105.
6. M. Kreitner, K.-H. Wagner, G. Alth, R. Ebermann, H. Foissy and I. Elmadfa, Finding optimal photosensitisers for the decontamination of foods by the photodynamic effect, *Forum Nutr.*, 2003, **56**, 367–369.
7. P. Meisel and T. Kocher, Photodynamic therapy for periodontal diseases: State of the art, *J. Photochem. Photobiol. B*, 2005, **79**, 159–170.
8. M.R. Hamblin and T. Hasan, Photodynamic therapy: A new antimicrobial approach to infectious disease?, *Photochem. Photobiol. Sci.*, 2004, **3**, 436–450.
9. C.E. Millson, W. Thurrell, G. Buonaccorsi, M. Wilson, A.J. Macrobert and S.G. Bown, The effect of low-power laser light at different doses on gastric mucosa sensitized with methylene blue, haematoporphyrin derivative or toluidine blue, *Lasers Med. Sci.*, 1997, **12**, 145–150.
10. M.R. Hamblin, J. Viveiros, C. Yang, A. Ahmadi, R.A. Ganz and M.J. Tolkoff, *Helicobacter pylori* accumulates photoactive porphyrins and is killed by visible light, *Antimicrob. Agents Chemother.*, 2005, **49**, 2822–2827.
11. M. Wilson, Lethal photosensitisation of oral bacteria and its potential application in the photodynamic therapy of oral infections, *Photochem. Photobiol. Sci.*, 2004, **3**, 412–418.
12. T.N. Demidova and M.R. Hamblin, Photodynamic therapy targeted to pathogens, *Int. J. Immunopathol. Pharmacol.*, 2004, **17**, 245–254.
13. I.M. Stender, *Treatment of human papilloma virus*, in *Photodynamic Therapy*, M.P. Goldman (ed), Elsevier Saunders, Philadelphia, 2005, pp. 77–88.
14. F. Wierrani, A. Kubin, R. Jindra, M. Henry, K. Gharehbaghi, W. Grin, J. Soltz-Szotz, G. Alth and W. Grunberger, 5-aminolevulinic acid-mediated photodynamic therapy of intraepithelial neoplasia and human papillomavirus of the uterine cervix – A new experimental approach, *Cancer Detect. Prev.*, 1999, **23**, 351–355.
15. E.-S. Abdel-Hady, P. Martin-Hirsh, M. Duggan-Keen, P.L. Stern, J.V. Moore, G. Corbitt, H.C. Kitchener and I.N. Hampson, Immunological and viral factors associated with the response of vulval intraepithelial neoplasia to photodynamic therapy, *Cancer Res.*, 2001, **61**, 192–196.
16. Z. Smetana, Z. Malik, A. Orenstein, E. Mendelson and H.-E. Ben, Treatment of viral infections with 5-aminolevulinic acid and light, *Lasers Surg. Med.*, 1997, **21**, 351–358.

17. T.G. Smith and K.C. Cain, Inactivation of *Plasmodiumfalciparum* by photodynamic excitation of hemecycle intermediates derived from aminolevulinic acid, *J. Infect. Dis.*, 2004, **190**, 184–191.

18. R.F. Donnelly, P.A. McCarron, J.M. Lightowler and A.D. Woolfson, Bioadhesive patch-based delivery of 5-aminolevulinic acid to the nail for photodynamic therapy of onychomycosis, *J. Control. Rel.*, 2005, **103**, 381–392.

19. P.G. Calzavara-Pinton, M. Venturini and R. Sala, A comprehensive overview of photodynamic therapy in the treatment of superficial fungal infections of the skin, *J. Photochem. Photobiol. B*, 2005, **78**, 1–6.

20. A.S. Morales, Antimalarial activity of 5-aminolevulinic acid-induced protoporphyrin IX photodynamic therapy in rodent model, M.Sc. Thesis, Queen's University, Kingston, Canada, 1995.

21. M. Korbelik and G.J. Dougherty, Photodynamic therapy-mediated immune response against subcutaneous mouse tumours, *Cancer Res.*, 1999, **59**, 1941–1946.

22. L.M. Skivka, O.B. Gorobets, V.V. Kutsenok, M.O. Lozinsky, A.N. Borisevich, A.G. Fedorchuk, V.V. Kholin and N.F. Gamaleya, 5-aminolevulinic acid mediated photodynamic therapy of Lewis lung carcinoma: A role of tumor infiltration with different cells of immune system, *Exp. Oncol.*, 2004, **26**, 312–315.

23. P.C. Kousis, S. Schneider, B.W. Henderson and S.O. Gollnick, Effects of photodynamic therapy (PDT) dose, Proceedings of the 10th World Congress of the International Photodynamic Association, Munich, Germany, 2005, abstract S13.06, 117.

24. E.F. Gudgin Dickson, J.C. Kennedy and R.H. Pottier, *Photodynamic therapy using 5-aminolevulinic acid-induced protoporphyrin IX*, in *Photodynamic Therapy*, T. Patrice (ed), The Royal Society of Chemistry, Cambridge, 2003, pp. 81–103.

25. F.D. Sheski and P.N. Mathur, Diagnosis and treatment of early lung cancer: As it stands today, *Sem. Respir. Crit. Care Med.*, 2004, **25**, 387–397.

26. K. Moghissi, K. Dixon, J.A.C. Thorpe, C. Oxtoby and M.R. Stringer, Photodynamic therapy (PDT) for lung cancer: The Yorkshire Laser Centre experience, *Photodiagn. Photodyn. Ther.*, 2004, **1**, 253–262.

27. M. Triesscheijn, M. Aalders, S. Burgers, F. Stewart and P. Baas, Fluorescence detection of pleural malignancies using 5-aminolevulinic acid, Proceedings of the 10th World Congress of the International Photodynamic Association, Munich, Germany, 2005, abstract S04.06, 67.

28. P. Babilas, S. Karrer, A. Sidoroff, M. Landthaler and R.-M. Szeimies, Photodynamic therapy in dermatology – An update, *Photodermatol. Photoimmunol. Photomed.*, 2005, **21**, 142–149.

29. J. Bartosik, I.-M. Stender, T. Kobayasi and M.S. Agren, Ultrastructural alteration of tape-stripped normal human skin after photodynamic therapy, *Eur. J. Dermatol.*, 2004, **14**, 91–95.

30. J.-Y. Fang, W.-R. Lee, S.-C. Shen, Y.-P. Fang and C.-H. Hu, Enhancement of topical 5-aminolaevulinic acid delivery by erbium: YAG laser and microdermabrasion: A comparison with iontophoresis and electroporation, *Br. J. Dermatol.*, 2004, **151**, 132–140.

31. A. Orenstein, G. Kostenich, L. Roitman, H. Tsur, D. Katanick, J. Kopolovic, B. Ehrenberg and Z. Malik, Photodynamic therapy of malignant lesions of the skin mediated by topical application of 5-aminolevulinic acid in combination with DMSO and EDTA, *Lasers Life Sci.*, 1996, **7**, 49–57.

32. S.L. Marcus, R.S. Sobel, A.L. Golub, R.L. Carrol, S. Lundahl and D.G. Shulman, Photodynamic therapy and photodiagnosis using endogenous photosensitisation

induced by 5-aminolevulinic acid: Current clinical and development status, *J. Clin. Laser Med. Surg.*, 1996, **14**, 59–66.

33. K.F.M. Fan, C. Hopper, P.M. Speight, G. Buonaccorsi, A.J. MacRobert and S.G. Bown, Photodynamic therapy using 5-aminolevulinic acid for premalignant and malignant lesions of the oral cavity, *Cancer*, 1996, **78**, 1374–1383.

34. A. Salim, J.A. Leman, J.H. MacColl, R. Chapman and C.A. Morton, Randomized comparison of photodynamic therapy with topical 5-fluorouracil in Bowen's disease, *Br. J. Dermatol.*, 2003, **148**, 539–543.

35. R. Waidelich, H. Stepp, R. Baumgartner, E. Weninger, A. Hofstetter and M. Kriegmair, Clinical experience with 5-aminolevulinic acid and photodynamic therapy for refractory superficial bladder cancer, *J. Urol.*, 2001, **165**, 1904–1907.

36. D.J. Robinson, H.S. de Bruijn, N. van der Veen, M.R. Stringer, S.B. Brown and W.M. Starr, Fluorescence photobleaching of ALA-induced protoporphyrin IX during photodynamic therapy of normal hairless mouse skin: The effect of light dose and irradiance and the resulting biological effect., *Photochem. Photobiol.*, 1998, **67**, 140–149.

37. H. Messmann, P. Mlkvy, G. Buonaccorsi, C.L. Davies, A.J. MacRobert and S.G. Bown, Enhancement of photodynamic therapy with 5-aminolaevulinic acid-induced porphyrin photosensitisation in normal rat colon by threshold and light fractionation studies, *Br. J. Cancer*, 1995, **72**, 589–594.

38. A. Curnow, B.W. McIlroy, M.J. Postle-Hacon, A.J. MacRobert and S.G. Bown, Light dose fractionation to enhance photodynamic therapy using 5-aminolaevulinic acid in the normal rat colon, *Photochem. Photobiol.*, 1999, **69**, 71–76.

39. D.J. Robinson, H.S. de Bruijn, W.M. Star and H.J. Sterenborg, Dose and timing of the first light fraction in two-fold illumination schemes for topical ALA-mediated photodynamic therapy of hairless mouse skin, *Photochem. Photobiol.*, 2003, **77**, 319–323.

40. F. Tomaselli, A. Maier, H. Pinter, H. Stranzl and F.M. Smolle-Jüttner, Photodynamic therapy enhanced by hyperbaric oxygen in acute endoluminal palliation of malignant bronchial stenosis (clinical pilot study in 40 patients), *Eur. J. Cardio-Thorac. Surg.*, 2001, **19**, 549–554.

41. J. Hanania and Z. Malik, The effect of EDTA and serum on endogenous porphyrin accumulation and photodynamic sensitisation of human K562 leukemic cells, *Cancer Lett.*, 1992, **65**, 127–131.

42. B. Ortel, A. Tanew and H. Honigsmann, Lethal photosensitisation by endogenous porphyrins of PAM cells – Modification by desferrioxamine, *J. Photochem. Photobiol. B*, 1993, **17**, 273–278.

43. N. Rebeiz, C.C. Rebeiz, S. Arkins, K.W. Kelley and C.A. Rebeiz, Photodestruction of tumor cells by induction of endogenous accumulation of protoporphyrin IX: Enhancement by 1,10-phenanthroline, *Photochem. Photobiol.*, 1992, **55**, 431–435.

44. N. Rebeiz, S. Arkins, C.A. Rebeiz, J. Simon, J.F. Zachary and K.W. Kelley, Induction of tumor necrosis by δ-aminolevulinic acid and 1,10-phenanthroline photodynamic therapy, *Cancer Res.*, 1996, **56**, 339–344.

45. A. Curnow, B.W. McIlroy, M.J. Postle-Hacon, J.B. Porter, A.J. MacRobert and S.G. Bown, Enhancement of 5-aminolaevulinic acid induced photodynamic therapy using iron chelating agents, *Br. J. Cancer*, 1998, **78**, 1278–1282.

46. G. Bhasin, H. Kausar and M. Arthar, Protoporphyrin-IX accumulation and cutaneous tumour regression in mice using a ferrochelatase inhibitor, *Cancer Lett.*, 2002, **187**, 9–16.

47. Y. Akita, K. Kazaki, A. Nakagawa, T. Saito, S. Ito, Y. Tamada, S. Fujiwara, N. Nishikawa, K. Uchida, K. Yoshikawa, T. Noguchi, O. Miyaishi, K. Shimozato, S. Saga, Y. Matsumoto, Cyclooxygenase-2 is a possible target of treatment approach in conjunction with photodynamic therapy for various disorders in skin and oral cavity, *Br. J. Dermatol.* 2004, 151, 472–480.

48. Q. Peng, T. Warloe, J. Moan, A. Godal, F. Apricena, K.-E. Giercksky and J.M. Nesland, Antitumor effect of 5-aminolaevulinic acid-mediated photodynamic therapy can be enhanced by the use of a low dose of photofrin in human tumor xenografts, *Cancer Res.*, 2001, **61**, 5824–5832.

49. H. Takahashi, S. Nakajima, I. Sakata, A. Ishida-Yamamoto and H. Iizuka, Photodynamic therapy using a novel photosensitizer ATX-S10 (Na): Comparative effect with 5-aminolevulinic acid on squamous cell carcinoma cell line SCC15, ultraviolet B-induced skin tumour, and phorbol ester-induced hyperproliferative skin, *Arch. Dermatol. Res.*, 2005, **296**, 496–502.

50. R.L. Goyan and D.T. Cramb, Near-infrared two-photon excitation of protoporphryin IX: Photodynamics and photoproduct generation, *Photochem. Photobiol.*, 2000, **72**, 821–827.

51. M. Fournier, C. Pépin, D. Houde, R. Ouellet and J.E. van Lier, Ultrafast studies of the excited-state dynamics of copper and nickel phthalocyanine tetrasulfonates: Potential sensitizers for the two-photon photodynamic therapy of tumors, *Photochem. Photobiol. Sci.*, 2004, **3**, 120–126.

52. A. Orenstein, G. Kostenich, Y. Kopolovic, T. Babushkina and Z. Malik, Enhancement of ALA-PDT damage by IR-induced hyperthermia on a colon carcinoma model, *Photochem. Photobiol.*, 1999, **69**, 703–707.

53. J.C.M. Bremner, G.E. Adams, J.K. Pearson, J.M. Sansom, I.J. Stratford, J. Bedwell, S.G. Bown, A.J. MacRobert and D. Philips, Increasing the effect of photodynamic therapy on the RIF-1 murine sarcoma, using the bioreductive drugs RSU1069 and RB6145, *Br. J. Cancer*, 1992, **66**, 1070–1076.

54. C.M. Peterson, J.-G. Shiah, Y. Sun, P. Kopečková, T. Minko, R.C. Straight and J. Kopeček, HPMA copolymer delivery of chemotherapy and photodynamic therapy in ovarian cancer, *Adv. Exp. Med. Biol.*, 2003, **519**, 101–123.

55. N. Umegaki, R. Moritsugu, S. Katoh, K. Harada, H. Nakano, K. Tamai, K. Hanada and M. Tanaka, Photodynamic therapy may be useful in debulking cutaneous lymphoma prior to radiotherapy, *Clin. Exp. Dermatol.*, 2004, **29**, 42–45.

56. A. Nanashima, H. Yamaguchi, S. Shibasaki, N. Ide, T. Sawai, T. Tsuji, S. Hidaka, Y. Sumida, T. Nakagoe and T. Nagayasu, Adjuvant photodynamic therapy for bile duct carcinoma after surgery A preliminary study, *J. Gastroenterol.*, 2004, **39**, 1095–1101.

57. R.R. Allison, G.H. Downie, R. Cuenca, X.-H. Hu, C.J.H. Childs and C.H. Sibata, Photosensitizers in clinical PDT, *Photodiagn. Photodyn. Ther.*, 2004, **1**, 27–42.

# Appendix

## A.1. Manufacturers of Light Sources and Applicators for FD and PDT

The following table provides a non-exhaustive list of manufacturers of light sources and applicators suitable for ALA-PDT and ALA-FD. The authors give no warranty for completeness or appropriateness of the company profile, address, *etc.*

| Name and Location | Website | Products | Application | Contact |
|---|---|---|---|---|
| Applied Optronics, South Plainfield, NJ | www.applied-optronics.com | Laser Diode Systems 635 nm to NIR wavelengths | PDT or FD applications | Tel: +1 908 753-6300 Fax: +1 908 753-4041 |
| Biocam GmbH, Regensburg, Germany | www.biocam.de | Fluorescence imaging system | FD of skin cancer | Tel: +49 941 78 53 98-0 Fax: +49 941 78 53 98-10 |
| Biolitec AG, Jena, Germany | www.biolitec.com | Diode laser 635 nm, up to 4 W, light applicators | PDT | Tel: +49 3641 508 550 Fax: +49 3641 508 599 |

*(continued)*

| Name and Location | Website | Products | Application | Contact |
|---|---|---|---|---|
| Biospec, Moscow, Russia | www.biospec.ru | Laser/light systems and dosimetry tools | Light sources, PDT dosimetry, software | Tel: +7 095 248-7352<br>Fax: +7 095 132-8200 |
| Curalux GbR, Munich, Germany | www.curalux.de | Power-beam splitter, fiber applicators, consulting | PDT | Tel: +49 89 70954884<br>Fax: +49 89 70954864 |
| Diomed, Cambridge, UK | www.diomedinc.com | Diode laser 630 nm, cylindrical fiber applicators | PDT, laser developed for Photofrin-PDT | Tel: +44 1223 729300<br>Fax: +44 1223 729329<br>Tel: +1 978 475 7771<br>Fax: +1 978 475 8488 |
| DUSA Pharmaceuticals, Inc., Wilmington, MA | www.dusapharma.com | Blu-U PDT-lamp Levulan Kerastick (ALA) | PDT of skin lesions: actinic keratosis; drug and light source | Tel: +1 978 657-7500<br>Fax: +1 978 657-9193 |
| High-power Devices, North Brunswick, NJ | www.hpdinc.com | Diode Lasers, 635, 655, 670, and 690 nm | PDT and FD applications | Tel: +1 732 249-2228<br>Fax: +1 732 249-8139 |
| Laserscope, San Jose, CA | www.laserscope.com | PDT-laser, tunable | PDT | Tel:+1 408 943 0636<br>Fax: +1 408 428 0512 |
| Lightguideoptics GmbH, Rheinbach, Germany | www.lgoptics.de | Radial diffusers, bare fibers | PDT | Tel: +49 2226 15850<br>Fax: +49 2226 158520 |

| Company | Website | Technology | Application | Contact |
|---|---|---|---|---|
| Lightsciences corp., Snoqualmie, WA | www.lightsciences.com | Linear LED array, non-ALA PDT-drug | PDT, mainly interstitial | Tel: +1 425 369-2800 Fax: +1 425 369-2801 |
| Lumacare, Newport Beach, CA | www.lumacare.com | PDT-lamp | PDT, mainly in dermatology | Tel: +1 949 422 1963 |
| Lumenis Inc., Santa Clara, CA | www.lumenis.com | "Intense pulsed light (IPL)", flashlamp (red and IR) and blue light source (Clearlight) | rejuvenation of photodamaged skin, Acne treatment | Tel: +1 408 764-3000 Fax: +1 408 764-3999 |
| Medlight SA, Ecublens, Switzerland | www.medlight.com | Fiber-based PDT-applicators | PDT with light sources that can be coupled to fibers | Tel: +41 21 697 0775 Fax: +41 21 697 0779 |
| Olympus Deutschland GmbH, Hamburg, Germany | www.olympus-owi.de | Fluorescence excitation light source, fluorescence cystoscopes, specialized camera | Fluorescence cystoscopy | Tel: +49 40 237 73 –0 Fax: +49 40 237 73 -57 26 |
| PerkinElmer Optoelectronics, Fremont, CA | www.perkinelmer.com | "Intense pulsed light (IPL)" flashlamp | rejuvenation of photodamaged skin | Tel: +1 510 979-6500 Fax: +1 510 687-1140 |
| Pharmacyclics Inc., Sunnyvale, CA | www.pharmacyclics.com | Texaphyrin-based PDT and Radiation sensitizers Light sources and dosimetry | PDT & Radiation Sensitizers | Tel: +1 408 774-0330 Fax: +1 408 774-0340 |

(*continued*)

| Name and Location | Website | Products | Application | Contact |
|---|---|---|---|---|
| PhotoCure ASA, Oslo, Norway | www.photocure.com | metvix (see Galderma) hexvix (ALA-esters) Aktilite LED-light source for PDT | PDT of skin and bladder, Drug and light source | Tel: +47 22 06 22 10 Fax: +47 22 06 22 18 |
| Photo-therapeutics Ltd, Altrincham, UK | www.phototherapeutics.com | LED-based PDT-lamp (Omnilux) | PDT in dermatology | Tel: +44 161 9255610 Fax: +44 161 9255628 |
| Pioneer Optics Company, Windsor Locks, CT | Contact Ron Hille: ronhille@aol.com | Customer bare end and diffuser fibers for PDT Cylindrical Diffuser Fibers Microlens Fibers | PDT | Tel: +860 292 8705 Fax: +860 292 8706 |
| PRP Optoelectronics Ltd, Towcester, UK | www.prpopto.com | LED-based PDT-lamp | PDT in dermatology | Tel: +44 1327 359135 Fax: +44 1327 359602 |
| Somta Ltd, Riga, Latvia | www.somta.lv | Fiber light guides, spherical and cylindrical diffusers | PDT | Tel: +371 7228249 Fax: +371 7820113 |
| SpectraCure AB, Lund, Sweden | www.spectracure.se | PDT-system with dosimetry-monitoring | PDT of skin lesions (basal cell ca), interstitial approach | Tel: +46 46 286 37 72 Fax: +46 46 286 37 79 |

| Company | Website | Product | Application | Contact |
|---|---|---|---|---|
| StockerYale Inc., Salem, NH | www.stockeryale.com (http://www.stockeryale.com/i/leds/app/pdt.htm) | LED-based lamp | PDT in dermatology | Tel: +1 603 893 8778<br>Fax: +1 603 893 5604 |
| Karl Storz GmbH & Co. KG, Tuttlingen, Germany | www.karlstorz.de | Fluorescence excitation light source (D-Light), fluorescence endoscopes, specialized cameras | Endoscopic FD | Tel: +49 7461 708-0<br>Fax: +49 7461 708-105 |
| Herbert Waldmann GmbH & Co. KG, Villingen-Schwenningen, Germany | www.waldmann-medizintechnik.com | FD-system, PDT-lamps (bulbs and LED-based) | FD and PDT in dermatology | Tel: +49 7720 601 0<br>Fax: +49 7720 601 290 |
| Richard Wolf GmbH, Knittlingen, Germany | www.richard-wolf.de | Fluorescence excitation light source, fluorescence cystoscopes, specialized camera | Fluorescence cystoscopy | Tel: +49 70 43 35-0<br>Fax: +49 70 43 35-300 |
| Saalmann GmbH, Herford, Germany | www.saalmann.net | FD&PDT-system (Wood's-lamp and green light) | FD and PDT in dermatology | Tel: +49 5221/20 44<br>Fax: +49 5221/272 35 |

## A.2. Manufacturers of ALA, ALA-Derivatives and Porphyrins

The following table provides a non-exhaustive list of manufacturers of ALA, ALA-derivatives and porphyrins independent on whether approval has been obtained or not. The authors give no warranty for completeness or appropriateness of the company profile, address, *etc.* ALA and some porphyrins can also be purchased with big suppliers of chemical compounds.

| Name and Location | Website | Products | Application | Contact |
|---|---|---|---|---|
| DUSA Pharmaceuticals, Inc., Wilmington, MA | www.dusapharma.com | Levulan Kerastick (ALA), Blu-U PDT-lamp | PDT of skin lesions: actinic keratosis; drug and light source | Tel: +1 978 657-7500 Fax: +1 978 657-9193 |
| Frontier Scientific Inc., Logan, UT | www.frontiersci.com | Porphyrin Products | ALA, PpIX for scientific purpose | Tel: +1 435 753-1901 Fax: +1 435 753-6731 |
| Fudan-Zhangjiang Bio-Pharmaceutical Ltd., Shanghai, China | www.fd-zj.com | ALA | FD and PDT Clinical studies | Tel: +86 21 58953355 Fax: +86 21 58553990 |
| GALDERMA Laboratorium GmbH, Duesseldorf, Germany | www.galderma.de | Metvix-Creme (see PhotoCure) | PDT of skin lesions: actinic keratoses and basal cell carcinoma | Tel: +49 211 5 86 01-04 Fax: +49 211 45 44 008 |
| medac GmbH, Hamburg, Germany | www.medac.de | ALA | Clinical studies on FD and PDT of bladder cancer (see also photonamic) | Tel: +49 4103 8006 0 Fax: +49 4103 8006 100 |

| Company | Website | Product | Purpose | Contact |
|---|---|---|---|---|
| NIOPIK (Federal State Unitary Enterprise State Scientific Center of Organic Intermediates and dyes), Moscow, Russia | www.niopik.ru | ALA (Alasens) | FD and PDT clinical studies | Tel: +7 095 2513100 Fax: 7 095 2541200 |
| PhotoCure ASA, Oslo, Norway | www.photocure.com | Metvix, (see Galderma) Hexvix (ALA-esters) Aktilite LED-light source for PDT | PDT of skin and bladder, drug and light source | Tel: +47 22 06 22 10 Fax: +47 22 06 22 18 |
| Photoderma, Ecublens, Switzerland | www.photoderma.com | PDT-procedure | Experimental PDT in dermatology: hair removal, acne, warts etc., procedure | Tel: +41 21 311 9041 Fax: +41 21 311 90 42 |
| photonamic GmbH & Co. KG, Hamburg, Germany | www.photonamic.de | ALA | FD and PDT: clinical studies for FD and PDT of brain tumours (malignant glioma) | Tel: +49 4103 8006-701 Fax: +49 4103 8006-710 |
| Porphyrin Systems GbR, Lübeck, Germany | www.porphyrin-systems.de | Porphyrin products | ALA, PpIX for scientific purpose | Tel.: +49 451 2903 310 Fax: +49 451 2903 319 |

# Subject Index